Smart Village Infrastructure and Sustainable Rural Communities

Mohammad Ayoub Khan
University of Bisha, Saudi Arabia

Bhumika Gupta
Govind Ballabh Pant Institute of Engineering and Technology, India

Agya Ram Verma
Govind Ballabh Pant Institute of Engineering and Technology, India

Pushkar Praveen
Govind Ballabh Pant Institute of Engineering and Technology, India

Cathryn J. Peoples
Ulster University, UK

A volume in the Advances in Electronic Government, Digital Divide, and Regional Development (AEGDDRD) Book Series

IGI Global
PUBLISHER of TIMELY KNOWLEDGE

Published in the United States of America by
IGI Global
Engineering Science Reference (an imprint of IGI Global)
701 E. Chocolate Avenue
Hershey PA, USA 17033
Tel: 717-533-8845
Fax: 717-533-8661
E-mail: cust@igi-global.com
Web site: http://www.igi-global.com

Library of Congress Cataloging-in-Publication Data

Names: Khan, Mohammad Ayoub, 1980- editor.
Title: Smart village infrastructure and sustainable rural communities / edited by Mohammad Ayoub Golf Khan, Bhumika Gupta, Agya Ram Verma, Pushkar Praveen, Cathryn Jennifer Peoples.
Description: Hershey, PA : Information Science Reference, [2023] | Includes bibliographical references and index. | Summary: "The objective of this book is to collect the state-of-art and innovative ideas on empowering rural communities through the concept of smart village and Industry 5.0 from the expert in the domain of agriculture process improvement, sustainability, irrigation system, transportation, smart education, sanitation, safe drinking water, internal road, tree plantation, health, and many other factors. This book will be able to create interest in the reader to fill the gap between the smart city infrastructure and suitability in the village. This will provide an opportunity for the researcher and developed to think about how a village could get benefit from smart technologies and achieve sustainability"-- Provided by publisher.
Identifiers: LCCN 2022047447 (print) | LCCN 2022047448 (ebook) | ISBN 9781668464182 (h/c) | ISBN 9781668464229 (s/c) | ISBN 9781668464199 (eISBN)
Subjects: LCSH: Smart cities. | Rural development. | Sustainable urban development.
Classification: LCC TD159.4 .S688 2023 (print) | LCC TD159.4 (ebook) | DDC 307.1/412--dc23/eng/20221202
LC record available at https://lccn.loc.gov/2022047447
LC ebook record available at https://lccn.loc.gov/2022047448

This book is published in the IGI Global book series Advances in Electronic Government, Digital Divide, and Regional Development (AEGDDRD) (ISSN: 2326-9103; eISSN: 2326-9111)

British Cataloguing in Publication Data
A Cataloguing in Publication record for this book is available from the British Library.

All work contributed to this book is new, previously-unpublished material.
The views expressed in this book are those of the authors, but not necessarily of the publisher.

For electronic access to this publication, please contact: eresources@igi-global.com.

Advances in Electronic Government, Digital Divide, and Regional Development (AEGDDRD) Book Series

ISSN:2326-9103
EISSN:2326-9111

Zaigham Mahmood
University of Derby, UK & North West University, South Africa

MISSION

The successful use of digital technologies (including social media and mobile technologies) to provide public services and foster economic development has become an objective for governments around the world. The development towards electronic government (or e-government) not only affects the efficiency and effectiveness of public services, but also has the potential to transform the nature of government interactions with its citizens. Current research and practice on the adoption of electronic/digital government and the implementation in organizations around the world aims to emphasize the extensiveness of this growing field.

The Advances in Electronic Government, Digital Divide & Regional Development (AEGDDRD) book series aims to publish authored, edited and case books encompassing the current and innovative research and practice discussing all aspects of electronic government development, implementation and adoption as well the effective use of the emerging technologies (including social media and mobile technologies) for a more effective electronic governance (or e-governance).

COVERAGE

- Urban Development, Urban Economy
- Current Research and Emerging Trends in E-Government Development
- E-Government in Developing Countries and Technology Adoption
- Citizens Participation and Adoption of E-Government Provision
- Online Government, E-Government, M-Government
- Electronic Government, Digital Democracy, Digital Government
- Knowledge Divide, Digital Divide
- Emerging Technologies within the Public Sector
- E-Citizenship, Inclusive Government, Connected Government
- Frameworks and Methodologies for E-Government Development

IGI Global is currently accepting manuscripts for publication within this series. To submit a proposal for a volume in this series, please contact our Acquisition Editors at Acquisitions@igi-global.com or visit: http://www.igi-global.com/publish/.

The Advances in Electronic Government, Digital Divide, and Regional Development (AEGDDRD) Book Series (ISSN 2326-9103) is published by IGI Global, 701 E. Chocolate Avenue, Hershey, PA 17033-1240, USA, www.igi-global.com. This series is composed of titles available for purchase individually; each title is edited to be contextually exclusive from any other title within the series. For pricing and ordering information please visit http://www.igi-global.com/book-series/advances-electronic-government-digital-divide/37153. Postmaster: Send all address changes to above address.

Titles in this Series

For a list of additional titles in this series, please visit: http://www.igi-global.com/book-series/

AI, IoT, and Blockchain Breakthroughs in E-Governance
Kavita Saini (Galgotias University, India) A. Mummoorthy (Malla Reddy College of Engineering and Technology, India) Roopa Chandrika (Department of Information Technology, Malla Reddy College of Engineering and Technology, India) and N.S. Gowri Ganesh (Saveetha Engineering College, India)
Information Science Reference • © 2023 • 243pp • H/C (ISBN: 9781668476970) • US $240.00

Glocal Policy and Strategies for Blockchain Building Ecosystems and Sustainability
Gülsün Kurubacak (Anadolu Üniversitesi Açıköğretim Fakültesi, Turkey) Ramesh Chander Sharma (Dr. B.R. Ambedkar University, Delhi, India) and Hakan Yıldırım (Eskişehir Osmangazi University, Turkey)
Information Science Reference • © 2023 • 332pp • H/C (ISBN: 9781668441534) • US $270.00

The COVID-19 Pandemic and the Digitalization of Diplomacy
Floribert Patrick Calvain Endong (University of Dschang, Cameroon)
Information Science Reference • © 2023 • 263pp • H/C (ISBN: 9781799883944) • US $235.00

Handbook of Research on Current Advances and Challenges of Borderlands, Migration, and Geopolitics
Rui Alexandre Castanho (WSB University, Poland)
Information Science Reference • © 2023 • 444pp • H/C (ISBN: 9781668470206) • US $270.00

Handbook of Research on Cyber Approaches to Public Administration and Social Policy
Fahri Özsungur (Mersin University, Turkey)
Information Science Reference • © 2022 • 692pp • H/C (ISBN: 9781668433805) • US $315.00

IGI Global
PUBLISHER of TIMELY KNOWLEDGE

701 East Chocolate Avenue, Hershey, PA 17033, USA
Tel: 717-533-8845 x100 • Fax: 717-533-8661
E-Mail: cust@igi-global.com • www.igi-global.com

Table of Contents

Preface..xviii

Chapter 1
Smart Village Infrastructure and Rural Communities 1
Pushkar Praveen, Govind Ballabh Pant Institute of Engineering and Technology, India
Ayoub Khan, University of Bisha, Saudi Arabia
Agya Ram Verma, Govind Ballabh Pant Institute of Engineering and Technology, India
Manoj Kumar, Govind Ballabh Pant Institute of Engineering and Technology, India
Cathryn Peoples, University of Ulster, UK

Chapter 2
Artificial Intelligence and Machine Learning-Based Agriculture 16
Sakshi Rajput, Maharaja Surajmal Institute of Technology, Guru Gobind Singh Indraprastha University, India
Lititia Khanna, Maharaja Surajmal Institute of Technology, Guru Gobind Singh Indraprastha University, India
Priti Kumari, National Institute of Technology, Patna, India

Chapter 3
Importance of Drone Technology in Indian Agriculture, Farming......................35
Jyoti Yadav, Galgotias University, India
Usha Chauhan, Galgotias University, India
Divya Sharma, Galgotias University, India

Chapter 4
Advancement in Agricultural Techniques With the Introduction of Artificial Intelligence and Image Processing....................47
Mritunjay Rai, Shri Ramswaroop Memorial University, India
Nalini Tyagi, Delhi College of Technology and Management, India
Padmesh Tripathi, KCC Institute of Technology and Management, India
Nitendra Kumar, Amity Business School, Amity University, Noida, India
Priti Kumari, National Institute of Technology, Patna, India

Chapter 5
Role of Image Processing in Agricultural Practices 69
Agrima Saxena, Department of Computer Science Engineering, Galgotias University, Greater Noida, India
Ayush Bharadwaj, Department of Computer Science Engineering, Galgotias University, Greater Noida, India
Usha Chauhan, Department of Electrical, Electronics and Communications, Galgotias University, Greater Noida, India
Divya Sharma, Department of Electrical, Electronics and Communications, Galgotias University, India
Priti Kumari, National Institute of Technology, Patna, India

Chapter 6
Applications of Machine Learning in Agriculture....................99
Padmesh Tripathi, KCC Institute of Technology and Management, India
Nitendra Kumar, Amity Business School, Amity University, Noida, India
Mritunjay Rai, Shri Ramswaroop Memorial University, India
Pushpendra Kumar Shukla, Vindhya Institute of Technology and Science, India
Kailash Nath Verma, IIMT College of Engineering, India

Chapter 7
Achieving Sustainable Agriculture: An AI-Based Approach 119
Syed Asif Ali, SMI University, Karachi, Pakistan

Chapter 8
An Outlook Over Smart Irrigation System for Sustainable Rural Development134
R. Muthuminal, SNS College of Technology, India
R. Mohana Priya, SNS College of Technology, India

Chapter 9
Artificial Intelligence Technology in Different Fields......................................161
Sakshi Rajput, Maharaja Surajmal Institute of Technology, India
Deepak Sarangi, Maharaja Surajmal Institute of Technology, India
Preeti Sehrawat, Maharaja Surajmal Institute of Technology, India

Chapter 10
Revolutionizing Agriculture With Cloud-Connected Irrigation Technology 184
Abhishek Kumar Singh, Govind Ballabh Pant Institute of Engineering and Technology, India
Pushkar Praveen, Govind Ballabh Pant Institute of Engineering and Technology, India
Devansh Tripathi, Govind Ballabh Pant Institute of Engineering and Technology, India
Ved Prakash Pandey, Govind Ballabh Pant Institute of Engineering and Technology, India
Pratibha Verma, Rajkiya Degree College, Puranpur, India

Chapter 11
COVID-19 Diagnosis and Treatment Challenges in Rural India.......................208
Shanti Chandra, Indian Institute of Information technology, India
Shivaji Kanojiya, Indian Institute of Technology (BHU), Varanasi, India
Agya Ram Verma, Govind Ballabh Pant Institute of Engineering and Technology, India
Yatendra Kumar, M.J.P. Rohilkhand University, India

Chapter 12
Air Pollution Effects on the Habitat of Insects and Strategic Control to Improve Productivity 216
Loveneet Mishra, Department of DEECE, Galgotias University, India
Usha Chauhan, Department of DEECE, Galgotias University, India
S. P. S. Chauhan, SCSE, Galgotias University, India
Harshit Singh, SCSE, Galgotias University, India

Chapter 13
Automation of Street Light of Adarsh Gram With Manual Intervention and Automatic Luminosity Control With Traffic Flow 228
Rahul Kumar Singh, Indian Institute of Information Technology, Design, and Manufacturing, Jabalpur, India
Anju Bala, Dr. Bhim Rao Ambedkar Engineering College of Information Technology, India
Pratibha Verma, Rajkiya Degree College, Puranpur, India

Chapter 14
Advanced Detection: Fungi-Bacterial Diseases in Plants Using Modified Deep Neural Network and DSURF 248
Pooja Singh, SCSE, Galgotias University, India
Usha Chauhan, DEECE, Galgotias University, India
S. P. S. Chauhan, SCSE, Galgotias University, India
Harshit Singh, SCSE, Galgotias University, India

Chapter 15
Optimal Control Strategies for Mitigating Frequency Deviation in Microgrid-Based Smart Village Systems 269
Sudheeksha Misra, Govind Ballabh Pant Institute of Engineering and Technology, India
Bhola Jha, Govind Ballabh Pant Institute of Engineering and Technology, India
V. M. Mishra, Govind Ballabh Pant Institute of Engineering and Technology, India
Surjeet Singh Patel, Govind Ballabh Pant Institute of Engineering and Technology, India

Chapter 16
Solar-Powered Pesticide-Spraying RFID Robot 283
K. Raju, St. Joseph Engineering College, India
Roshan Rajeshwar Tulaskar, St. Joseph Engineering College, India
Canute Sherwin, St. Joseph Engineering College, India

Compilation of References 308

About the Contributors 338

Index 341

Detailed Table of Contents

Preface...xviii

Chapter 1
Smart Village Infrastructure and Rural Communities 1
Pushkar Praveen, Govind Ballabh Pant Institute of Engineering and Technology, India
Ayoub Khan, University of Bisha, Saudi Arabia
Agya Ram Verma, Govind Ballabh Pant Institute of Engineering and Technology, India
Manoj Kumar, Govind Ballabh Pant Institute of Engineering and Technology, India
Cathryn Peoples, University of Ulster, UK

Rural communities around the world face significant challenges in accessing basic services such as healthcare, education, and sanitation, as well as in promoting economic growth and environmental sustainability. In short, smart village infrastructure has emerged as a potential solution to the challenges faced by rural communities around the world. By leveraging technology and innovation, smart village infrastructure has the potential to increase access to basic services, promote economic growth and environmental sustainability, and improve the overall quality of life in rural communities. However, addressing the challenges of funding, skilled labor, and community engagement requires a holistic approach that involves collaboration between multiple stakeholders. The policy recommendations provided in this paper can help promote smart village infrastructure and sustainable rural communities.

Chapter 2
Artificial Intelligence and Machine Learning-Based Agriculture 16
Sakshi Rajput, Maharaja Surajmal Institute of Technology, Guru Gobind Singh Indraprastha University, India
Lititia Khanna, Maharaja Surajmal Institute of Technology, Guru Gobind Singh Indraprastha University, India
Priti Kumari, National Institute of Technology, Patna, India

According to the Food and Agriculture Organization of the United Nations, by 2050, there will be an additional nine billion people living on the planet, yet there will only be an additional 4% of land under cultivation. In such a situation, more effective agricultural practices can be achieved using recent technical developments and fixes for the industry's current bottlenecks. Direct use of artificial intelligence and machine learning in the agricultural industry could represent a paradigm shift in the way farming is now carried out. A farmer may accomplish more with fewer resources using AI-powered farming solutions, which also improve quality and ensure speedy GTM (go-to-market) strategies for crops. This chapter offers a perspective of how AI might fuel the various agricultural sectors. Additionally, it looks into future AI-powered concepts and potential challenges.

Chapter 3
Importance of Drone Technology in Indian Agriculture, Farming......................35
Jyoti Yadav, Galgotias University, India
Usha Chauhan, Galgotias University, India
Divya Sharma, Galgotias University, India

The population is quickly growing, as is the need for food. The farmers' traditional tactics are insufficient to achieve these goals. As a result, new automated methods (such as drone technology) emerged. These innovative techniques met food demands while also offering employment opportunities for billions of people. Drone technology saves water, pesticides, and herbicides, protects soil fertility, and allows for more efficient use of labour to boost productivity and quality. This article will look at the usage of drones in agricultural applications. According to the literature, drones may be utilised for a range of agricultural reasons. The authors took the approach of doing a thorough review of past studies in this topic. This study describes the current state of agricultural drone technology, including crop health monitoring and farm activities such as weed control, evapotranspiration estimation, and spraying. The study concludes by recommending more farmers to invest in drone technology to boost agricultural output.

Chapter 4
Advancement in Agricultural Techniques With the Introduction of Artificial Intelligence and Image Processing..47
Mritunjay Rai, Shri Ramswaroop Memorial University, India
Nalini Tyagi, Delhi College of Technology and Management, India
Padmesh Tripathi, KCC Institute of Technology and Management, India
Nitendra Kumar, Amity Business School, Amity University, Noida, India
Priti Kumari, National Institute of Technology, Patna, India

Food is the most fundamental requirement of human beings, and the only way this requirement can be satisfied is through agricultural production. The agricultural sector not only fulfills the need for food but also contributes to any country's economy. It also provides employment on a large scale. Agriculture is the backbone of any developing country due to its help in employment and economic sectors. This field is broadly divided into three parts: pre-harvest, harvest, and post-harvest. Advancement in various technologies like machine learning, image processing, the internet of things, and machine vision helps to reduce risk in farming and also increase profit. Using these techniques, a farmer can gain a brief understanding of the crop throughout its life cycle with greater precision and accuracy. This chapter provides a quick overview of different technologies along with their advantages and disadvantages. These technologies make farming more productive, accurate, and cost-effective, since they require less human effort and are less expensive.

Chapter 5
Role of Image Processing in Agricultural Practices 69
Agrima Saxena, Department of Computer Science Engineering, Galgotias University, Greater Noida, India
Ayush Bharadwaj, Department of Computer Science Engineering, Galgotias University, Greater Noida, India
Usha Chauhan, Department of Electrical, Electronics and Communications, Galgotias University, Greater Noida, India
Divya Sharma, Department of Electrical, Electronics and Communications, Galgotias University, India
Priti Kumari, National Institute of Technology, Patna, India

Agriculture is considered necessary to sustain life on this planet. With a growing population, we need agricultural productivity to increase significantly to meet the demands. Image processing has proven to be a powerful tool for analysis in many fields and applications. The agricultural sector, where parameters such as canopy cover, yield, and product quality were important for farmers. It is often the case that professional advice is not affordable, and that the availability of an expert and their services may take time. The situation could be greatly improved if image processing was combined with a reliable communication network, as this would eliminate the need to obtain expert advice within a limited time frame. This can be achieved because image processing is a powerful tool for parameter analysis.

Chapter 6
Applications of Machine Learning in Agriculture..99
Padmesh Tripathi, KCC Institute of Technology and Management, India
Nitendra Kumar, Amity Business School, Amity University, Noida, India
Mritunjay Rai, Shri Ramswaroop Memorial University, India
Pushpendra Kumar Shukla, Vindhya Institute of Technology and Science, India
Kailash Nath Verma, IIMT College of Engineering, India

Today's epoch christened as the modern epoch, is the epoch of information and technologies. No field is untouched with the emergence of technologies. In different segments of agriculture, numerous technologies have been employed extensively. Machine learning has extensively been employed in every field today. It is composed of high-performance computing and big data technologies. Machine learning (ML) and its variants have been employed in different domains of agriculture like yield prediction, weed detection, disease detection, water management, soil management, livestock production, etc. Machine learning techniques have emerged as a boon to agriculture and have increased the quality and quantity of agricultural products. In this chapter, some applications of machine learning technologies in agriculture have been explored.

Chapter 7
Achieving Sustainable Agriculture: An AI-Based Approach 119
Syed Asif Ali, SMI University, Karachi, Pakistan

Sustainable agriculture means an integrated agricultural system (IAS), which is a prolonged or long-term support system that fulfils the requirements of agricultural products, improves the environment (by controlling, improving, and monitoring natural biological cycles), and boosts the economy. Crop product quality and quantity have increased significantly over the last half-century, owing in large part to organic chemistry. Organic chemistry, however, remains a key component of crop protection and helps to create techniques to boost crop output. In the modern era of technology, artificial intelligence approaches are used for the analysis of soil nutrients and their optimal management on the basis of supervised learning, depending on behaviour analysis and prior experience. Soil nutrients are an essential component of soil fertility and environmental sustainability. This chapter will cover soil nutrient measurements, water irrigation, and predator control through artificial intelligence techniques such as decision making, artificial neural networks, and machine learning.

Chapter 8
An Outlook Over Smart Irrigation System for Sustainable Rural Development134
R. Muthuminal, SNS College of Technology, India
R. Mohana Priya, SNS College of Technology, India

The process of applying water to the soil artificially by means of systems such as pumps, tubes, sprays, pipes is called irrigation. As irrigation plays a vital role in the growth of plants or cultivating species to satisfy the basic needs of human beings, it is necessary to manage it in a proper way. Usually, irrigation is done at the regions which are subjected to irregular rain or dry or drought prone areas in order to manage the crop yield along with the consideration of quality. Since the mode of applying the water in a uniform manner is the main task in irrigation, several irrigation systems are adopted. The growth of the agricultural crops, maintaining landscapes, reducing weeds, and also to regenerate the disturbed soils in dry areas or also during the season of less rainfall are dependent on irrigation for their prosperity. Hence, in this chapter, the authors detail the background of irrigation, how it impacts the crop yield and demand, smart technologies available, growth analysis, and intelligent and smart irrigation systems with its benefits and necessities.

Chapter 9
Artificial Intelligence Technology in Different Fields 161
Sakshi Rajput, Maharaja Surajmal Institute of Technology, India
Deepak Sarangi, Maharaja Surajmal Institute of Technology, India
Preeti Sehrawat, Maharaja Surajmal Institute of Technology, India

The chapter consists of information regarding different usage of artificial intelligence systems in the health sector and the transportation sector in different types of ways which will help to maintain a good experience for the users. The chapter includes the data that shows the sections where AI is used in healthcare and transportation sectors like smart devices used in hospitals such as chatbots for patients, virtual nurse assistants, robotic surgeries, smart traffic control system which will help the peoples to avoid unnecessary long trips and save their time, drone systems, etc. Along with this, the authors have data about growth in these sectors by using AI systems, which make it cost-efficient with better performance than before.

Chapter 10
Revolutionizing Agriculture With Cloud-Connected Irrigation Technology 184
Abhishek Kumar Singh, Govind Ballabh Pant Institute of Engineering and Technology, India
Pushkar Praveen, Govind Ballabh Pant Institute of Engineering and Technology, India
Devansh Tripathi, Govind Ballabh Pant Institute of Engineering and Technology, India
Ved Prakash Pandey, Govind Ballabh Pant Institute of Engineering and Technology, India
Pratibha Verma, Rajkiya Degree College, Puranpur, India

The current global water crisis, caused by factors such as population growth, urbanization, and climate change, has a significant impact on agriculture, which consumes 70% of the world's freshwater. One potential solution to this problem is the use of intelligent irrigation systems. Cloud-based smart irrigation systems offer a more efficient and cost-effective approach to water consumption management. This study examines the concept of cloud-based smart irrigation systems, their advantages, and their potential application in agriculture. To begin, the authors detail the various components of a cloud-based smart irrigation system, including sensors, controllers, and cloud computing. They then explore the benefits of cloud-based systems, which include scalability, cost-effectiveness, and real-time data processing capabilities. They also delve into the different types of sensors used in smart irrigation systems, such as soil moisture, weather, and crop sensors, as well as their respective advantages and limitations. Furthermore, they discuss the various controllers utilized in smart irrigation systems, such as scheduling, sensor-based, and weather-based controllers. They elaborate on how these controllers function and the advantages they offer in terms of water conservation and crop yield optimization. Lastly, they examine the practical application of cloud-based smart irrigation systems in agriculture, including how to create, implement, and tailor the system to suit the specific requirements of a particular farm. They also consider potential barriers to adoption, such as cost and technological complexity, and how to overcome them. In short, cloud-based smart irrigation systems present a promising solution to the water crisis in agriculture. These systems can significantly reduce water waste, increase crop yield, and save costs by utilizing cloud computing technology to manage irrigation. With careful planning, installation, and configuration, cloud-based smart irrigation systems have the potential to be an effective tool for farmers in improving their productivity and sustainability.

Chapter 11
COVID-19 Diagnosis and Treatment Challenges in Rural India.......................208
Shanti Chandra, Indian Institute of Information technology, India
Shivaji Kanojiya, Indian Institute of Technology (BHU), Varanasi, India
Agya Ram Verma, Govind Ballabh Pant Institute of Engineering and Technology, India
Yatendra Kumar, M.J.P. Rohilkhand University, India

COVID-19 was first identified in Wuhan, China in December 2019. India is the second most affected country of the world. About 516 million people have been affected worldwide by it so far, and approximately 6 million deaths have been caused. In India, more than 40 million people have been infected according to official data released by the government. Early detection of COVID-19 infection is the first part to deal with this pandemic. However, detection of COVID-19 infection is the biggest problem in rural India because of bad connectivity and poor medical facilities. After a certain period of time there are waves of increasing numbers of cases because of its variants. Rural India faced the problem of shortage of medical resources because of a spike in the number of cases in a region. The goal of this study is to develop a deep learning model that can help in detecting COVID-19 cases using chest x-rays, thus leading to inexpensive and fast early detection of COVID-19 to prevent its spread.

Chapter 12
Air Pollution Effects on the Habitat of Insects and Strategic Control to Improve Productivity 216
Loveneet Mishra, Department of DEECE, Galgotias University, India
Usha Chauhan, Department of DEECE, Galgotias University, India
S. P. S. Chauhan, SCSE, Galgotias University, India
Harshit Singh, SCSE, Galgotias University, India

Urbanization has significantly worsened the ecosystem by releasing harmful substances and gases into the air through air pollution, which has an impact on the population demographics of insect herbivores. A pest outbreak and the formation of new pest biotypes are caused by the careless application of wide spectrum pesticides, which decimate the population of important pollinators and foragers. The potential efficiency of phloem-feeding homopteran insects such aphids is increased by pollutants like sulphur dioxides, oxides of nitrogen, fluorides, and heavy metals in the environment. Industrial melanism in moths is caused by a decline in the light-coloured lichens as a result of pollution. However, trophic amplification caused by high toxicant concentrations reduces the susceptibility of the parasitise & predators. The population growth of invading alien pest species & tropical disease vectors is amplified by air pollution and the acceleration of global warming. However, steps have been taken to reduce air pollution and boost agricultural output. The fabric filtration method uses filter cakes to remove air contaminants in this manner. Successful electrostatic precipitation strategies remove charged droplets of solid or liquid from syngas where the suspended particles exist. However, methods like thermal incineration & catalytic oxidation, which essentially use a catalyst activator to transform organic chemicals in waste gas streams into CO2, liquid water vapor, as well as inorganic-gases at extreme temps, are in use. Additional to these methods Due to their low operating and capital expenses, biofilters could be employed in conjunction with the other treatment technologies. Thus, successful reduction of air pollution can restore the health of agricultural plants and lessen food shortage.

Chapter 13
Automation of Street Light of Adarsh Gram With Manual Intervention and Automatic Luminosity Control With Traffic Flow 228
Rahul Kumar Singh, Indian Institute of Information Technology, Design, and Manufacturing, Jabalpur, India
Anju Bala, Dr. Bhim Rao Ambedkar Engineering College of Information Technology, India
Pratibha Verma, Rajkiya Degree College, Puranpur, India

Sixty-five percent of India's population still lives in the villages. Even after years of independence, the condition of villages are not good for living. The basic facilities and amenities like drinking water, hospital, and electricity are not available in most villages in India. The streets of villages are mostly dark either due to no light or malfunctioning lights. This leads women and children facing problems at night while going outside. The authors try to overcome this problem with the concept of electrification of street lights with minimum power consumption by automatic control along with intensity control with simple, cheap, durable, and easily reparable components with and without manual intervention and further expanding into solar power.

Chapter 14
Advanced Detection: Fungi-Bacterial Diseases in Plants Using Modified Deep Neural Network and DSURF 248
Pooja Singh, SCSE, Galgotias University, India
Usha Chauhan, DEECE, Galgotias University, India
S. P. S. Chauhan, SCSE, Galgotias University, India
Harshit Singh, SCSE, Galgotias University, India

In the present scenario, due to climate change, farmers crops get fungi due to bacteria since soil temperatures change very rapidly according to sudden climate changes for which the crop is getting spoiled. At this advanced era, disease can be detected early so that crops are safe. Different types of fungi-bacterial disease will be detected and prevented by machine learning-based predicted deterministic probabilistic and artificial technology-based CNN for colour changes in plants. This chapter described machine learning techniques and proposed modified algorithms to identify and classify plant diseases. Deep neural network (DNN) models and algorithms are used to improve object accuracy and entropy to reduce the complexity of computational processes and improve the features during deep learning processes (e.g., modified deep neural network [MDNN]). Additionally, they support dynamic feature extraction DSURF and classifier combinations for creating image clusters with the help of clustering and deterministic probability.

Chapter 15
Optimal Control Strategies for Mitigating Frequency Deviation in Microgrid-Based Smart Village Systems 269
Sudheeksha Misra, Govind Ballabh Pant Institute of Engineering and Technology, India
Bhola Jha, Govind Ballabh Pant Institute of Engineering and Technology, India
V. M. Mishra, Govind Ballabh Pant Institute of Engineering and Technology, India
Surjeet Singh Patel, Govind Ballabh Pant Institute of Engineering and Technology, India

Frequency deviation is a common issue in microgrids, especially in smart village systems where the energy demand varies significantly due to changing loads and intermittent renewable energy sources. This chapter aims to propose optimal control strategies that effectively mitigate frequency deviation in microgrid-based smart village systems. The research will explore the variation of frequency by using different control techniques such as conventional PI, PI plus fuzzy, and PI plus ANFIS. The proposed strategies will be evaluated using simulation models. The results shows that the PI plus ANFIS shows the best results, which will provide valuable insights for designing reliable and efficient microgrid-based smart village systems, addressing the problem of frequency deviation and improving overall system performance.

Chapter 16
Solar-Powered Pesticide-Spraying RFID Robot 283
K. Raju, St. Joseph Engineering College, India
Roshan Rajeshwar Tulaskar, St. Joseph Engineering College, India
Canute Sherwin, St. Joseph Engineering College, India

In this chapter, a solar powered pesticide-spraying RFID robot has been designed and developed for its use in different types of fruits and vegetable fields. The robot is equipped with DC motors, DC pump, a solar panel, and an ultrasonic sensor. The robot's navigation has been automated using RFID (radio frequency identification) technology. An ultrasonic sensor is used for sensing any obstacle in the path of the robot's navigation. A solar panel of 30W is used for charging the batteries that are used for driving the DC motors and the DC pump. The wheels of the robot are powered by DC motors, and the DC pump is used for spraying pesticides through the nozzles. The height of the nozzles can be varied depending on the size of the crop. The robot has been designed for spraying pesticides automatically thereby controlling pests, improving productivity, and reducing the health hazards of farmers posed by pesticides.

Compilation of References 308

About the Contributors 338

Index 341

Preface

Rural areas are often overlooked and underserved when it comes to infrastructure development and sustainability initiatives. However, the COVID-19 pandemic has highlighted the importance of rural areas and the need to invest in their development to ensure that they are resilient and able to withstand future shocks. The concept of a smart village is one that can help to address this issue. Rural communities have always been an essential part of every country's growth and development. However, despite their significance, rural areas often lag behind urban areas in terms of access to basic infrastructure, technology, and resources. This gap creates a significant challenge for sustainable development and hinders the potential of rural communities.

To overcome these challenges, the concept of smart village infrastructure and sustainable rural communities has emerged. Smart village infrastructure involves the use of innovative technology and sustainable practices to improve the quality of life in rural areas. Sustainable rural communities are those that prioritize environmental, social, and economic sustainability, ensuring that they can thrive for generations to come.

Smart Village Infrastructure and Sustainable Rural Communities aims to provide readers with a comprehensive understanding of the various aspects of smart village infrastructure and its role in creating sustainable rural communities. The book is a compilation of research, case studies, and expert insights that showcase the various opportunities and challenges that come with implementing smart village infrastructure and provides an in-depth analysis of the concept of smart village infrastructure and sustainable rural communities. It aims to contribute to the understanding and promotion of sustainable development in rural areas. It covers topics such as renewable energy, waste management, water conservation, agriculture, and education. It also discusses the challenges and opportunities associated with implementing sustainable practices in rural areas.

The book's target audience includes policymakers, researchers, academics, students, and professionals interested in sustainable development, rural development, and infrastructure planning. It provides an excellent resource for those looking to understand the key elements of sustainable rural development and how to implement them in practice. The book's chapters are written by experts from various fields, including engineering, environmental science, social sciences, and economics.

The diversity of perspectives ensures that the book covers a wide range of topics and provides a comprehensive understanding of smart village infrastructure and sustainable rural communities.

In summary, this book presents a compelling case for the importance of sustainable rural development and smart village infrastructure. It provides a valuable resource for those looking to understand the key elements of sustainable rural development and implement them in practice. The book's insights and recommendations have the potential to transform rural communities and create a more sustainable future for all.

ORGANIZATION OF THE BOOK

This book contains 16 chapters. A brief description of each of the chapters is as follows:

Chapter 1, titled "Smart Village Infrastructure and Rural Communities," presents the concept of smart village infrastructure and its potential to promote sustainable development in rural communities facing challenges in accessing basic services such as healthcare, education, and sanitation, as well as in promoting economic growth and environmental sustainability. To address these challenges, the chapter recommends a holistic approach involving collaboration between local communities, governments, NGOs, and the private sector, and provides several policy recommendations, including investing in basic infrastructure, promoting ICT access and literacy, providing financial and technical support to rural entrepreneurs, and promoting community engagement and participation.

Chapter 2, titled "Artificial Intelligence and Machine Learning-Based Agriculture," presents the concept of artificial intelligence (AI) and its various subsets, including machine learning (ML), which allows software applications to predict outcomes more accurately based on previous data. It then goes on to highlight the importance of agriculture in the global economy, noting the need for increased productivity to meet growing demand due to population growth. This article suggests that AI and ML have become essential technologies in the agricultural sector, allowing for smarter approaches and more efficient farming methods and also mentioned various computer-based systems developed for yield detection, weed detection, and harvest quality, which have upgraded real-time monitoring, crop production, processing, harvesting, and advertising.

Chapter 3, titled "Importance of Drone Technology in Indian Agriculture, Farming," presents the growing population and the need for food, which traditional farming methods cannot meet. The article suggests that new automated methods, including drone technology, have emerged to meet the increasing food demands while also providing employment opportunities. The use of drones in agriculture can save water, pesticides, herbicides, protect soil fertility, and increase productivity and quality. The article reviews past studies on the use of drones in agriculture and discusses various agricultural applications such as crop health monitoring, weed control, evapotranspiration estimation, and spraying. The study recommends that more farmers invest in drone technology to boost agricultural output.

Chapter 4, titled "Advancement in Agricultural Techniques With the Introduction of Artificial Intelligence and Image Processing," presents the importance of agriculture, which is essential for human existence and economic development. However, farmers are reluctant to adopt new technologies; making this field the most neglected one globally. This is despite the fact that human food demand is expected to increase by up to 70% by 2040. Agricultural advancements such as the use of hybrid seeds, improved irrigation techniques, and new types of fertilizer have contributed to an increase in agricultural production. A new agricultural revolution is being fueled by data and connection, with machine learning, image processing analytics, and networked sensors contributing to increased productivity. Technological advancements allow for the monitoring of crops at various stages and the adoption of automated solutions in the industry.

Chapter 5, titled "Role of Image Processing in Agricultural Practices," presents the importance of agriculture to sustain life and meet the growing population's demands. The increasing need for agricultural productivity and the challenges farmers face in obtaining professional advice in a timely and affordable manner are discussed. Image processing is identified as a powerful tool for analyzing important parameters such as canopy cover, yield, and product quality in the agricultural sector. By combining image processing with a reliable communication network, farmers can eliminate the need for expert advice within a limited time frame, which could greatly improve the situation. The article suggests that this can be achieved because image processing is a powerful tool for parameter analysis.

Chapter 6, titled "Application of Machine Learning in Agriculture," presents the widespread use of information and technologies in various fields, including agriculture. Machine learning, which emerged in the 1940s, has become an essential technology in agriculture. It involves high-performance computing and big data technologies, and it has been widely used in yield prediction, weed and disease detection, water and soil management, and livestock production. Machine learning has greatly enhanced the quality and quantity of agricultural products, making it a boon for the agricultural industry. This article explores some of the applications of machine learning in agriculture.

Chapter 7, titled "Achieving Sustainable Agriculture: An AI-Based Approach," presents the concept of sustainable agriculture, which involves an Integrated Agricultural System (IAS) that meets the requirements of agricultural products while also improving the environment and boosting the economy. Organic chemistry has played a significant role in improving crop quality and quantity in the last 50 years, and it continues to be a crucial component of crop protection. The article also highlights the use of artificial intelligence (AI) in soil nutrient analysis and management through techniques like decision making, artificial neural networks, and machine learning. Soil nutrients are essential for soil fertility and environmental sustainability, and AI can help in optimal management of these nutrients, as well as in water irrigation and predator control.

Chapter 8, titled "An Outlook Over Smart Irrigation System for Sustainable Rural Development," presents a modern technique for watering crops that employs remotely controlled equipment, connected sensors, and other technology. This enables farmers and their staff to automate their workflow and monitor the irrigation operation in real-time. With the help of smart irrigation, farmers can anticipate potential hazards, make irrigation plans based on weather updates and moisture levels, and makes use of data from previous irrigation sessions. As a result, farming becomes a more efficient, less wasteful, and more predictable process. By drawing on their past experiences and knowledge, farmers can adapt to changes and plan for the future with greater confidence.

Chapter 9, titled "Artificial Intelligence Technology in Different Fields," presents the growing demand for automation in today's fast-paced world and the potential benefits of using Artificial Intelligence (AI) and Machine Learning (ML) to perform various tasks more efficiently. The article traces the origins of AI and credits British logician and computer pioneer Alan Mathison Turing for its earliest substantial work in the mid-20th century. The chapter highlights the significant impact of AI and ML on various sectors, including IT, Agriculture, Healthcare, Transportation, Defense Services, and Industries. The article then focuses on the applications of AI and ML in the Healthcare and Transportation sectors.

Chapter 10, titled "Revolutionizing Agriculture With Cloud-Connected Irrigation Technology," presents the benefits of automatic water management systems, also known as irrigation systems, for farmers in managing their crops efficiently. The manual irrigation process has the risk of crops being either overwatered or under watered due to human error or absence. The article proposes the "Cloud-Based Smart Irrigation System" as a solution to this problem, which uses Arduino UNO and various sensors, including the Capacitive Moisture Sensor, 5v Relay Module, and Wi-Fi module. The system records soil moisture levels and initiates the watering process if it falls below a threshold value. The data is sent to a cloud-based monitoring system and plotted in graph format. The article suggests adding a humidity sensor to take atmospheric humidity into account and programming the system to water crops based on both humidity and moisture levels for even more efficiency.

Chapter 11, titled "COVID-19 Diagnosis and Treatment Challenges in Rural India," discusses the emergence of the COVID-19 pandemic in December 2019 in Wuhan, China, which has affected a significant number of individuals worldwide and caused a large number of deaths. With the outbreak's continuous waves, many countries face a shortage of medical resources to handle the spike in cases. To tackle this issue, the study aims to develop a Deep Learning Model that can detect COVID-19 cases through chest X-rays, allowing for early detection and prevention of its spread at a low cost and faster pace.

Chapter 12, titled "Air Pollution, Its Effects on the Habitat of Insects, and Strategic Control to Improve Productivity and the Environment in the Future," discusses how urbanization has led to the release of harmful substances and gases into the air, causing air pollution that has negative effects on insect herbivores and their demographics. The careless use of wide-spectrum pesticides has led to pest outbreaks and the formation of new pest biotypes, affecting important pollinators and foragers. Certain pollutants in the environment, such as sulphur dioxide, oxides of nitrogen, fluorides, and heavy metals, increase the efficiency of phloem-feeding homopteran insects like aphids. Pollution has also led to industrial melanism in moths, caused by a decline in light-colored lichens. However, high toxicant concentrations reduce the susceptibility of parasitoids and predators. Air pollution has also accelerated global warming, leading to the growth of invading alien pest species and tropical disease vectors. The article also explores various methods to reduce air pollution, such as the fabric filtration method, electrostatic precipitation strategies, thermal incineration, catalytic oxidation, and biofilters. Successful reduction of air pollution can help restore the health of agricultural plants and lessen food shortages.

Chapter 13, titled "Automation of Street Light of Adarsh Gram With Manual Intervention and Luminosity Control With Traffic Flow," presents the Smart Village (Adarsh Gram) initiative, based on Mahatma Gandhi's concept of comprehensive rural development, aims to transform villages in India into model villages. The initiative is driven by local governments, and the main objective is to work on social and cultural development to create awareness among people in villages and transform them into models under the Sansad Adarsh Gram Yojana guidelines. The initiative covers various spheres like Personality, Sanitation, Plantation, Water supply, Electrification, Internal roads, Health, Financial, Human, Social, and Economic development. The initiative has been widely accepted and recognized across the country, with various local bodies, countries, global organizations, business tycoons, celebrities, and others coming forward to adopt villages and promote sustainable development. One of the challenges faced by villages is the non-availability of street lights, which is particularly difficult for women and children during the night. To overcome this challenge, the initiative proposes the electrification of street lights using simple, cheap, durable, and easily repairable components, with minimal power consumption, and without manual intervention. In the absence of electricity, solar-powered panels can be used as an alternative.

Chapter 14, titled "Advancement Detection: Fungi-Bacterial Diseases in Plants Using Modified Deep Neural Network and DSURF," presents the current scenario of climate change is affecting farmers as their crops are getting infected by fungi and bacteria due to sudden changes in soil temperature. However, with advanced technology, diseases can be detected early to prevent crop spoilage. This manuscript proposes machine learning techniques and modified algorithms to identify and classify plant diseases using Deep Neural Network (DNN) models and algorithms. The proposed techniques improve object accuracy and reduce computational complexity through dynamic feature extraction and classifier combinations using Clustering, Deterministic Probability, and AI-based strategies for better accuracy and robustness.

Chapter 15, titled "Optimal Control Strategies for Mitigating Frequency Deviation in Microgrid-Based Smart Village," focuses on addressing frequency deviation issues in microgrid-based smart village systems due to changing loads and intermittent renewable energy sources. The paper proposes optimal control strategies, including conventional PI, PI plus fuzzy, and PI plus ANFIS, and evaluates their effectiveness using simulation models. Results show that the PI plus ANFIS strategy provides the best results, offering valuable insights for designing reliable and efficient microgrid-based smart village systems while improving system performance.

Chapter 16, titled "Solar-Powered Pesticide-Spraying RFID Robot," describes the development of a prototype of a solar-powered pesticide spraying robot, designed for use in different types of fruits and vegetable fields. The robot is equipped with DC motors, a DC pump, a solar panel, and an ultrasonic sensor for navigation. RFID technology is used to automate the robot's navigation, while the ultrasonic sensor detects obstacles in the path. The robot is powered by a 30-watt solar panel, which charges the batteries used to drive the motors and pump. The robot's wheels are powered by the motors, while the pump sprays pesticides through the nozzles, with the height adjustable depending on the size of the crop. The robot has been designed to spray pesticides automatically, reducing the health hazards of farmers and controlling pests while improving productivity.

Mohammad Ayoub Khan
University of Bisha, Saudi Arabia

Bhumika Gupta
Govind Ballabh Pant Institute of Engineering and Technology, India

Agya Ram Verma
Govind Ballabh Pant Institute of Engineering and Technology, India

Pushkar Praveen
Govind Ballabh Pant Institute of Engineering and Technology, India

Cathryn J. Peoples
Ulster University, UK

Chapter 1
Smart Village Infrastructure and Rural Communities

Pushkar Praveen
Govind Ballabh Pant Institute of Engineering and Technology, India

Ayoub Khan
University of Bisha, Saudi Arabia

Agya Ram Verma
Govind Ballabh Pant Institute of Engineering and Technology, India

Manoj Kumar
Govind Ballabh Pant Institute of Engineering and Technology, India

Cathryn Peoples
University of Ulster, UK

ABSTRACT

Rural communities around the world face significant challenges in accessing basic services such as healthcare, education, and sanitation, as well as in promoting economic growth and environmental sustainability. In short, smart village infrastructure has emerged as a potential solution to the challenges faced by rural communities around the world. By leveraging technology and innovation, smart village infrastructure has the potential to increase access to basic services, promote economic growth and environmental sustainability, and improve the overall quality of life in rural communities. However, addressing the challenges of funding, skilled labor, and community engagement requires a holistic approach that involves collaboration between multiple stakeholders. The policy recommendations provided in this paper can help promote smart village infrastructure and sustainable rural communities.

DOI: 10.4018/978-1-6684-6418-2.ch001

INTRODUCTION

The past few years have witnessed an upsurge in interest in the creation of smart village infrastructure and the promotion of sustainable rural communities. The reason for this is that rural areas face a plethora of challenges, including inadequate access to basic services, lack of employment opportunities, and environmental deterioration. Smart villages are a response to these difficulties, aiming to use technology and innovation to enhance the quality of life of rural communities. This article delves into the concept of smart village infrastructure and its part in advancing sustainable rural communities.

Smart village infrastructure pertains to the implementation of innovative solutions that tackle the problems faced by rural areas. Its goal is to enhance access to fundamental services such as education, healthcare, and clean water, as well as promote economic development and environmental sustainability. Digital technologies, such as internet connectivity, mobile devices, and renewable energy sources, are the typical foundations of smart village infrastructure. Renewable energy sources such as solar, wind, and hydro power are essential components of this infrastructure. By providing electricity to rural areas, they enhance the quality of life for rural communities and decrease dependence on fossil fuels, which contribute significantly to climate change. Another critical component is the use of digital technologies to enhance access to fundamental services such as healthcare, education, and financial services through telemedicine, e-learning platforms, and mobile applications.

Sustainable rural communities are inextricably linked to the creation of smart village infrastructure. These communities can satisfy the needs of the present generation without jeopardizing the ability of future generations to satisfy their needs. This necessitates a focus on environmental sustainability, economic development, and social equity. A primary issue encountered by rural communities is the shortage of economic opportunities. Inclusive and environmentally sustainable economic development is required for sustainable rural communities. This includes supporting local businesses, promoting sustainable agriculture, and using renewable energy sources. Sustainable rural communities must prioritize environmental sustainability. This involves protecting natural resources such as forests and water sources, as well as decreasing greenhouse gas emissions by utilizing renewable energy sources. Finally, social equity is a fundamental element of sustainable rural communities, ensuring that all members of the community have access to basic services such as healthcare, education, and clean water, regardless of their socio-economic status.

The establishment of smart village infrastructure is essential for promoting sustainable rural communities. By leveraging technology and innovation, we can enhance access to basic services, promote economic development, and protect the environment. Sustainable rural communities must prioritize inclusive and environmentally sustainable economic development, safeguard natural resources, and guarantee social equity. As the world confronts issues such as climate change and rural-urban migration, the concept of smart villages will continue to play an increasingly vital role in promoting sustainable rural development.

According to Amit Degada et al. (2021), almost half of the global population, around 46%, lives in rural areas. In order to foster economic development, enhance agriculture, and improve healthcare and education, smart villages have become increasingly necessary alongside smart cities. The concept of a smart village involves the use of digital technology to modernize traditional rural practices. This digital transformation is built on a foundation of heterogeneous digital technologies, with the Internet of Things (IoT) serving as a central pillar. As a result, there are numerous research opportunities to create a secure, cost-effective, and efficient technical infrastructure for smart villages which is shown in Figure 1. This article highlights some of the key areas where IoT can be applied in the context of smart villages.

Figure 1. Smart village
Source: Degada (2021)

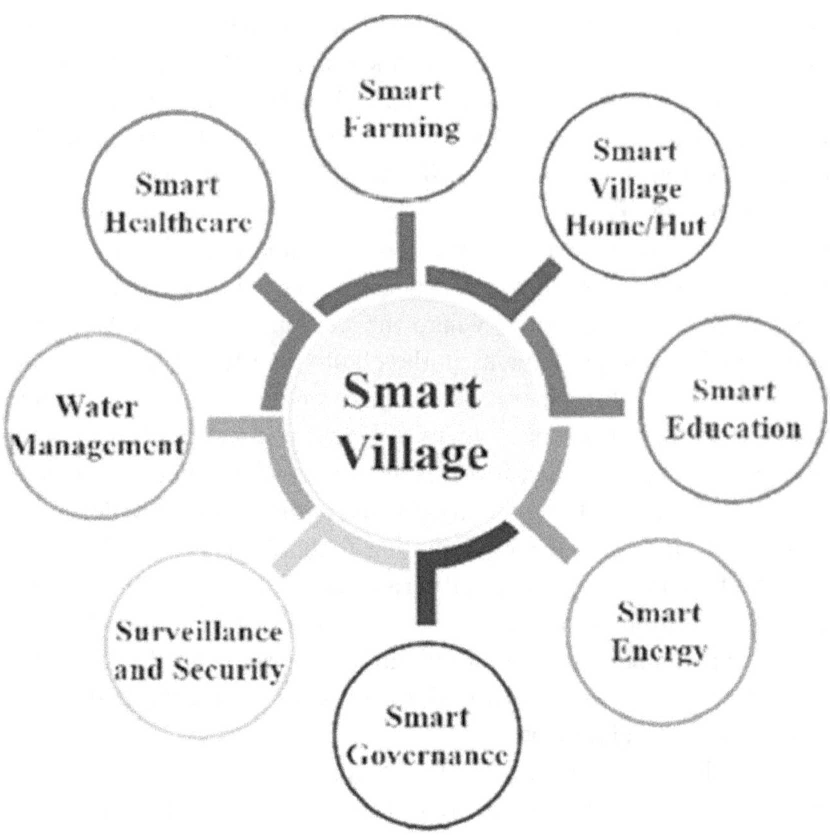

Ajay Yadav et al. (2022) discuss the problems faced by rural villages and proposes hybrid solutions based on the Internet of Things (IoT) to address these issues. The authors have done an excellent job of highlighting the key challenges faced by rural communities, such as lack of access to basic amenities, poor infrastructure, and limited healthcare facilities. They argue that these issues can be addressed through the deployment of IoT-based solutions, which can provide real-time monitoring and data analytics to improve service delivery and infrastructure management. Author presents a comprehensive analysis of various IoT-based solutions that can be implemented to address these challenges. The authors propose a hybrid solution that combines several IoT technologies, including wireless sensors, cloud computing, and machine learning algorithms, to create a robust system that can provide real-time monitoring and analysis of various parameters. The authors also present a case study of a smart village project that has been implemented in India, highlighting the benefits of using IoT-based solutions to address the challenges faced by rural communities. The case study provides valuable insights into the practical implementation of IoT-based solutions and the challenges that need to be addressed to ensure their successful deployment.

The authors provide a thorough discussion on the concept of Village 5.0, which is a modern approach to rural development that leverages advanced technologies to provide better infrastructure, access to education, healthcare, and other essential services to the rural population (Singh et al., 2022). The paper presents an in-depth analysis of the enabling technologies that are critical to the success of Village 5.0, including the Internet of Things (IoT), artificial intelligence (AI), blockchain, and cloud computing. The paper goes on to describe the various applications of these enabling technologies in the development of smart villages, including smart agriculture, smart healthcare, and smart education. The authors present case studies from different parts of the world, where the implementation of Village 5.0 has resulted in significant improvements in the quality of life of the rural population. One of the strengths of this paper is its emphasis on the importance of community engagement in the development of smart villages. The authors highlight the need for a collaborative approach that involves all stakeholders, including government, private sector, and the rural community, to ensure the sustainable development of smart villages. Overall, "Village 5.0: Enabling Technologies and its Applications in Development of Smart Village" is a well-written and informative paper that provides valuable insights into the role of enabling technologies in rural development. The authors present a compelling case for the implementation of Village 5.0, and their analysis and case studies serve as a useful reference for researchers and practitioners in the field of rural development.

Muhammad Ilham Alhari et al. (2013) focuses on developing an IT strategic transformation framework for the Smart Village Concept in Indonesia. The authors highlight the need for technology-enabled solutions to address the challenges faced by Indonesian villages and provide a comprehensive overview of the Smart Village Concept. The paper presents a detailed analysis of the current state of the Smart Village Concept and proposes a framework for IT strategic transformation to achieve the desired outcomes. The authors also discuss the importance of community participation in the process and emphasize the need for capacity building and training programs to ensure successful implementation. One of the strengths of this paper is the extensive literature review conducted by the authors, which provides a solid foundation for the proposed framework. The study also uses a mixed-method approach, combining qualitative and quantitative research methods to obtain a holistic understanding of the problem. Overall, the authors have successfully presented a well-researched and well-structured paper that provides valuable insights into the development of IT strategic transformation for the Smart Village Concept in Indonesia. The paper is relevant to researchers, policymakers, and practitioners in the field of information technology and community development.

Susy Ella et al. (2019) presented the importance of information and communication technology (ICT) in the development of smart villages in Indonesia. The authors begin by providing a brief introduction to the concept of smart villages and their potential to revolutionize rural development. They go on to explain the benefits of using ICT in building smart villages, such as improving access to education, healthcare, and market information. The authors also provide examples of successful smart village initiatives in Indonesia that have utilized ICT to great effect. One of the strengths of this article is the detailed discussion of the challenges that must be overcome to implement smart village initiatives. The authors note that lack of infrastructure, low digital literacy rates, and limited funding can all hinder the success of such initiatives. However, they also provide strategies for addressing these challenges, such as public-private partnerships and the development of digital literacy programs. Overall, "Utilization of ICT in Building a Smart Village Model for Village Development in Indonesia" is a valuable contribution to the literature on smart village development. The authors provide a thoughtful analysis of the benefits and challenges of using ICT in this context and offer practical recommendations for policymakers and practitioners. The article is also well-written and structured, making it accessible to a wide audience.

Why Smart Village

Smart village design models aim to create sustainable, livable, and technologically advanced rural communities. These models incorporate various aspects such as smart infrastructure, renewable energy, and digital connectivity to create a self-sufficient and prosperous community.

Here are some key elements of a smart village design model:

- **Smart Infrastructure:** The village should have access to smart infrastructure, including efficient transportation systems, modern healthcare facilities, and reliable water and sanitation services.
- **Renewable Energy:** The village should be powered by renewable energy sources like solar, wind, or hydro power. This will reduce dependence on fossil fuels and increase energy efficiency.
- **Digital Connectivity:** High-speed internet access and other digital technologies can connect the village with the outside world and provide access to information, education, and job opportunities.
- **Sustainable Agriculture:** Smart village design models prioritize sustainable agriculture practices that reduce the environmental impact of farming while increasing productivity and profitability.
- **Waste Management:** Proper waste management practices can help keep the village clean and healthy while reducing the environmental impact of waste.
- **Education and Training:** Education and training programs can help build skills and capacity within the community, enabling them to take advantage of new technologies and opportunities.
- **Access to Technology:** Technology is used to enhance various aspects of life in the village, such as healthcare, education, agriculture, and communication. For example, telemedicine can be used to provide healthcare services to remote areas, while digital platforms can be used to facilitate online learning and training.
- **Community Participation:** Community participation is key to the success of any smart village design model. Local residents should be involved in the planning and implementation of initiatives to ensure that they meet the needs of the community.
- **Entrepreneurship:** Smart villages encourage entrepreneurship and support the development of local businesses. This helps to create jobs and promote economic development in the community.

Examples of Smart Village

A smart village is a rural area that uses innovative technology to improve the quality of life for its residents while preserving its natural resources.

1. **Piplantri, Rajasthan, India:** In Piplantri, the villagers have implemented a unique initiative to promote the conservation of the environment and empower their female population. For every girl child born in the village, the community

plants 111 trees in her name. This initiative has led to the planting of over 350,000 trees in the village, which has not only helped to mitigate climate change but has also created a sustainable source of income for the villagers through the sale of wood and other forest products. In addition to this, Piplantri has implemented a range of other sustainable initiatives, including the use of solar power to generate electricity, the construction of eco-friendly homes made from locally sourced materials, and the creation of a community garden to promote organic farming (Nair, 2017).

Table 1. Smart village goal

Traditional Village	Smart Village	Issues Addressed
Agriculture	Smart Farming Smart Irrigation Smart dairy Smart agriculture waste management	• Keeping track of climate conditions and predicting future changes • Implementing an irrigation and fertilizer management system • Monitoring the health of crops and using predictive analysis to prevent issues • Tracking the health and performance of individual animals and analyzing the collective herd • Forecasting agriculture production and meeting market demand • Reducing agricultural waste through effective recycling and waste management (Degada et al., 2021)
Public Activity	Smart Education Cultural activity Security Smart Healthcare	• Accessibility, personalization and engagement to all students • Preservation of cultural heritage • Cyber security, physical security and emergency response • Healthcare accessible and personalize to each patients.
House hold Resources	Smart Huts Smart Food Management Smart Energy & Power distribution Smart drinking water management Smart domestic waste management	• Smart huts having all the sensor based facility • Smart storage and distribution of food product • Supply of drinking water to everywhere in the village • Reduce and recycle the domestic waste effectively
Connectivity	Smart Transportation Digital Connectivity	• Accessible and convenient connectivity to other parts of the country • Digitally connectivity and internet service available to each and everyone
Business	Entrepreneurship Smart agriculture market Online product delivery	• Focus on self employment and entrepreneurship • Market or online platform for agriculture product selling • Cloud based product delivery service for quick and fast delivery of agriculture product

2. **Punsari, India:** Punsari, a quaint village in India, has undergone a remarkable transformation thanks to technology. This small community has been elevated into a “smart village” with the help of various technological advancements. Among these advancements include free Wi-Fi, CCTV cameras for added security, a telemedicine center, a digital school, and a water ATM. The integration of these technological innovations has helped to modernize the village and make it more efficient and sustainable. Punsari now serves as a prime example of how technology can positively impact rural communities and enhance the quality of life for its residents (Joshi, 2019).
3. **Nyamata, Rwanda:** Nyamata is a rural village in Rwanda that has implemented a solar-powered irrigation system, allowing farmers to grow crops year-round. Nyamata is a rural village located in Rwanda that has implemented several sustainable and innovative initiatives to improve the quality of life for its residents. One of these initiatives is a solar-powered irrigation system that enables farmers to grow crops all year round, increasing food security and reducing the reliance on rain-fed agriculture. The village has also installed a water purification system that provides clean drinking water to its inhabitants, which has reduced the incidence of waterborne diseases. Additionally, Nyamata has established a digital education center that provides access to educational resources and promotes digital literacy. This center has been especially helpful for children who may not have access to traditional educational resources. Furthermore, the village has a health clinic that provides medical care to its residents. The clinic is staffed with trained medical personnel and has basic medical equipment to diagnose and treat common ailments. This has greatly improved the health and well-being of the village’s residents. Overall, Nyamata’s sustainable initiatives have transformed the village into a thriving community that is self-sufficient and prosperous. The village’s success serves as an inspiration to other communities that are looking to improve their living conditions while protecting the environment (Morgenstern, 2019).
4. **Kamikatsu, Japan:** Kamikatsu is a small town located in Japan, known for its unique waste management system. The town has set a goal to achieve zero waste by 2020 and has been actively implementing practices to achieve this. The town’s waste management system involves separating waste into 45 different categories, such as paper, plastic, metal, and glass. Residents are responsible for properly sorting their waste into the appropriate category, and the town has designated collection areas for each type of waste. Kamikatsu also has a recycling center that allows residents to drop off items that are not collected curbside, such as electronics and furniture. The center is staffed by volunteers who help sort the items and ensure they are recycled properly. The town has become a popular tourist destination for those interested in sustainable living

and waste reduction. Visitors can learn about the town's waste management practices and participate in workshops to learn how to reduce their own waste. Kamikatsu's unique waste management system has gained international recognition, and the town has received numerous awards for its efforts towards sustainability (Morgenstern, 2019).

5. **Santander, Spain:** Santander, Spain is an example of a smart village, which aims to use technology to improve the lives of its residents. In Santander, there are a number of initiatives that have been implemented to create a more sustainable and efficient community. One such initiative is the use of sensors throughout the village, which collect data on a variety of factors such as air quality, noise levels, and traffic flow. This data is then analyzed to identify areas where improvements can be made, such as reducing congestion or improving pedestrian safety. Another initiative is the creation of a "smart park" which uses renewable energy sources to power its lighting and irrigation systems. The park also features a bike sharing program, which encourages residents to use sustainable modes of transportation. Santander also offers a variety of digital services to its residents, such as a mobile app that allows them to access information on public transportation, parking availability, and local events. This app also includes a feature that allows residents to report problems such as potholes or broken streetlights, making it easier for the village to address these issues. Overall, the smart village of Santander demonstrates the potential of technology to create more sustainable and efficient communities, while improving the quality of life for residents (Evers, 2022).
6. **Ljubljana, Slovenia:** Ljubljana is the capital city of Slovenia and has implemented a number of smart city initiatives. The city has installed a smart parking system, which allows drivers to find available parking spots using a mobile app. It has also implemented a smart waste management system, which uses sensors to monitor waste levels in garbage bins and optimizes waste collection routes (Zavratnik et al., 2019).

Scope of Smart Village and Sustainable Rural Communities Worldwide

The concept of smart villages and sustainable rural communities has gained global attention as a means of addressing the challenges of rural development through technology and innovation. Smart villages aim to promote sustainable rural development by leveraging the benefits of digital technologies, while empowering local communities to achieve their goals. In many parts of the world, smart villages can have a significant impact on agriculture and rural livelihoods, healthcare, education, infrastructure, and governance. By using digital technologies, farmers and rural communities can access information on weather patterns, crop yields, and market prices, which can help improve their agricultural practices and increase their income. Smart villages can also improve access to healthcare by providing telemedicine and other digital health services to remote and rural areas.

In terms of education, digital technologies can be used to provide access to quality education in rural areas, through online learning platforms and other innovative solutions. Additionally, smart villages can improve infrastructure in rural areas, such as through the use of renewable energy sources, smart water management, and waste management solutions. Moreover, the use of digital technologies can improve governance in rural areas, by providing access to information, enabling citizen participation, and improving transparency and accountability. However, there are also challenges to implementing smart villages and sustainable rural communities, such as infrastructure limitations and the need for community engagement and capacity building.

Overall, the scope of smart villages and sustainable rural communities is vast and has the potential to transform rural areas worldwide by providing innovative solutions to address their development challenges. It requires a multi-stakeholder approach involving governments, private sector, and local communities to ensure their success and sustainability.

Role of IoT and Machine Learning in Development of Smart Village

The combination of the Internet of Things (IoT) and machine learning (ML) has enormous potential to transform various industries, such as healthcare, agriculture, energy, and transportation, leading to a more sustainable and efficient society, particularly in rural areas. This concept has given rise to the development of "smart villages," which leverage these technologies to improve the standard of living and economic prospects of rural communities. IoT in smart villages can offer several benefits, such as remote monitoring and control of various systems. For example, placing sensors on farms to measure soil moisture, temperature, and other variables, and analyzing the data using ML algorithms can optimize irrigation and fertilization schedules, resulting in higher crop yields and reduced water consumption. In the same way, smart grids can help manage energy consumption and distribution, reducing waste and lowering costs. The integration of wearable devices and telemedicine can improve healthcare in rural areas, particularly for early detection and prevention of serious illnesses. Additionally, smart traffic management systems can optimize traffic flow, reducing congestion and travel times. Moreover, online learning platforms and e-commerce platforms can provide educational resources and economic opportunities to rural areas. However, the implementation of IoT and ML in smart villages faces several challenges such as lack of infrastructure, including reliable internet connectivity and power supply, and cultural barriers. Therefore, governments, NGOs, and private organizations must make a concerted effort to invest in infrastructure development and provide education and training to rural communities. Ultimately, leveraging IoT and ML can revolutionize rural communities' operations and provide significant benefits to smart villages, leading to a more sustainable and efficient society.

Nina Cvar et al. (2020) discusses the implementation of the Internet of Things (IoT) technology in smart cities and smart villages as shown in Figure 2. The paper explores the similarities and differences between the two concepts and their future prospects.

Figure 2. Internet of things (IoT) architecture components
Source: Cvar et al. (2020)

The authors note that smart cities and smart villages share common goals of improving the quality of life for citizens, reducing resource consumption, and increasing efficiency. However, they differ in their scale, population density, and technological infrastructure. Smart cities typically have larger populations and more advanced technological infrastructure, while smart villages are more rural and may have limited technological resources. The paper discusses the potential benefits of implementing IoT technology in both smart cities and smart villages, such as improved public services, enhanced sustainability, and increased economic development. However, the authors also note that there are challenges to implementing IoT technology in these contexts, such as data privacy and security concerns, infrastructure limitations, and unequal access to technology.

Overall, the paper provides a comprehensive overview of the use of IoT technology in smart cities and smart villages, highlighting both the similarities and differences between the two concepts. The authors suggest that further research is needed to address the challenges associated with implementing IoT technology in these contexts and to fully realize the potential benefits for citizens.

Smart Village Design: Challenges and Opportunities

Smart village refers to a rural area that has implemented advanced technologies to improve the quality of life of its inhabitants. While the idea of a smart village is promising, there are several challenges that make its implementation difficult. Firstly, there is a lack of basic infrastructure in many rural areas, such as electricity, water supply, and transportation. Without these basic amenities, it is difficult to implement advanced technologies and create a smart village. Providing basic infrastructure in rural areas requires significant investment and planning. Secondly, there is a lack of awareness and understanding of technology in rural communities. Many people in rural areas have limited exposure to technology and may not understand the benefits that it can bring. As a result, there may be resistance to implementing new technologies or difficulty in adopting them. Thirdly, there is a lack of skilled manpower to implement and maintain the technologies required for a smart village. Rural areas often have limited access to education and training, which makes it difficult to find skilled workers to implement and maintain the necessary technologies. Finally, there is a lack of funding for implementing and maintaining smart village technologies. While there are many potential benefits to implementing smart village technologies, the costs associated with these projects can be high. Finding funding sources and maintaining financial sustainability can be a major challenge.

Challenges for Smart Village Design

- **Lack of Infrastructure:** Rural areas often have limited access to basic infrastructure such as roads, electricity, and water supply. This can make it challenging to implement Smart Village Design as it requires the deployment of various technologies that depend on reliable infrastructure.

- **Limited Resources:** Rural communities often have limited financial resources, which can make it difficult to invest in expensive technologies. This can limit the scope of Smart Village Design projects and make it challenging to achieve their full potential.
- **Digital Divide:** Access to digital technologies is often limited in rural areas. This can make it challenging to implement Smart Village Design, which relies heavily on digital solutions such as sensors, data analytics, and remote monitoring.
- **Cultural Differences:** Rural communities have unique cultural identities that may not align with the goals of Smart Village Design. This can make it challenging to introduce new technologies and solutions without disrupting local culture and traditions.

Opportunities in Smart Village

- **Sustainable Development:** Smart Village Design provides an opportunity to promote sustainable development in rural areas. It involves the use of renewable energy sources, waste management solutions, and eco-friendly practices, which can help preserve the environment and promote long-term development.
- **Inclusive Development:** Smart Village Design aims to promote inclusive development, which involves ensuring that all members of the community benefit from development initiatives. This can help reduce inequalities and promote social cohesion.
- **Digital Empowerment:** Smart Village Design provides an opportunity to bridge the digital divide in rural areas. By introducing digital technologies and solutions, rural communities can gain access to new opportunities and improve their livelihoods.
- **Preserving Local Culture:** Smart Village Design can be designed to preserve local culture and traditions. By involving the community in the design process, Smart Village projects can be tailored to fit the unique needs and preferences of each community.

Smart Village Design presents both challenges and opportunities. Overcoming the challenges requires a holistic approach that involves addressing the underlying issues of limited infrastructure, limited resources, digital divide, and cultural differences. At the same time, Smart Village Design provides opportunities to promote sustainable and inclusive development, bridge the digital divide, and preserve local culture. To ensure the success of Smart Village Design, it is essential to involve the community in the design process and ensure that their unique needs and preferences are taken into account.

CONCLUSION

In recent years, there has been an increasing focus on smart villages as an important area of research. This chapter provides an overview of the different components that make up a smart village, including smart infrastructure, smart farming, smart transportation, smart energy, smart healthcare, smart technology, smart governance, smart education, and smart citizens. It is important to note that the level of implementation of these components can vary depending on the vision of the specific smart village. However, there are many challenges associated with building smart villages, including the complexity and diversity of the issues involved. One important factor is the need to keep design costs low in order to make smart rural development economically feasible. This will also help to promote sustainability and reduce the burden on the country as a whole. Machine learning and data analytics are also crucial technologies that must be considered in conjunction with other strategies for building smart villages.

REFERENCES

Alhari, M. I. (2022). Developing IT Strategic Transformation of Smart Village Concept for Indonesian Village Model. *2022 International Conference Advancement in Data Science, E-learning and Information Systems (ICADEIS)*. IEEE. 10.1109/ICADEIS56544.2022.10037570

Cvar, N. (2020). *The Use of IoT Technology in Smart Cities and Smart Villages: Similarities, Differences, and Future Prospects*. MDPI. www.mdpi.com/journal/sensors

Degada, A., Thapliyal, H., & Mohanty, S. P. (2021). Smart Village: An IoT Based Digital Transformation In *2021 IEEE 7th World Forum on Internet of Things (WF-IoT)*. IEEE. 10.1109/WF-IoT51360.2021.9594980

Ella, S., & Andari, R. N. (2019). Utilization of ICT in Building a Smart Village Model for Village Development in Indonesia. *In 2019 International Conference on ICT for Smart Society (ICISS)*. IEEE. 10.1109/ICISS48059.2019.8969820

Evers, M. (2022). Urban Planning Goes Digital in Spanish 'Smart City'. *Spiegel*. https://www.spiegel.de/international/world/santander-a-digital-smart-city-prototype-in-spain-a-888480.html

Joshi, S. R. (2019). Emerging Model Villages in India: A Study of Punsari Village from the State of Gujarat (India). *Croatian and Comparative Public Administration, 19*(2). doi:10.31297/hkju.19.2.3

Morgenstern, J. (2019). Kamikatsu: The Zero Waste Town That Turned Its Whole Economy Sustainable. *Forbes*. https://www.forbes.com/sites/juliemorgenstern/2019/06/27/kamikatsu-the-zero-waste-town-that-turned-its-whole-economy-sustainable/?sh=4b4d31743794

Nair, S. M. (2017). Piplantri: The Village That Celebrates Girl Child with 111 Trees. *The Indian Express*. https://indianexpress.com/article/trending/trending-in-india/piplantri-the-village-that-celebrates-girl-child-with-111-trees-4928185/

Pruet, P., Ang, C. S., Farzin, D., & Chaiwut, N. (2015). Exploring the internet of "educational things" (IOET) in rural underprivileged areas. In *12th International Conference on Electrical Engineering/Electronics, Computer, Telecommunications and Information Technology (ECTI-CON)*. IEEE.

Singh, K. B., Sengar, N., Das, D., & Misra, S. C. (2022). Village 5.0: Enabling Technologies and its Applications in Development of Smart Village. In *2022 IEEE International Symposium on Smart Electronic Systems (iSES)* (pp.556-561). IEEE. 10.1109/iSES54909.2022.00122

Yadav, A., Rashmi, A. V., Goyal, V., & Mukherjee, R. (2022). Smart Village: Problems and Hybrid Solutions from IoT Perspective. In *2022 2nd International Conference on Innovative Sustainable Computational Technologies (CISCT)*, (pp.1-5). IEEE. 10.1109/CISCT55310.2022.10046527

Zavratnik, V., Kos, A., & Duh, E. S. (2019). Smart Villages in Slovenia: Examples of Good Pilot Practices, In Smart Villages in the EU and Beyond (Emerald Studies in Politics and Technology). Emerald Publishing Limited. doi:10.1108/978-1-78769-845-120191010

Chapter 2
Artificial Intelligence and Machine Learning-Based Agriculture

Sakshi Rajput
Maharaja Surajmal Institute of Technology, Guru Gobind Singh Indraprastha University, India

Lititia Khanna
Maharaja Surajmal Institute of Technology, Guru Gobind Singh Indraprastha University, India

Priti Kumari
National Institute of Technology, Patna, India

ABSTRACT

According to the Food and Agriculture Organization of the United Nations, by 2050, there will be an additional nine billion people living on the planet, yet there will only be an additional 4% of land under cultivation. In such a situation, more effective agricultural practices can be achieved using recent technical developments and fixes for the industry's current bottlenecks. Direct use of artificial intelligence and machine learning in the agricultural industry could represent a paradigm shift in the way farming is now carried out. A farmer may accomplish more with fewer resources using AI-powered farming solutions, which also improve quality and ensure speedy GTM (go-to-market) strategies for crops. This chapter offers a perspective of how AI might fuel the various agricultural sectors. Additionally, it looks into future AI-powered concepts and potential challenges.

DOI: 10.4018/978-1-6684-6418-2.ch002

INTRODUCTION

Artificial intelligence (AI) is based upon the concept that human intelligence can be described in such a way that machines can easily emulate and perform both simple and complex tasks. Artificial intelligence goals include learning, recognition, logical thinking and reasoning. Specific applications of artificial intelligence include natural language processing, expert systems, computer vision and speech recognition. Some examples are visual recognition systems in self-driving vehicles, recommendation systems that suggests products you may like based upon past purchases, language and voice recognition for Siri, virtual assistant in Apple iPhone.

Machine learning (ML) deals with the study of various different computer algorithms that are automatically improved using experience and data. It is a subset of artificial intelligence that allows software applications to predict outcomes more accurately without explicit planning. ML algorithms use previous data as input values to predict the output. A computer program is said to learn if its performance P improves with experience E on a given task T. E.g., Task (T) is teaching alphabets to a kid. Show the kid the many images of different alphabets and then pronounce those alphabets. The training of a kid is his/her experience (E). The training data has two parts: main data (images of the alphabet) and supervisory information (speak the alphabet). After some time, the kid gains experience and finally after n times of training evaluation is done for the kid. During evaluation, simply show the kid only the image of the alphabet and the kid is asked to pronounce it. As the kid speaks, compare the two and then calculate the performance (P).

Figure 1. Artificial Intelligence subsets

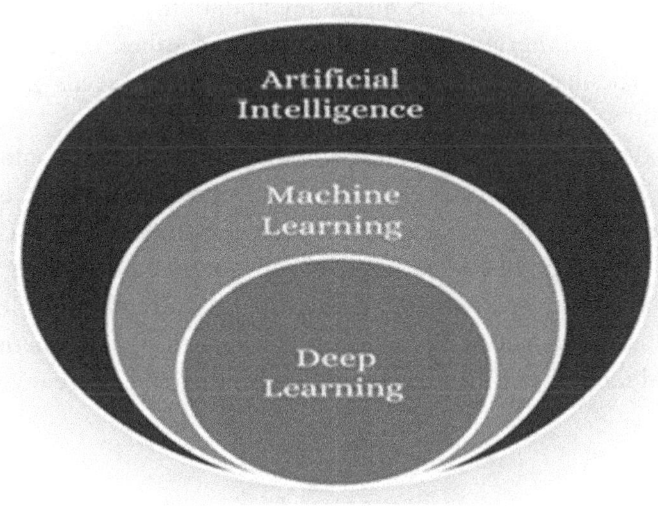

Farming and Agriculture are one of the most important and oldest occupations in the world. It plays a major role in increasing the economy. Agriculture is a $5 trillion industry worldwide. With the world's population projected to exceed 9 billion by 2050, there is need to increase agricultural production by 70 percent to meet the requirements. As the world's population grows, land, water and other resources will not be sufficient to continue the supply and demand chain. Maximising productivity therefore requires smarter approaches and more efficient farming. Therefore, in order to meet these needs and create great employment opportunities for many people in this sector, some new automated methods are introduced and being developed.

Artificial Intelligence has become one of the most essential technologies in everything from robotics, healthcare, agriculture, banks, and more. In agricultural sector, it plays a very important role in modifying the agricultural industry. Artificial Intelligence protects the agricultural sector from a variety of factors, including population growth, climate change, food security and agricultural sector employment issues. Nowadays farming system has reached another level with Artificial Intelligence and Machine Learning. Artificial Intelligence has upgraded real time monitoring, crop production, processing, harvesting and advertising. Various computer-based high-tech systems have been developed to find out various key parameters such as yield detection, weed detection and harvest quality.

CHALLENGES FACED DURING AGRICULTURE USING TRADITIONAL METHODS

The challenges faced during agriculture using the traditional methods are listed below:

1. Various meteorological factors such as precipitation, humidity and temperature play a major role in agriculture. Pollution can change the climate rapidly, making it really difficult for farmers to properly harvest, sow seeds and prepare soil for farming.
2. For a good and healthy harvest, the soil must be fertile and contain necessary nutrients such as nitrogen, phosphorus and potassium. Poor quality crops result, if essential nutrients are ineffectively present in the soil. However, this soil quality is difficult to determine by conventional methods.
3. Crops need to be protected from weeds in the agricultural life cycle. Otherwise, it can increase production costs and also absorb nutrients from the soil. Thus, traditional methods of identifying and preventing weeds are not efficient.

PROCESS OF AGRICULTURE

The levels involved in lifecycle of agriculture are listed below:

- Soil preparation:
  Farmers prepare the soil for sowing in the early stages of farming. During this process, large clods are broken up and debris such as stones, sticks and roots are removed. Also, the addition of fertilisers and organic matter varies according to the type of cultivation, in order to create ideal growing conditions for plants.

- Sowing seeds:
  At this time, attention should be paid to the distance between the 2 seeds and the depth at which the seeds are planted. Meteorological conditions such as humidity, precipitation and temperature play a critical role in this stage.

- Adding Fertiliser:
  Maintaining soil fertility is an essential factor in keeping farmers growing nutritious and healthy plants. Farmers rely on fertilisers as these substances contains phytonutrients such as nitrogen (N), potassium (K) and phosphorus (P). Fertilisers are planted nutrients sprayed in agricultural fields to add-on the necessary elements naturally present in the soil. This step also determines the harvest quality.

- Irrigation:
  This step helps in dampening of soil and maintaining moisture levels. Overwatering or underwatering can hinder crop growth and damage crops if not done properly.

- Weed Control:
  Weeds are the unwanted plants that grows around the harvest or at boundaries of the vegetation. Weed control is one of the most critical step as weeds affect yields, increase production costs, reduce crop quality and inhibit harvest.

- Harvesting:
  It is a method of collecting ripened crops from the field. This activity is labour intensive as it requires a lot of manpower. This step involves post-harvest handling that includes washing, cooling, sorting and storing.

- Storage:
  In this stage of the post-harvest, produce is stored in a manner that ensures food security other than the farming season. This also includes packing and transporting the harvest. (Pravar Jain, Analytics Vidhya, 2021)

Figure 2. Process of agriculture

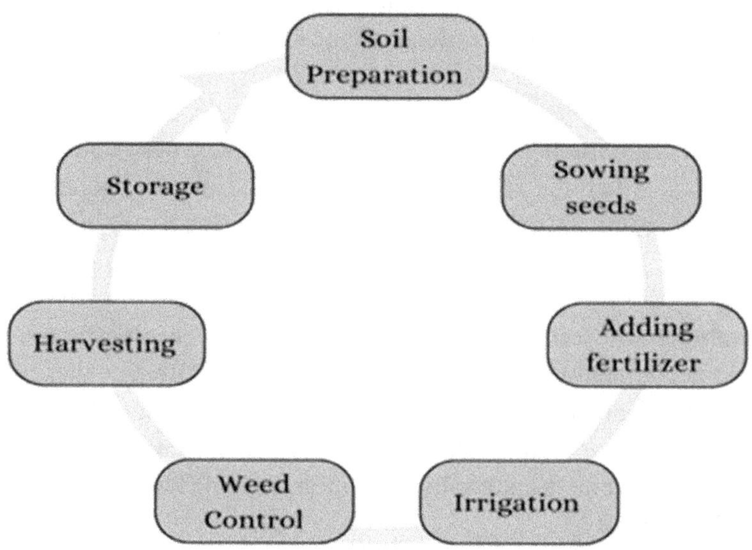

APPLICATIONS OF ARTIFICIAL INTELLIGENCE IN AGRICULTURE

Using traditional farming, there are numerous challenges faced by farmers. Artificial Intelligence is widely used in this field to solve these challenges. Industries enable healthier crops, monitoring soil and crop growth conditions, controlling pests, organising farmer's data, assisting workloads, and performing a variety of agricultural chores throughout the food supply chain using Machine Learning and Artificial Intelligence.

Crop Health and Soil Monitoring System

Soil type and soil nutrients play an essential role in the type of crops grown and the crop quality. Increase in pollution and deforestation is deteriorating the soil and crop quality, which is difficult to determine.

Figure 3. Artificial Intelligence in agriculture

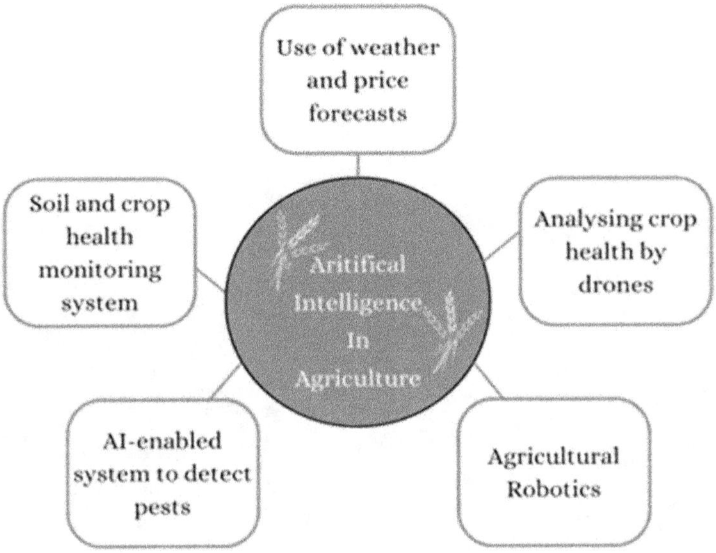

Germany-based technology start-up PEAT, created an Artificial Intelligence based app called "Plantix" that is used to detect nutrient deficiency in the soil, as well as diseases and pests in crops. This allows farmers to know how to use fertilisers and contributes to improvement in crop quality. This application uses image recognition technology. Farmers take snapshots of crops with smartphones. This application also provides soil restoration methods with guidelines and various solutions through some small videos.

Trace Genomics, another Machine Learning based company which helps farmers perform analysis of soil. This application helps in monitoring crop health and soil to produce productive and healthier plants.

Use of Weather and Price Forecasts

Increasing environmental pollution and changing climatic conditions make it really difficult for the farmers to decide the correct time to sow seeds. With the help of machine learning and artificial intelligence, farmers use weather forecasts to analyse climatic conditions to help them plan which crops to grow and when to sow. Price forecasting gives farmers a more accurate picture of crop prices in the coming weeks, helping them maximise their profits.

Figure 4. Weather forecast for a farmland

Analysing Crop Health Using Drones

SkySquirrel technologies unveils Aerial imaging solution using drones to monitor crop health. In this method, a drone collects data from the field, which is then transferred from the drone to a computer via a USB drive for analysis by an expert.

The company uses various algorithms to analyse recorded photographs and provide detailed reports on the recent condition of farms. Helping farmers identify pests and bacteria in time to take the necessary steps to provide farmers with appropriate guidance.

Nowadays, applications and tools are developed to assist farmers in sustainable agriculture.

These Artificial Intelligence applications in agriculture can help reduce inaccuracies and mismanagement by providing farmers with appropriate advice on crop rotation, water management, timely harvesting, types of crops to grow, optimal planting, pest infestation, and nutritional management.

Using machine learning approaches combined with imagery clicked by drones and satellites, AI-enabled technology analyses crop sustainability, predicts weather conditions and examine data such as solar radiation, wind speed, temperature and rainfall which is used to assess farms for detecting presence of on-farm pests and diseases and insufficient crop nutrition.

Nowadays, without physical connectivity farmers can reap the benefits of Artificial Intelligence with a SMS-enabled phone and simple tools like Seeding and Farm maintenance apps. Simultaneously, farmers with Wi-Fi access could use the Artificial Intelligence application to receive continuous Artificial Intelligence specific plans for land and farms. With such Internet of Things and Artificial Intelligence-driven solutions will help farmers meet the global demand for more food by continuously increasing production and income without exhausting natural resources.

Weeds can be easily detected with an Artificial Intelligence sensor, and the area affected by weeds can also be detected. If such areas are found, they can be applied in a targeted manner to reduce herbicide use, saving time and harvesting. There are various Artificial Intelligence companies building drones and robots with Artificial Intelligence and computer vision that can spray weeds with precision. Artificial Intelligence sprayers significantly decrease the number of synthetics used in the field, improving crop quality and saving money.

In the future, Artificial Intelligence and Machine Learning transform farmers into agricultural engineers, using resultant data to optimise yields down to separate rows of crops.

Figure 5. Drones analysing crop health and spraying pesticides

Agricultural Robotics

Artificial Intelligence and Machine Learning companies are manufacturing robots that are used to carry out various different tasks easily in the large agricultural farms. These types of robots are prepared to fight weeds and harvest more and more crops at a higher rate than humans.

These type of robots are trained to observe and monitor harvest quality and identify weeds while simultaneously harvesting and packing the crop. These robots also tackle the challenges of forced labour.

AI-Based Systems to Detect Pests

Pests are the farmer's worst foes when it comes to damaging crops. The Artificial Intelligence system uses satellite imagery and compares it with previous data using Artificial Intelligence approaches to detect whether insects have landed and the types of insects such as locusts, grasshoppers, etc. Take the necessary precautions and apply the necessary pest control to help farmers get rid of and fight pests using ML and AI.

AI START-UPS IN AGRICULTURE

Few examples of Popular Start-Ups that improved various different factors of agriculture using Machine Learning and Artificial Intelligence are listed below:

1. **Fasal:** Use of Artificial Intelligence based agriculture is increasing every day in different parts of the world. Still, the number of farms per farmer is lower in poor areas than in wealthy areas. This is useful for automated monitoring as it requires less bandwidth and smaller size devices to capture complete farm data. Indian start-up Fasal is working in this field. Provide farmers with real-time data and insights using affordable sensors and Artificial Intelligence tools. This allows farmers to benefit from real-time information related to everyday operations on the field. The company devices are easier to deploy in smaller venues. They are creating AI-based systems to make precision farming reachable to all farmers.
2. **Prospera:** An Israeli start-up founded in 2014. The company develops smart solutions for effective farming. The company build cloud based solutions that are used to collect data from the field such as aerial, water, soil photography, etc. and combines it with devices in the field. This device is well known as

"Prospera" and gains insight from the collected data. This device makes use of several Artificial Intelligence technologies, tools and sensors.

3. **Blue River Technology:** Blue River Technology is a Californian start-up started in 2011. This company is developing next-generation agricultural equipment that utilises Artificial intelligence, computer vision, and robotics technology. This device uses computer vision to identify individual plants, Machine Learning to determine actions, and robotics to carry out actions. This saves farmer's money and synthetics in farming.
4. **Farmbot:** It is an open source CNC software package and machine precision farming which is designed to help anyone grow plants in their own space. The full product "Farmbot" is priced at $4000 and allows anybody to do achieve farming starting from seed planting to weed identification on their own using physical bots and software systems. Also provides web app that could be downloaded to any computer system or smartphone helping farmers manage farming at any time, at any place.
5. **OneSoil:** It is an application developed to help farmers making good decisions. The application uses machine learning algorithms and computer vision for precision agriculture. Remotely monitor crops, identify field problems, check weather forecasts, calculate nitrogen, potassium and phosphorus fertiliser ratios, and more.

BENEFITS OF USING ARTIFICIAL INTELLIGENCE IN AGRICULTURE

1. Artificial Intelligence provides satisfactory decisions
   Predictive analytics is a real boon for agribusiness. This helps farmers to address key agricultural challenges such as analyse market demand, price forecasts, and find the best times to sow and harvest crops. In addition, Artificial Intelligence machines can calculate crop and soil health, monitor weather, make fertiliser recommendations and even evaluate crop quality. All these benefits of Artificial intelligence in agriculture will help farmers make better decisions and farm more efficiently.

2. Artificial Intelligence saves cost
   Precision agriculture with Artificial Intelligence enabled equipment helps farmers grow more and more plants with limited costs and resources. Artificial Intelligence provides farmers with real-time recognition to help them make better decisions at every stage of farming. With the right decisions, time and money both can be used efficiently with less wastage of products and chemicals. In

addition, farmers can identify specific areas in need of irrigation, fertilisation and pesticide treatment, which reduces the extensive use of synthetics during harvest. All of this adds up to less herbicide use, better crop quality, and higher returns with less resources.

3. Artificial Intelligence alleviates labour shortages
   Agriculture has always had a labour shortage problem. Artificial Intelligence can solve this problem through agricultural automation. Artificial Intelligence and automation will help farmers get the job done without adding headcount. Examples include self-driving tractors, vertical farming software, intelligent fertilisation and irrigation systems, precision farming, intelligent spraying and AI robots for harvesting. AI-controlled equipments and systems are faster and more precise than human farm workers. (Pravar Jain, Analytics Vidhya, 2021)

CHALLENGES FACED DUE TO ARTIFICIAL INTELLIGENCE IN AGRICULTURE

1. Unfamiliar with artificial intelligence machines
   Although there are many benefits to using Artificial Intelligence in agriculture, people in most parts of the world are unfamiliar with using AI-enabled solutions and devices. To solve the problem, Artificial Intelligence companies need to provide farmers with basic equipment and, as they become familiar with them, farmers should be provided with intelligent machines.

2. Lack of experience with New Technologies:
   Introducing Artificial Intelligence and new technologies into agriculture in developing countries could become a daunting task. It is really hard to sell such technology in areas where such technology is not used. In such areas, farmers need someone's help to use these techniques and tools.

3. Security and Privacy Issues:
   There are no explicit regulations or guidelines regarding the use of Artificial intelligence which could raise various legal issues. Software and Internet usage can also create security and privacy issues, such as data leaks and cyberattacks. All of these issues might pose major problems for farm owners and farmers.

ARTIFICIAL INTELLIGENCE IN INDIAN AGRICULTURE

Agriculture is most common productive industries for Machine Learning and Artificial Intelligence. Artificial Intelligence and Internet of Things sensors feeding algorithms with real-time input are improving crop yields, increasing agricultural efficiency and reducing production costs. Global spending on various intelligent agricultural techniques and machines, such as Machine Learning and Artificial Intelligence, is expected to increase by three times till 2025, reaching $15.3 billion. IoT-based agriculture monitoring (IoTAg) is the fastest growing technology segment in smart and connected agriculture, predicted to reach $4.5 billion by 2025, estimated by PwC.

The Government of India will provide funds allocated to the state of Rs. 17.563 billion and Rs. 24.227 billion for introducing new technologies and tools that includes drones, artificial intelligence and remote sensors, block chain and GIS in agriculture. In addition, the government has allocated ICAR (Indian Agricultural Research Institute) with 73.025 billion INR and 79.0818 million INR for 2020-21 and 2021-22 respectively to conduct agricultural groundwork and development to develop new technologies and demonstrate these technologies to farmers for the introduction of new technologies.

Figure 6. Artificial intelligence in Indian agriculture

In addition to the government's focus on improving service delivery and promoting market access for farmers, it is also focusing on decreasing transaction costs and encouraging farmer producer organisations (FPO) to increase bargaining power. Great attention has also been paid to developing infrastructure to secure better links between agricultural producers and domestic farmers and international markets. Low-cost, high-yield, disease and pest resistant, climate tolerant crop, horticultural, animal and fisheries varieties and technologies developed by ICAR will improve production and productivity, reduce production costs and increase farmer's income. Adoption of the farming system model created by ICAR has enabled farmers to improve incomes and economic status. In addition, the State Farmer Income Enhancement Strategy provided to the State by ICAR will help increase farmer's income.

Below are some of the areas where the integration of artificial intelligence has the greatest potential to improve agriculture:

- Cognitive computing become the ground-breaking technology in agriculture due to its ability to understand, learn and interact with diverse environment to maximise production. Microsoft is recently working in Andhra Pradesh with 175 farmers to provide field, fertiliser and agriculture related advisory services. The initiative has increased average yield per hectare by 30% over the past year. A pilot project was completed with an agricultural AI application that communicates deadlines, soil preparation, soil test-based fertilisation, optimal application depth, seed treatment and more. Additionally, field sensors and mobile agricultural robots, laser scanners and multidisciplinary cameras are used for non-measurable assets and radiation areas.
- Image-Based Precision Agriculture, Remote Sensing, Internet of Things, and Proximity Sensing, relates to historic meteorology, soil reports, current studies, weather, insect infestation, and drone imagery used for field surveys, intelligent data integration, harvest monitoring and depth field analysis.
- Use of image recognition with intelligent algorithms to identify plant and cations, pest infestation and disease prognosis is becoming more popular. Animal and human violations are identified by monitoring real-time video feeds of each crop field using machine learning AI-based surveillance systems. Sending alerts immediately can be very helpful in preventing crop damage.
- Yield mapping to observe pattern in larger datasets and figure out their orthogonality in irrigation and real-time system optimization to rate the effectiveness of frequent irrigation for crop planning.
- Shortage of Farm workers nowadays, and smart AI tractors, agri-bots and robotics based on Artificial Intelligence and machine learning have become viable options for many remote farms struggling to find workers. The robots

are able to harvest very fast, trace weeds and perform weeding more precisely, and decrease operating cost and labour reliance. Nowadays, Chatbots assist farmers by answering their questions and providing guidance and advice on specific agricultural and yield related issues.

- Remove barriers in bringing fresher and safer crops to market and improve tracking and traceability of agricultural supply chains, resulting in greater transparency and control across the supply chain and reduced inventory depletion.

The use of Artificial intelligence in agriculture is promising, but not without challenges. AI systems demand for larger amount of data to train these machines to make precise predictions. Spatial data is easy to collect, but temporary data for large agricultural areas is difficult to find. Developing powerful machine learning models takes time, as the data infrastructure must be mature. This is one reason why Artificial Intelligence is used for agronomic products such as fertilisers, pesticides and seeds rather than in field solutions. Other critical drawback is the excessive cost of various AI systems accessible in the agricultural market. Solutions should be open source and affordable so that the technologies could be accessed even at the farmer's level. With continuous effort and extensible innovation from both the private sectors, these technological interventions entirely transformed agriculture and changed farmer's lives for the better. (SIRU, 2022)

RESEARCH IN AGRICULTURE USING ARTIFICIAL INTELLIGENCE AND MACHINE LEARNING

1. IoT driven application
   Huge amounts of data are created every day in unstructured and structured formats. This data relates to rainfall, ground report, new research, weather pattern, susceptibility to pest infestation, and images by cameras and drones. IoT techniques for detecting and providing intelligent solutions to increase yields.

2. Image based precision agriculture
   In the present world scenario, precision agriculture is one of the most specialised fields in agriculture. Drone imaging is useful for rigorous field analysis, crop monitoring, and field scanning. A fusion of IoT, drone data and computer vision technology will help farmers act faster. Commercial drone manufacturers such as Aerialtronics are adopting the IBM Watson IoT platform and Visual Recognition API for real-time application for analysis of the images.

3. Field management
   Real-time estimates during crop season by making field maps and discovering different parts of field where crops need pesticides, water and fertilisers using high-resolution imagery from drone and helicopter systems.

4. Disease Detection
   Image acquisition and analysis reliably divides images of plant leaves into surface areas, including leaf background, diseased areas, and disease-free areas. The diseased or infected area is clipped and sent to the laboratories for further prognosis that helps to identify pests and detect nutritional deficiencies.

Figure 7. Disease detection
("Artificial Intelligence in Agriculture: An Emerging Era of Research," 2019)

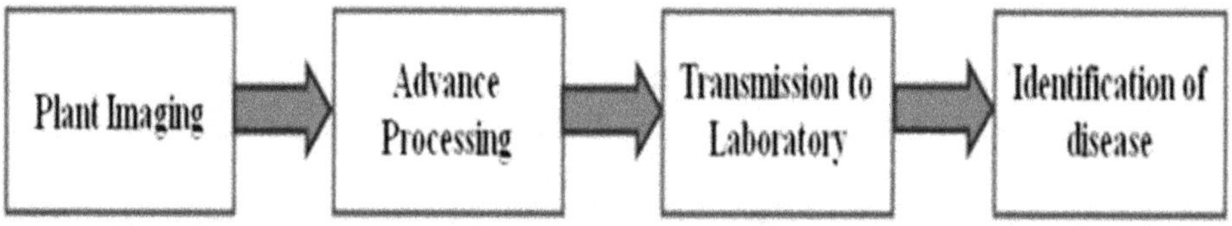

5. Plant Stress recognition using Artificial Intelligence and Machine Learning
   Detect different stress levels in plants via high-resolution Artificial Intelligence powered images and data from multiple sensors. This entire dataset generated from various sources should be used as input data for Artificial Intelligence and Machine Learning models. This allows the combination of this data to include specific parameters for plant stress detection. The developed Artificial Intelligence and Machine Learning model is trained on different plant images and can detect different stress levels in crops. This whole approach is broken down into different steps of detection, classification, analysis and prediction to provide suitable and improved decisions.

6. Identification of readiness of crops
   To determine how ripe the green fruits are, several photographs of crops are taken using white and UVA light. Farmers could determine different levels for the fruit or crop category's readiness using this approach. Before putting them on the market, add them to sorted batches.

Figure 8. Plant stress recognition using Artificial Intelligence
("Artificial Intelligence in Agriculture: An Emerging Era of Research," 2019)

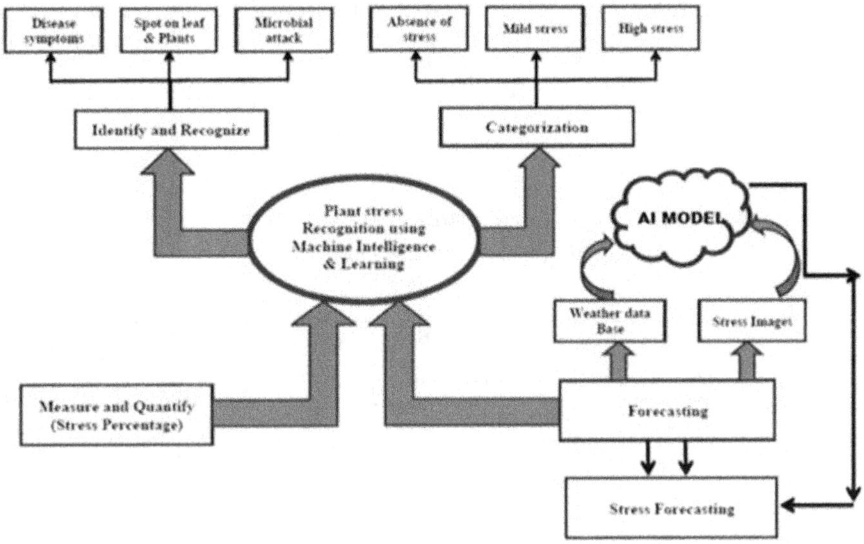

7. Health monitoring system for crops
   In addition to hyperspectral imaging and 3D laser scanning, remote sensing (RS) techniques are critical for building yield indicators on thousands of acres of farmland. In terms of time and effort, it could revolutionize the way farmers monitor their fields. This technology is also used to monitor the entire crop life cycle, including generating reports in case of anomalies.

8. Calculate the number of agronomic products used during farming
   Cognitive solutions recommend the best crop and hybrid seed selections for farmers based on various parameters such as soil conditions, weather forecast, seed type and pest prevalence in a particular area. Personalised recommendations based on farm requirements, local conditions and past successful farming history. Other external factors such as market trends, crop prices, requirements, consumer needs and aesthetics can also be considered to help farmers make knowledgeable decisions.

9. Robotics in precision agriculture
   Artificial Intelligence and Machine Learning models used to detect optimum sowing period for different seasons, statistical climate data, real-time Moisture Adequacy Data (MAI) from daily precipitation and soil moisture statistics to create forecast charts and also carter inputs on best sowing time to farmers.

To predict possible pest attacks, Microsoft is working with United Phosphorus Limited to develop a Pest risk prediction Application Programming Interface (API). This API offers the strategic advantage of Artificial Intelligence and machine learning to proactively notify potential pest attacks. Depending on weather conditions and crop growth stage, pest infestation is expected to be high, medium or low. (Khandelwal and Chavhan., 2019).

Figure 9. Robotics in precision agriculture
("Artificial Intelligence in Agriculture: An Emerging Era of Research," 2019)

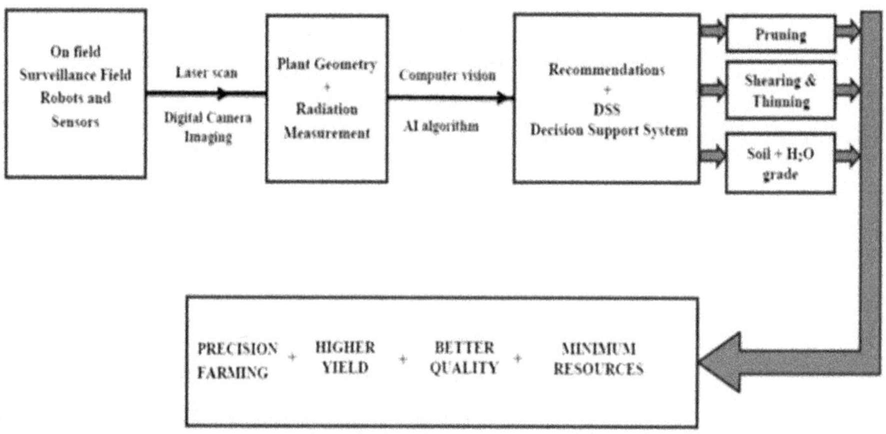

CONCLUSION

Farming is becoming more precise because of artificial intelligence, which not only helps farmers automate their operations but also produces higher yields and better-quality crops with fewer resource consumption.

In the future, companies will work to improve artificial intelligence and machine learning-based goods and service such as drones, training data for automated system manufacturing, etc will advance the technology and uncover more useful applications to assist the world in resolving issues related to food production for a expanding population.

REFERENCES

Java. (n.d.). *Artificial Intelligence in agriculture - javatpoint.* Java. https://www.javatpoint.com/artificial-intelligence-in-agriculture#:~:text=AI%20enables%20better%20decision%2Dmaking&text=Moreover%2C%20AI%2Dpowered%20machines%20can,decisions%20and%20do%20efficient%20farming

Badia Melis, R., Qian, J. P., Fan, B. L., Hoyos-Echevarria, P., Ruiz-García, L., & Yang, X. T. (2016). Artificial neural networks and thermal image for temperature prediction in apples. *Food and Bioprocess Technology*, *9*(7), 1089–1099. doi:10.100711947-016-1700-7

Balleda, K. (2014). Agpest: An efficient rule-based expert system to prevent pest diseases of rice & wheat crops. In *Proc. Intelligent Systems and Control (ISCO)-*. IEEE. (Original work published 2014) doi:10.1109/ISCO.2014.7103957

Strategic Investment Research Unit (SIRU). (2022, June 17). *Artificial Intelligence in Indian agriculture.* Investment Promotion and Facilitation Agency. https://www.investindia.gov.in/team-india-blogs/artificial-intelligence-indian-agriculture

E-Agriculture. (n.d.). *Can artificial intelligence help improve agricultural productivity?* Can Artificial Intelligence help improve agricultural productivity? FAO. https://www.fao.org/e-agriculture/news/can-artificial-intelligence-help-improve-agricultural-productivity

Hopfield, J. J. (1982). Neural Networks and Physical Systems with Emergent Collective Computational Abilities. *Proceedings of the National Academy of Sciences of the United States of America*, *79*(8), 2554–2558. doi:10.1073/pnas.79.8.2554 PMID:6953413

Jain, P. (2021, July 23). *Artificial Intelligence in agriculture: Using modern day AI to solve traditional farming problems.* Analytics Vidhya. https://www.analyticsvidhya.com/blog/2020/11/artificial-intelligence-in-agriculture-using-modern-day-ai-to-solve-traditional-farming-problems/

Jha, K., & Shah, M. (2019, June 17). *A comprehensive review on automation in agriculture using artificial intelligence.* Artificial Intelligence in Agriculture. https://www.sciencedirect.com/science/article/pii/S2589721719300182#bbb0025

Karmokar, B. C. (2015). Tea leaf diseases recognition using neural network ensemble. *International Journal of Computer Applications*, *114*(17), 27–30. doi:10.5120/20071-1993

Khandelwal, P. M., & Chavhan, H. (2019). *Artificial Intelligence in Agriculture: An Emerging Era of Research*. Kavikulguru Institute of Technology and Science.

Polya, G. (2004). *How to Solve It: A New Aspect of Mathematical Method*. Princeton University Press.

Sladojevic, S., Arsenovic, M., Anderla, A., Culibrk, D., & Stefanovic, D. (2016). Deep neural networks based recognition of plant diseases by leaf image classification. *Computational Intelligence and Neuroscience*, *2016*, 1–11. doi:10.1155/2016/3289801 PMID:27418923

Temple, J. (2020, April 2). *How artificial intelligence is helping farmers and babies in the developing world*. MIT Technology Review. https://www.technologyreview.com/s/613210/how-artificial-intelligence-is-helping-farmers-and-babies-in-the-developing-world/

Dharmaraj, V. & Vijayanand, C. (2018). Artificial Intelligence (AI) in Agriculture. *International Journal of Current Microbiology and Applied Sciences, 7*(12)

Walch, K. (2019, July 9). How ai is Transforming Agriculture. *Forbes*. https://www.forbes.com/sites/cognitiveworld/2019/07/05/how-ai-is-transforming-agriculture/#5442e7ad4ad1

Zha, J. (2022). Artificial Intelligence in Agriculture. *Journal of Physics*.

Chapter 3
Importance of Drone Technology in Indian Agriculture, Farming

Jyoti Yadav
Galgotias University, India

Usha Chauhan
Galgotias University, India

Divya Sharma
Galgotias University, India

ABSTRACT

The population is quickly growing, as is the need for food. The farmers' traditional tactics are insufficient to achieve these goals. As a result, new automated methods (such as drone technology) emerged. These innovative techniques met food demands while also offering employment opportunities for billions of people. Drone technology saves water, pesticides, and herbicides, protects soil fertility, and allows for more efficient use of labour to boost productivity and quality. This article will look at the usage of drones in agricultural applications. According to the literature, drones may be utilised for a range of agricultural reasons. The authors took the approach of doing a thorough review of past studies in this topic. This study describes the current state of agricultural drone technology, including crop health monitoring and farm activities such as weed control, evapotranspiration estimation, and spraying. The study concludes by recommending more farmers to invest in drone technology to boost agricultural output.

DOI: 10.4018/978-1-6684-6418-2.ch003

INTRODUCTION

Despite the fact that India is heavily reliant on agriculture, it is still a long way from using cutting-edge technologies to ensure a strong crop. UAVs are already being used in precision agriculture, photogrammetry, and remote sensing in developed countries. It is extremely rapid, and it has the potential to minimise a farmer's workload. UAVs are generally equipped with cameras and microphones for crop monitoring, as well as pesticide sprayers. Previously, several UAV models were utilized for both military and civilian purposes (Aditya & Kulkarni, 2016). A scientific evaluation of UAVs in precision agriculture looks at their value in agricultural tasks such crop monitoring, crop height estimate, pesticide spraying, soil and field analysis, and so on. Their hardware implementations, on the other hand, are fully reliant on crucial factors such as weight, range of flight, payload, configuration, and cost.

The Unmanned Aerial Vehicle (UAV), sometimes known as a drone, is a robotic vehicle that can operate remotely for a variety of purposes. Initially designed primarily for military purposes, it gradually expanded to other parts of society. According to www.dronedeploy.com, approximately 25 million acres of land across seven continents and 160 nations is being used for one purpose or another by drone technology (Rahman, 2014). Agriculture is not immune. In an era of labour shortages and the enormous relevance of precision agriculture, the agricultural world has warmly adopted drone technology. Only a few years ago, drone technology appeared to be an unattainable possibility for the impoverished agricultural community, but the situation has steadily improved, and it is now an easily obtainable technology with minimal financial investment. According to research, drone technology would generate more than 70 billion Euros (Ve oustraete 2015) in the world economy between 2015 and 2025.

Types of Drone

According to Watts et al., there are numerous basis for categorising the d ones the varieties of drones (2012). Drones have been classified into seven categories in this study: Micro Air Vehicles (MAV), Vertical take-off and landing (VTOL), Low Altitude- Short Endurance (LASE), Low Altitude- Short Endurance (LASE), LASE Close, Low Altitude - Long Endurance (LALE), Medium Altitude - Long Endurance (MALE) and High Altitude- Long Endurance (HALE).

1. **Micro Air Vehicles (MAV):** Because of their small size, they are sometimes referred to as Nano-Air aircraft, which allows military models of these air craft to be carried in soldiers' backpacks and provides discreet monitoring capabilities in confined locations. Such aircraft typically fly at lower altitudes

(330 m), with battery life constraints leading to flying periods of 5-30 minutes. However, their commercial development is still in its infancy.

2. **Vertical take-off and landing (VTOL):** Because these aircrafts do not require a take-off and landing flight, they are typically utilised in situations where topographical limits necessitate this advanced capability. They fly at various heights depending on the mission profile, but generally at lower altitudes since the high battery power consumption for flight hovering limit flying midframes. VTOLs have the significant benefit of being portable to distant area networks without the requirement for route complexes.
3. **Low Altitude Short Endurance (LASE):** Miniature unmanned aerial systems frequently reduce the need for aeroplane runways built for rapid zone deployment and transfer. To launch an aeroplane using a catapult, it should weigh between 2 and 5 kg and have a wing span of less than three metres. Weight reductions tend to limit the range of strength and connectivity to 1—2 hours, but only within a few kilometres of ground stations.
4. **LASE Close:** This category defines small unmanned aircraft that require larger runways with better capabilities. They have a reasonably long flying time and can operate at an altitude of 1500 m, and these limits were addressed significantly by intentionally modified "record-breaker" aircraft.
5. **Low Altitude - Long Endurance (LALE):** This type of drone was designated by the United States Federal Aviation Administration based on its capacity to carry input payloads to an extreme upper end with a high weight (kg) at comparatively higher altitude for extended periods of time.
6. **Medium Altitude - Long Endurance (MALE):** These aircrafts are often significantly larger than low-altitude UAV groups, flying at altitudes of up to 9,000 m for hundreds of kilometres from their base stations over several hours. These play an important role in making strategic choices for military defence services, and their use is increasing in certain civil applications.
7. **High Altitude- Long Endurance (HALE):** These are the largest and most complicated aircrafts, with planes larger than most manned planes in commercial aviation. These drones have the capacity to fly at 20,000 metres or higher for thousands of kilometres and have a flying time of more than 30 hours in space.

Role of Drone in Farming

1. **Field and Soil Assessment:** Before and after crop planting, data obtained by drones on soil analysis is useful in planning crop species to be sown, planting pattern, and establishing the quantity and timing of irrigation and fertiliser treatment. These farm-level management initiatives can boost overall agricultural

output (Bencini et al., 2009). This technique in agriculture allows for the adaptation of site-specific management practises, such as precision farming.

2. **Plant Establishment:** Due to labour shortages, crop planting has become a costly and time-consuming endeavour that previously required a large amount of human labour. Drones have made large-scale agricultural sowing more precise and accurate in a short period of time. This method of planting with drones has reduced the cost of planting by up to 85% and eliminates the difficult labour associated with on-the-ground planting.
3. **Precision Crop Spraying:** Site-specific crop spraying may be accomplished utilising drones outfitted with sensors that scan the cultivated area in real time and guarantee a precise amount of liquid (such as insecticides and nutrients) is sprayed on that particular location (Yueling et al., 2011). According to experts, drones can perform aerial spraying up to five times faster than traditional spraying. It improves spraying accuracy while saving farmers time and money. It minimises pesticide contamination in groundwater indirectly.

Figure 1. Data on wings: A close look at drones in India

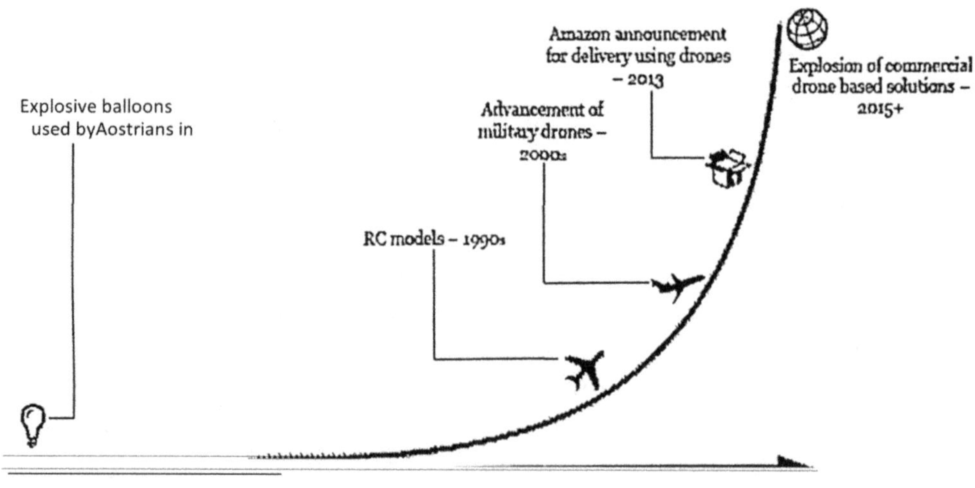

4. **The Unexpected Weather Extremes That Plague Crop Output:** The ease and efficiency of massive-scale observation of crops and agricultural land are the most significant advantages of deploying unmanned drones. Satellite images had been utilised to acquire a view for a long time assisting in the detection of possible difficulties in crop monitoring (Xia et al., 2016). At the moment, time series animations can deliver exact results cultivation of a crop and improved crop management.

5. **Irrigation Management:** Agricultural drones fitted with thermal sensing cameras may provide incredible insight into difficult regions of the field for irrigation application. Several insights were observed using these thermal digital cameras, ranging from low moisture strained to water laden conditions, allowing farmers to make irrigation management decisions depending on soil water status. It encourages the precise application of water in fields.
6. **Crop Health Evaluation:** Crop health evaluation is critical for detecting crop, bacterial, and fungal illnesses. Drones scan the crop using green visible light and near-infrared light to determine disease incidence in spatial and temporal changes based on crop reflectance (Niu, 2019). It is feasible to detect illness early on and intervene to protect the crop.
7. **Livestock Monitoring:** Drones offer a variety of possible uses in animal husbandry. Each animal is equipped with sensors or radio frequency identification (FRIDs) tags so that drones can monitor feeding activities and mobility. Using this, livestock tracking may be done much more often, with less time and investment in manpower. Remote sensing-fencing, virtual borders, or remote sensing-zoning fundamentally refers to the creation of a virtual barrier or security fence over a geographical region of intere5t, particularly in free range animal grazing

Figure 2. This block diagram explains the different types of uses of drone in agriculture

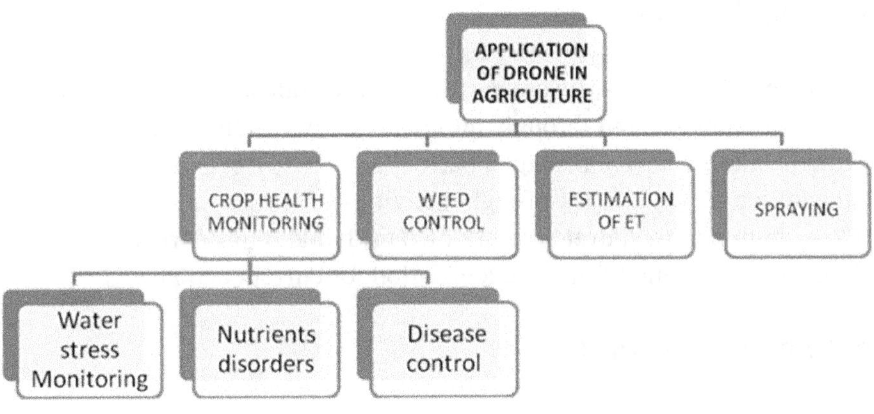

Agriculture Tasks Completed

One of the most visible applications of drone technology is automated crop phenotyping. Many earlier research on crop phenotyping utilising drones have been conducted (Grenzdorffer, 2019; Holman et al. 2016; Yang et al., 2017; Watanabe et al., 2017; Sugiura, et al., 2017). Image processing and drone technology are two concurrent technologies that are used in various investigations. Deep learning technologies, as well as aerial images captured by drones, have been utilised in a variety of occasions. Drone technology was used to identify weeds (Barrero et al., 2015; Bah, et al., 2018; Lopez, et al., 2016; David et al., 2016; Pena et al., 2015; Bah, et al., 2018; Stroppiana, et al., 2018).

Legislation Regarding Drone Technology in India

For numerous years, the Civil Aviation Ministry aimed to establish India as a global leader in the drone economy. Officials were worried about unregulated drone flights and the harm they pose, so restrictions were devised to prohibit illegal activities and ensure public safety (Sarghini & De Vivo, 2017). Toward the end, the development of worldwide standard drone law, including suitable rules for the industrial use of diverse drone technology, was necessary. All of this takes several years to create through a Civil Aviation Requirement (CAR) because: (1) drone technology advancements have advanced relatively swiftly; (2) different nations have been experimenting with their drone legislation and so no ICAO (The International Civil Aviation Organization)

The legislation establishes "No Drone Zones" along international borders, in and around airports in all states, key destinations, and other military and national services.

Drone Regulations

1. Drone hardware and software certification for safe and controlled flight
2. Airspace management through automated activities linked to an overarching framework in airspace management
3. In addition to visual track-of-sight activities
4. Contribution to the development of a worldwide set of norms
5. Suggestions for modifying the legislation for current and new CARs

Enforcement Measures

The legal proceedings include (a) the suspension or cancellation of a UAOP/UIN due to a violation of legislative requirements; (b) policies under sections of the Aircraft Acts 1934 and Aircraft Legislation in question, or for additional legislative provisions; and (c) sanctions under relevant IPCs (such as 287, 336, 337, 338, or any applicable clause of the IPC).

Agriculture's Challenges and Opportunities

The establishment of too stringent or even deficient laws and regulations governing the importation and use of UAVs may stymie the development of a high-growth business capable of attracting and employing educated professionals in rural areas. It is difficult to assess how they affect the use of drones in agricultural tasks (Sarghini & De Vivo, 2017). Nonetheless, several nations have implemented a federal ban on the import and use of UAVs. The following were the primary difficulties:

1. Drone technology, training integration, and deployment are regarded to be more expensive.
2. There is still some uncertainty regarding UAVs. Legislation is meagre, minimal, or a low interpretation of aviation regulations that do not properly match UAVs. As a result, regulations must be established to regulate the expanding UAV possibilities and application areas (Vardhan et al., 2014). Nations such as the United Kingdom, the United States, Spain, and Germany are setting the way in this manner by developing criteria for drone usage and identifying locations where they may be used. Nonetheless, other nations throughout the world lag far behind.
3. From a commercial standpoint, it is difficult to justify the profitability of drones in the agricultural industry. Although there is some debate over the use of drones, there may be long-term cost savings and claims regarding the particular running expenditures of UAVs (Colomina & Molina, 2014). Compensation might be proposed for the replacement of wrecked UAVs, the acquisition of high-resolution picture lenses, as well as the accompanying technical solutions and other maintenance expenditures. This makes it tough to sell to farmers and agricultural entrepreneurs.
4. System integration and data analytics of high performance drone systems enable drone control in the gathering of real-time pictures and transmission of information, as well as machine learning applications for helpful insights into drone operation (Reinecke & Prinsloo, 2017). It also demonstrates the technological aspects. The difficulties associated with the adaptation of drones in agriculture.
5. Many people are concerned that the use of unmanned aerial vehicles (UAVs) for tracking and surveillance would breach their privacy. A key difficulty is the absence of defined operational and technical norms for UAVs to function safely (Espinoza et al., 2017). Due to different internal security flaws in UAV operations, global location, tapping, and infiltration may occur.

6. The flying time of an unmanned aerial vehicle is determined by the power of its batteries. Most UAVs, particularly multi-rotors, have batteries that can only maintain a flight length of between 10 to 30 minutes and may be worse while flying at strong wind speeds (Huang et al., 2018). Battery technology must be developed, and a solution to employ bigger capacity batteries with a tiny footprint in UAVs must be discovered. Drones used for agricultural spraying are only effective in locations where other equipment is unavailable, such as in hilly terrain; they are significantly less dependable and far more expensive than a broader range of field-based spraying technologies.

The Rapidly Changing World of Agriculture

Agriculture used to entail a lot of guesswork: how would the weather patterns be? What kind of agricultural yield may be expected?

Farmers have found innovative ways to handle these concerns with new technology throughout the years. Agriculture drones are the next step in this process. Agriculture drones may help with everything from precision agriculture to appropriately distributing weed control or fertiliser to improving field management (Yallappa et al., 2017). As a result, operating costs are reduced, crop quality improves, and production rates increase.

Farming activities do not resemble those of even a few decades ago. Thanks to technological advancements, today's growers can optimise every element of their operations, from field spraying to grow cycles and crop health. Drones and other unmanned aerial vehicles (UAVs) have played an important part in this transformation (UAV) (Shah et al., 2020). When farmers deploy an agriculture drone, they gain from in-depth data analysis and mission planning, as well as new technology capable of managing physical labour. Whether you're a small-scale farmer or the CEO of a huge corporation, drone technology may help you meet your harvest goals and produce more crops with fewer resources (Simelli & Tsagaris, 2015). Crop spraying and field mapping are two of the most crucial applications for the right UAV and payload.

UAV Spraying a Field

Every grower struggles with making efficient use of the aforementioned. Spraying too much concentrated in one place increases costs and may degrade product quality (Sonaa et al., 2016). However, if the concentration is too low, your crops may get overrun with weeds, famished, or consumed by insects and other predators, resulting in a decrease in production rate. Pesticides, on the other hand, may be distributed equally and efficiently with the right agricultural drones and spraying payloads (Unpaprom et al., 2018).

Figure 3. Drone monitoring in agriculture

What were the results? Improved crop quality and increased output without the need for strenuous physical labour DJI drones may be used on practically every crop, including rice, wheat, corn, citrus trees, cotton, and many more.

Figure 4. Application of drone in agriculture

Drone spraying can have considerable environmental benefits in addition to covering more ground at a lower cost (Zheng et al., 2016). Drones can more efficiently disseminate pesticides, allowing you to use less for the same final outcome. UAV seeding can also assist you in better utilising your land (Sharma et al., 2022). With climate change and industrialisation expected to diminish the amount of land available for farming worldwide, prudent land use will be vital to preserving excellent outcomes in the coming years.

CONCLUSION

Drones have the potential to totally transform Indian agriculture. Drone manufacturing is projected to become more cost-effective as technology progresses. The hard effort and drudgery required in farming do not appeal to today's youth. Drones' ramifications may entice and drive young people to pursue jobs in agriculture. Drones provide real-time and high-quality aerial imagery over agricultural regions, as opposed to satellite photographs. Drones may also be used to pinpoint weeds and diseases, assess soil quality, detect vegetation fluctuations, and generate accurate elevation models. Drones will allow farmers to learn more about their farms. Farmers will be able to produce more food while using fewer pesticides as a consequence. Almost all farmers who have utilised drones have benefited in some way. They can make better use of their land, exterminate pests before they damage entire harvests, improve soil quality to stimulate growth in troubled areas, improve irrigation to heat-stressed plants, and track fires before they spread. As a consequence, drones may become a vital part of agriculture in the future, supporting farmers in better and more sustainable land and resource management.

REFERENCES

Aditya, S., & Kulkarni, S. C. (2016). Adoption and Utilization of Drones for Advanced Precision Farming: A Review. *International Journal on Recent and Innovation Trends in Computing and Communication, 4*(5), 563-565.

Bencini, L., Chiti, F., Collodi, G., Di Palma, D., Fantacci, R., Manes, A., & Manes, G. (2009). Agricultural Monitoring Based on Wireless Sensor Network Technology: Real Long-life Deployments for Physiology and Pathogens Control. In *Third International Conference on Sensor Technologies and Applications*. IEEE. 10.1109/SENSORCOMM.2009.63

Bing, C. (2014). The Design and Operation of City Agricultural Sightseeing System with the Internet of Things Technology. In *Sixth International Conference on Measuring Technology and Mechatronics Automation*. IEEE.

Colomina, I., & Molina, P. (2014). Unmanned aerial systems for photogrammetry and remote sensing: A review. *ISPRS Journal of Photogrammetry and Remote Sensing*, *92*, 79–97. doi:10.1016/j.isprsjprs.2014.02.013

Espinoza, C. Z., Khot, L. R., Sankaran, S., & Jacoby, P. W. (2017). High Resolution Multispectral and Thermal Remote Sensing Based Water Stress assessment. In Subsurface Irrigated Grapevines. *Remote Sensing (Basel)*, *9*(9), 961. doi:10.3390/rs9090961

Huang, Y., Reddy, K. N., Fletcher, R. S., & Pennington, D. (2018). 2017. UAV Low-Altitude Remote Sensing for Precision Weed Management. *Weed Technology*, *32*(1), 2–6. doi:10.1017/wet.2017.89

Niu, H. (2019). Estimating evapotranspiration with UAVs in agriculture: A review. In *2019 ASABE Annual International Meeting*. American Society of Agricultural and Biological Engineers, 2019.

Rahman, M. A. (2014).Enabling drone communications with WiMAX Technology. In *The 5th International Conference on Information, Intelligence, Systems and Applications* (pp 323-328). IEEE. 10.1109/IISA.2014.6878796

Reinecke, M., & Prinsloo, T. (2017). The influence of drone monitoring on crop health and harvest size. In *IEEE 1st International Conference in Next Generation Computing Applications (Next Comp)* (pp. 5-10). IEEE. 10.1109/NEXTCOMP.2017.8016168

Sarghini, F., & De Vivo, A. (2017a). Interference analysis of a heavy lift multi rotor drone flow field and transported spraying system. *Chemical Engineering Transactions*, *58*, 631–636.

Sarghini, F., & De Vivo, A. (2017b). Analysis of preliminary design requirements of a heavy lift multi rotor drone for agricultural use. *Chemical Engineering Transactions*, *58*, 625–630.

Shah, D., Dixit, R., Shah, A., Shah, P., & Shah, M. (2020). A comprehensive analysis regarding several breakthroughs based on computer intelligence targeting various syndromes. *Augment Hum Res*, *5*(1), 14. doi:10.100741133-020-00033-z

Sharma, U., Sharma, D., Saleem, S., Kumar, M., Singh, S.P. (2022). Artificial Intelligence-Based Sustainable Agricultural Practices. In *Artificial Intelligence Applications in Agriculture and Food Quality Improvement*. IGI Global. doi:10.4018/978-1-6684-5141-0.ch001

Simelli, I., & Tsagaris, A. (2015). The use of unmanned aerial systems (UAS) in agriculture. *7th International Conference on Information and Communication Technologies In Agriculture,* (pp. 730-736).

Sonaa, G., Passonia, D., Pintoa, L., Pagliaria, D., Masseroni, D., Ortuani, B., & Facchib, A. (2016). UAV multispectral survey to map soil and crop for precision farming applications. *The International Archives of the Photogrammetry, Remote Sensing and Spatial Information Sciences, XLI*(B1), 1023–1029. doi:10.5194/isprs-archives-XLI-B1-1023-2016

Unpaprom, Y., Dussadeeb, N., & Ramaraj, R. (2018). Modern Agriculture Drones the Development of Smart Farmers. Academic Press.

Vardhan, P.H., Deepak, S., Aditya, P.T., & Arul, S. (2014). Development of Automated Aerial Pesticide Sprayer. *International Journal of Engineering Science and Research Technology, 3*(4).

Xia, T., Kustas, W. P., Anderson, M. C., Alfieri, J. G., Gao, F., McKee, L., Prueger, J. H., Geli, H. M. E., Neale, C. M. U., Sanchez, L., Alsina, M. M., & Wang, Z. (2016). Mapping evapotranspiration with highresolution aircraft imagery over vineyards using one-and two-source modeling schemes. *Hydrology and Earth System Sciences, 2016*(4), 1523–1545. doi:10.5194/hess-20-1523-2016

Yallappa, D., Veerangouda, M., Maski, D., Palled, V., & Bheemanna, M. 2017. Development and evaluation of drone mounted sprayer for pesticide applications to crops. In *2017 IEEE Global Humanitarian Technology Conference (GHTC)*. IEEE. 10.1109/GHTC.2017.8239330

Yueling, S., Zengfeng, Z., & Hongsheng, W. (2011). Agricultural information technology development and innovation path. In *International Conference on Electronics, Communications and Control (ICECC)*. IEEE.

Zheng, H., Zhou, X., Cheng, T., Yao, X., Tian, Y., Cao, W., & Zhu, Y. 2016. Evaluation of a UAVbased hyperspectral frame camera for monitoring the leaf nitrogen concentration in rice. In *2016 IEEE International Geoscience and Remote Sensing Symposium (IGARSS)*. IEEE. 10.1109/IGARSS.2016.7730917

Chapter 4
Advancement in Agricultural Techniques With the Introduction of Artificial Intelligence and Image Processing

Mritunjay Rai
https://orcid.org/0000-0002-8911-4826
Shri Ramswaroop Memorial University, India

Nalini Tyagi
Delhi College of Technology and Management, India

Padmesh Tripathi
https://orcid.org/0000-0001-9455-1652
KCC Institute of Technology and Management, India

Nitendra Kumar
https://orcid.org/0000-0001-7834-7926
Amity Business School, Amity University, Noida, India

Priti Kumari
National Institute of Technology, Patna, India

DOI: 10.4018/978-1-6684-6418-2.ch004

ABSTRACT

Food is the most fundamental requirement of human beings, and the only way this requirement can be satisfied is through agricultural production. The agricultural sector not only fulfills the need for food but also contributes to any country's economy. It also provides employment on a large scale. Agriculture is the backbone of any developing country due to its help in employment and economic sectors. This field is broadly divided into three parts: pre-harvest, harvest, and post-harvest. Advancement in various technologies like machine learning, image processing, the internet of things, and machine vision helps to reduce risk in farming and also increase profit. Using these techniques, a farmer can gain a brief understanding of the crop throughout its life cycle with greater precision and accuracy. This chapter provides a quick overview of different technologies along with their advantages and disadvantages. These technologies make farming more productive, accurate, and cost-effective, since they require less human effort and are less expensive.

1. INTRODUCTION

Agriculture, despite being the most fundamental component of human existence, is the most neglected field in the world because farmers are reluctant to adopt new technologies. This is despite the fact that agriculture is the most essential component of human existence. Farmers are still resistant to adopting new technologies and using non-traditional agricultural practices (Tripathi, et al. 2022). We reside in a nation where 70% of the population is agriculturally dependent. Therefore, this field is significant not only for food but also from an economic standpoint. According to numerous estimates, human food demand will increase by up to 70% (Business Bliss Consultants FZE, 2018, Searchinger, et al., 2019, Vasudevan, 2016) by 2040 compared to current levels. Therefore, we should take concrete measures to eliminate food waste caused by human error. Due to the increase in machine development, this industry has undergone a dramatic transition during the past half century. These developments increase the size, speed, and productivity of farm machinery, allowing for the cultivation of more areas in a more efficient manner. The use of hybrid seeds, improved irrigation techniques, and new types of fertilizer have all contributed to a growth in agricultural production. Currently, a new agricultural revolution is being fueled by data and connection. Detecting weeds, enhancing irrigation techniques and other input efficiency, and promoting crop production's sustainability and resiliency could all contribute to a production increase. Machine learning (ML), image processing analytics, (Tyagi, et al. 2022), networked sensors, and other emerging technologies may contribute to a production increase. As we all know, information is power, and technological advancements allow for the monitoring of crops at various stages. These innovations have facilitated the adoption of automated solutions in numerous industries. As technology progresses, its applications grow increasingly diversified and widespread. These are the steps that are followed by every farmer for agricultural techniques that make farming easier, as shown in Figure 1.

1.1 Agricultural Practices

Practise 1: Seed selection
Practise 2: Soil preparation
Practise 3: Sowing
Practise 4: knowing requirement of crop (Manuring, irrigation, weeding etc.)
Practise 5: Harvesting
Practise 6: Post harvest Storage

Figure 1. Various agricultural practices [5]

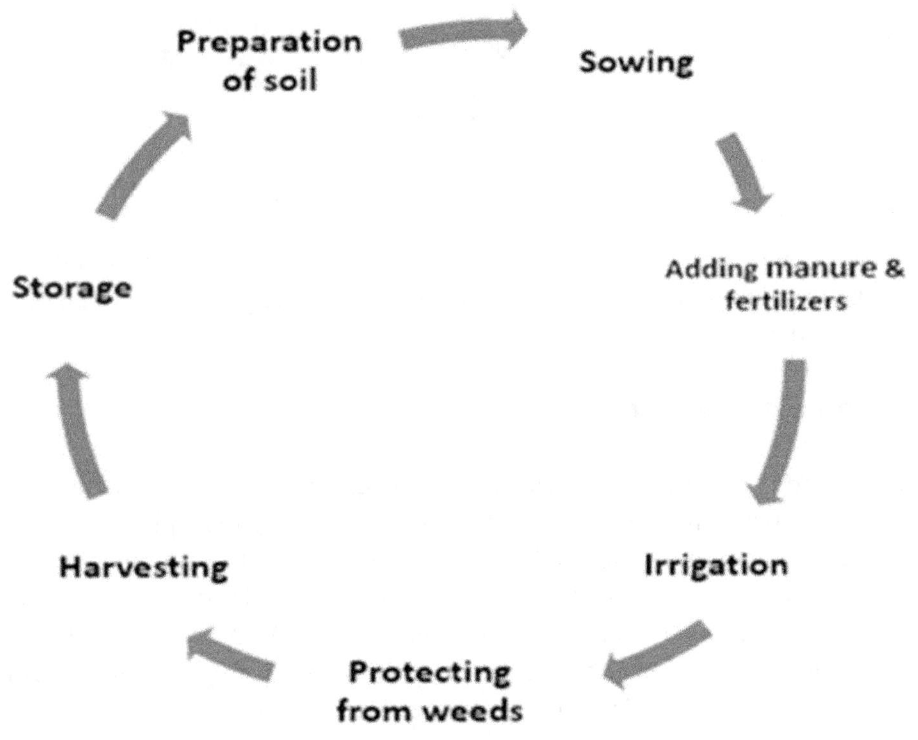

Broadly, the agricultural practices are divided into three major parts i.e. Pre-Harvesting, Harvesting, Post-Harvesting as illustrated in Fig. 2, and the further division is illustrated in Fig. 3.

Figure 2. Agriculture task categorization

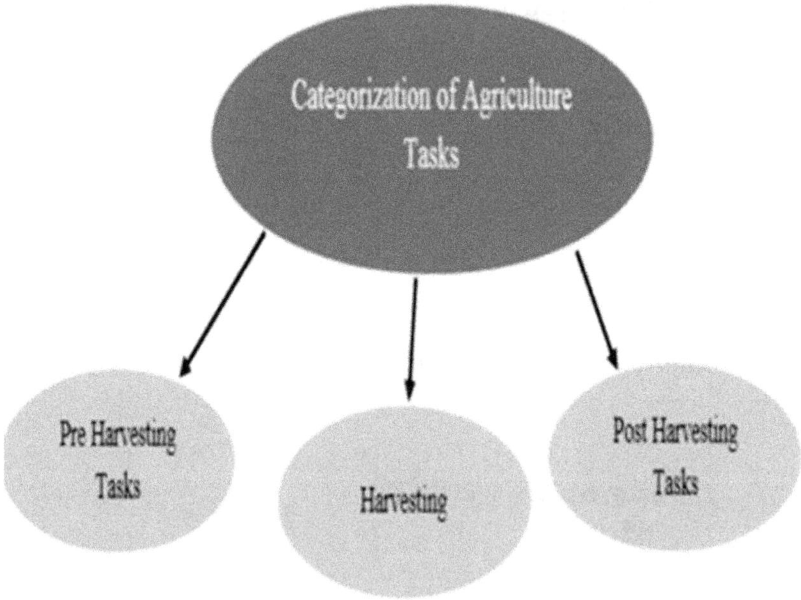

Multiple approaches exist for increasing the production of agricultural-based products with minimal effort. The different pre-requisites required before cultivation are discussed as below:

1.2 Land Management

Soil is a complicated and ambiguous natural resource that possesses a variety of characteristics. The effects of climate change on regional productivity can be determined by temperature. Algorithms based on machine learning examine evaporation processes, soil moisture, and temperature in order to obtain a clear insight into ecosystem dynamics and agricultural implications. ANN (Nahvi, et al., 2016) and ML algorithms can currently effectively manage forecasting and categorization problems. When machine learning techniques are used in agriculture, they make things much easier for domain experts.

Figure 3. Classification of agricultural stages

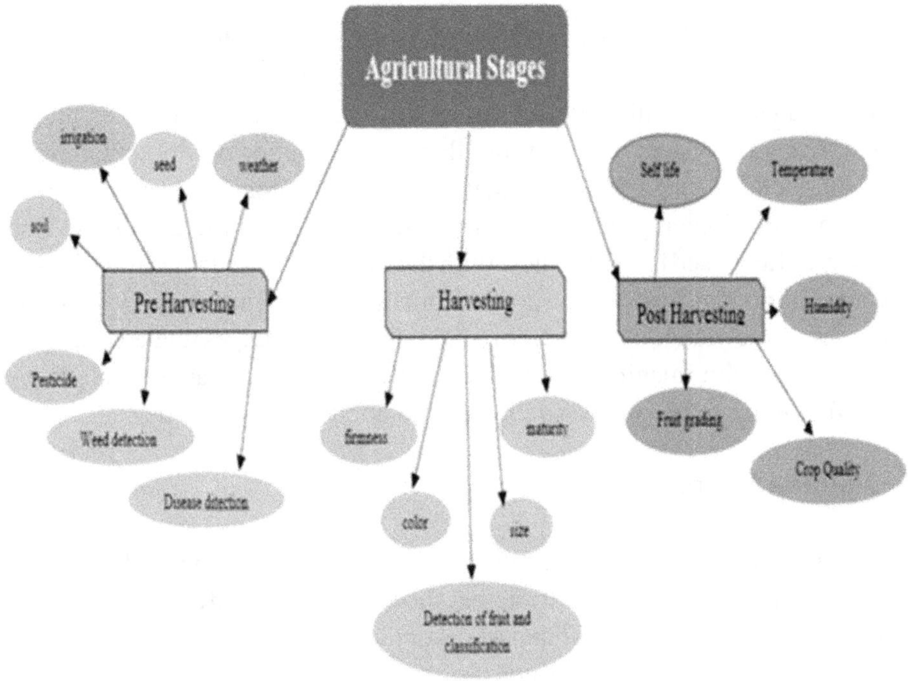

1.3 Breeding of Species

Species selection is a time-consuming process that involves looking for specific genes that affect how well plants use minerals and water, how well they adapt to climate change, how well they fight off disease, and whether they are healthier or taste better.

1.4 Recognition of Species

When classifying plants, people frequently compare the colour and shape of the leaves, but machine learning can produce more precise and immediate results by analysing the vein architecture of the leaves, which contains more information on leaf attributes. ML based methods like k-Nearest Neighbor (kNN), ANN and Support Vector Machines (SVM) are often used to identify plants (Malarvizhi, et al., 2021).

1.5 Water Management

The use of water in agriculture (irrigation) influences the hydrological, climatological, and agronomic equilibrium. The most advanced machine learning applications to date are those that estimate daily, weekly, or monthly evapotranspiration to enable more efficient irrigation system use and those that predict routine dew point temperature to assist in identifying probable weather events and estimating evapotranspiration and evaporation. Imagery data inputs from satellites, aircraft, and drones can be processed and analysed by AI engines. With the use of ML, particularly deep learning based algorithms, data can be interpreted from images and further analysis is done to identify some knowledge based pattern to address the challenges related to irrigation. Data combined with imaging of soil and plant based sensors can estimate the need of irrigation in real time and warn the farmers about potential issues.

1.6 Disease Diagnosis

The most frequent method of pest and disease control in both outdoor and a greenhouse setting is spraying pesticides evenly across the entire cultivated area. This technique necessitates the use of a large number of pesticides, resulting in significant financial and environmental expenses. Now, machine learning is an important part of precision agricultural management, in which agrochemical inputs are aimed at the right time, place, and plants (Latha et al., 2014).

1.7 Detection of Weeds

Besides illnesses, weeds pose the greatest threat to crop output. The main barrier to weed control is that it is difficult to recognise and differentiate the weed from crops. Historically, men were employed specifically for the purpose of weed detection. They will locate the weed by inspecting each and every part of the field. After that, weeds are removed manually by workers. In later years, with the development of technology, they began to use pesticides to eliminate weeds. In many regions of the world, however, weed detection is still conducted manually. Afterwards, a few technologies to automatically detect weeds emerged, but their lack of precision prevented them from reaching the public. Image processing, ML and computer vision based techniques can enhance the identification and categorization of weeds at a low cost and without adverse environmental impact. Similar methods will be used in the future to control robots that remove weeds and decrease the need for herbicides.

1.8 Yield Prediction

As it specifies yield mapping and estimation, management of crop, supply and demand matching of crop. One of the most important and well-known issues in precision agriculture is the prediction of yield. Computer vision technologies provide real-time data and thorough multidimensional analyses of crops, weather, and economic circumstances in order to increase efficiency for farmers and the wider population.

1.9 Quality of Crop

The accurate identification and categorization of the qualitative attributes of crops can lead to an increase in product value and a reduction in waste. Unlike human experts, machines can use seemingly unrelated data and connections to find and show new crop traits that improve their overall quality.

2. RELATED WORK

Numerous researchers have conducted a range of efforts to improve crop quality and yield. Table 1 discusses the contributions of researchers related to advancement in agriculture.

Table 1. Different researchers work on different techniques

Researcher	Methodology	Working area	Final result
Suchithra *et al.*, (2020)	Levenberg-Marquard artificial neural networks (*ANNs*)	Additionally, clay, silt, and sandy loam soils, carbon present in soil, and soil bulk density were used to anticipate soil fertility using partial least squares regression	techniques to estimate soil fertility
Kapoor, *et al.*, (2016)	IoT and Image Processing *IPT* of *MATLAB*	Find the cause of hindrance of growth of plants whether human made or environmental factor.	Combine image processing and *IoT*
Wilf *et al*, (2016)	Machine learning	Figure out what kind of leaf it is by its shape, texture, and colour, among other things.	The average recognition rate it demonstrates is relatively high, approaching *70%*.
Mohanraj *et al.*, (2016)	As main providers of data .NET with C# and XML are used and also Bluetooth for offline data transfer is used.	Reminder module that sends SMS reminders for reaping, application of fertilizers, pesticide treatment, and irrigation.	Offered a system involving a *TI CC3200* Launchpad networked with many sensors capable of measuring numerous facets of the examined land, which would also minimise labour costs.

continues on following page

Table 1. Continued

Researcher	Methodology	Working area	Final result
Gonzalez-de-Soto *et al.*, (2016)	AI-based approach for unmanned ground vehicles. Equipped with an internal combustion engine, a photovoltaic panel, batteries, and a hydrogen fuel cell.	Modifications to the commercial *CNHi* Boomer *3050 CVT* small tractor and a control system were proposed to give the vehicle autonomous driving capabilities.	Demonstrate an effective use of technology based on green energy sources, reducing toxic emissions by fifty percent.
Rokhmana *et al.*, (2015)	Unmanned Aerial Vehicles and *ML*	The deployment of their method resulted in the accurate calculation of stand per Hectare and individual palm trees using Orthophotos.	It demonstrates that a low-cost UAV equipped with a digital pocket camera may collect data used to count items and assist farms in calculating annual expenses. Simple to operate for farm employees.
Suksawat *et al.*, (2015)	different Image Processing	Develop a method wherein this could be discovered automatically by artificial intelligence	Utilizing technology as a supplement to increase the accuracy and efficiency of early disease detection on plants.
Marfatia *et al.*, (2013)	Explains about various techniques used to work for Varity of benefits such as *ML (LDC, K-NN), ANN*	The detection of insects for the purpose of pest control, the identification of weeds for the purpose of crop evaluation, the classification of fruits and vegetables for quality and grading, the harvesting of fruits and vegetables, nitrogen estimation for plants, and real-time object tracking.	Training neural networks achieves 75% to 96% accuracy, while fractal analysis and *CIELAB* produce 100% accuracy. The histogram often peaks between 85 and 97 percent when using conventional methods like colour mapping.
Mucherino *et al.*, (2009)	ML, K-nearest neighbours (*KNN*), One-R, and Apriori classifier algorithms)	Using phenotypic plant variables as inputs, the wheat yield was categorised and predicted.	Categorised the wheat yield as low, medium, and high using counter-propagation neural networks.
Mustafa *et al.*, (2008)	Machine vision	For finding ripeness of fruit.	Got success in finding size and quality of bananas also calculate the percentage of ripeness by using software.

3. METHODOLOGY

This section describes the remedies suggested by researchers and accepted by farmers to a satisfactory level.

3.1 Significance of Machine Learning in Agriculture

Agriculture is extremely important to India, which has established itself as the world's leader in the production of agricultural goods such as milk, wheat, cotton, vegetables, and fruits. There are a lot of uses for machine vision in agriculture in India, and they all show that it works. Diseases that affect the leaves, stems, fruits, and vegetables of an organism can be detected with the assistance of terrestrial and airborne sensors that transfer image data to equipment for evaluation and processing. In order to enhance the quality of fruits and vegetables, they must be classified.

The damped least-squares (DLS) method, also called the Levenberg-Marquardt (LMA or LM) method, is used to solve nonlinear least-squares problems. In the early days of ML techniques, soil fertility was estimated using the back-propagation approach in ANNs. Additionally, clay, silt, soil bulk density, soil carbon and sandy loam soils were used to predict soil fertility (Mustafa, et al., 2008). ML methods have been used to anticipate soil fertility, provide enough nutrients and water, and other agricultural soil challenges. Wheat yield was classified and predicted using the J48, K-nearest neighbours (KNN), One-R and Apriori classification algorithms using phenotypic plant variables as inputs (Kumar et al., 2016, De Paul et al., 2016; Goel, et al., 2022). For the assessment of soil organic carbon in eastern Australia, the random forest has been integrated with genetic algorithm feature selection (Schillaci, et al., 2017).. Organic carbon, soil pH, and cation exchange capacity were estimated using mid-infrared spectra (Wang, B. et al., 2018). The Bayesian network (Terhoeven-Urselmans, et al., 2010), is used to assess the soil nutrients (like phosphorus, nitrogen, potassium, copper, organic carbon, iron and zinc) and also the pH value to assess soil fertility. The portability of machine learning techniques to create wind speed has been studied. The soil and climate, which are directly associated with precision farming and crop growth, were analysed using, RF, ANN, Bayesian Networks, DT and SVM (Jia, et al., (2010).

According to studies, technology can be present and useful in a variety of facets of agriculture, including insect management for the detection of insects, crop evaluation for the identification of undesired weeds, quality and grading for the classification of fruits and vegetables, fruit and vegetable harvesting, estimation of plant nitrogen, and real-time object tracking. In addition, they offer a number of classifiers that are crucial for applying machine learning, such as the linear discriminant classifier (LDC), which aims to construct a boundary between image data and is used to extract product features. The K nearest classifier utilises fruit properties such as colour and shape to train the system using a Euclidean algorithm that seeks for the K shortest distance to the input fruit and assigns it to the K fruits nearest to its origin. Both linear and nonlinear data can be classified with SVM. It was used to identify weeds and determine in which areas pesticides needed to be applied. ANN is one of the most prevalent methods; it is inspired by a variety of well-known techniques and acts as a biological nervous system, making it robust when dealing with large data volumes. ANN has achieved remarkable results in classifying crops, such as apples, into categories (AA, A, B, C, D) that indicate their quality requirements. The Rule-Based System, also known as the fuzzy system, defines a set of rules for extracting data from photographs, extracting visual features, creating these rules, and identifying and classifying fruits. The table 2 presents the practices of various ML based approaches in agriculture.

By merging existing approaches and producing expressive outcomes, machine learning has proved its worth and efficacy in fruit grading and sorting.

Table 2. Machine learning in different practices

Part encountered	Features discussed	Algorithm to train the machine	Accuracy
Soil (Suchithra, & Maya, 2020)	Check soil fertility indices for different villages from available Organic Carbon, Soil Reaction, Boron, Potassium and Phosphorus	Extreme Learning Machine equipped with a variety of activation functions including, triangular basis, hyperbolic tangent Gaussian radial basis and sine-squared	up to *80%*
Seed (Maize) Sivakumar, et al., 2019)	Quality of seed by its color shape and size observation	*ML* techniques *KNN, SVM*, Googlenet, *VGG19*	*90%*
Haploid seeds of maize (Uzal, et al., 2018)	Morphology, texture, quality of seed	DeepSort, Support Vector Machine (*SVM*), *RF*, and Logistic Regression (*LR*)	Check whether true haploid or true diploid
Detection of pest and any dieses (Türkoğlu, & Hanbay, 2019)	Deep learning by classifying in *8* categories (*5* disease and *3* pest)	*ELM, SVM, KNN, VGG16, VGG19*	*98%*
Grading classification of cavendish banana for post harvesting (Ucat, et al., 2019)	Banana	Python OpenCV and Tensor flow	*90%*
Apple sorting system for bruise detection (Sofu, et al., 2016)	Automatic bruise detection	*K*- means, decision tree	*79%*
wheat yield (Romero, et al., 2013)	Categorised as low, medium, and high	counter-propagation neural networks and supervised Kohonen networks	Highly
Decide which insecticides to apply (Pantazi, et al., 2016)	Pest monitoring on kiwi fruit	Naive Bayes classifier, *DT, RF, SVM*, AdaBoost, and Logistic Regression (*LR*)	*80%*
Soil organic carbon (Hill, et al., 2014; Ritz, et al., 2015)	Forecast	An unbiased linear predictor Boosted regression trees	Quite accurate

3.2 Image Processing in Agriculture

Non-invasive techniques have improved in the last 10 years in terms of sensitivity, consistency, standardisation, high throughput, speed, and cost-effectiveness (Fabio, et al., 2017). The detection and real-time monitoring of plant diseases on a large scale and with high accuracy are made possible by spectroscopy-based, imaging-based, and adaptable remote sensing techniques. There are linkages between these coexisting and interacting non-invasive approaches. Spectroscopy-based processes have been formally acknowledged by the International Standards Committee (Elavarasan, et al., 2018). The most often used spectroscopic methods in plant disease research are electric impedance, visible-near infrared (VIS-NIR), and fluorescence spectroscopy (Borges, et al., 2012; Belasque, et al., 2008, Dhingra, et al., 2017). Consequently, due to their ability to deliver exact spatial information, imaging-based and RS approaches have gained a lot of attention. A range of imaging sensors, including digital [41, 42], fluorescent (Raza, et al., 2015), thermal (Sankaran, et al., 2013; Söderström, 2014), and multispectral or hyperspectral sensors (Wahabzada, et al., 2016; López-López, et al., 2016; Mahlein, et al., 2015), have been studied for the detection of symptomatic and asymptomatic plant diseases. These methods can be used for image segmentation, classification algorithm analysis, sensor development, and system construction (Al-Saddik, et al, 2017; Choudhary and Gulati, 2015),

The techniques for classifying any crop or identifying agricultural diseases can be broken down into the five processes depicted in Figure 4. To reduce noise, the image is pre-processed with techniques such as image clipping, smoothing, and enhancement before applying a grayscale filter to generate a histogram using a distribution function to enhance the image. Three layers of segmentation are involved such as:

- Boundary and spot detection, wherein a boundary detection method and eight connectivity of pixels are applied to an RGB image to produce an HSI model that detects the boundaries and spots of diseases.
- K-means clustering, in which the minimum distance between an object and a cluster is determined by searching a collection of leaf properties.
- The Otsu Threshold method converts grayscale images into binary format by assigning a value of 0 or 1 to pixels below or above a specified threshold.

Following these steps is feature extraction, which is where details like colour, texture, and edges are reconstructed in an image. Both neural networks and back propagation can be used to further refine the data after extraction.

Although it may contribute to several facets of the agriculture industry, image processing as an enabler would increase the rate of technological system adoption. Machine vision has been used successfully to detect diseases on fruits, insects, and for food grading. Weeds are harmful to farms because they compete for space, water, light, and nutrients with crops. Widespread is the development of numerous image processing-based systems that employ techniques such as the recognition of edges, colour, and other visual features (Zhang, 2022). In actuality, artificial visible examination (*Ruiz, et al., 2009*) remains the most basic and direct method. Nonetheless, this technique requires expert understanding of the pertinent plant phenotypic and plant pathological characteristics.

Figure 4. Classification of image processing

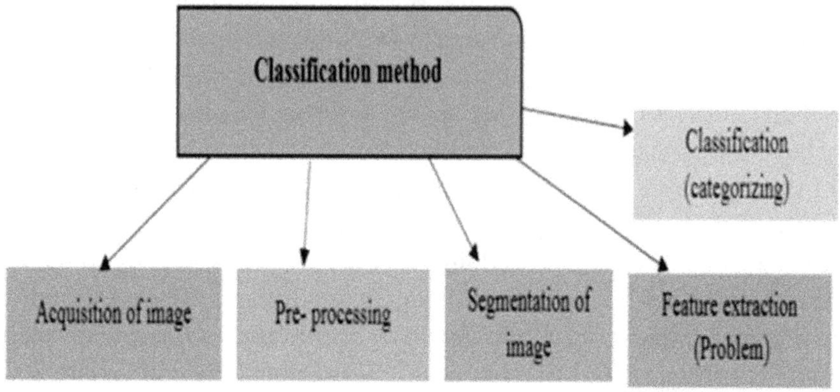

The biological molecular method (Sankaran, et al., 2010; Yvon, et al., 2009) is an additional common and straightforward technique for diagnosing plant diseases. Molecular biological technologies require precision sampling and intricate processing techniques. These procedures are more professional and systematic than artificial investigative techniques. These two operations are significant, efficient, and always involve manual monitoring and diagnosis of plant diseases. These manual operations are all expensive, labor-intensive, and time-consuming (Ali, et al., 2019). These constraints make it difficult to create and use artificial methods in agriculture on a large scale. These direct approaches for plant disease detection are frequently employed during the middle to later stages of infection, when symptoms are typically present. The importance of large-scale and real-time plant disease detection is developing in the context of modern facility agriculture and precision agriculture. This inevitable delay is not conducive to early identification.

3.3 Machine Vision

Due to natural human low visual abilities, they are unable to detect the initial signs of plant diseases as they occur at a microscopic level, necessitating the use of technology as a supplement to boost the accuracy and efficiently recognise plant diseases in their early stages (Pujari, et al., 2015; Tripathi, et al., 2022). The amount of labour necessary to manually identify plant diseases is the fundamental reason why it is necessary to build an artificial intelligence-based method for doing so (Khirade and Patil, 2015). This chapter aims to analyse further the methods and uses of the Agriculture sector, reviewing current suggestions that will bring humanity to a more technical approach aimed at a safer environment and highlighting the importance of adopting technologically based systems in our society. The use of machine vision to detect diseases on fruits, insects, and for food grading has proved successful. Widespread (Vibhute, et ak., 2012) is the development of machine vision-based systems that employ techniques such as the recognition of edges, shape, colour, and other weed features (Vibhute, et al., 2012; Tyagi, et al., 2012; Tripathi, et al., 2022; Rai, et al., 2022; Rai, et al., 2022). This chapter aims to analyse further the methods and uses of the agriculture sector, reviewing current suggestions that will bring humanity to a more technical approach aimed at a safer environment and highlighting the importance of adopting technologically based systems in our society.

It is worth to note that before processing the image, de-noising is vital in order to get correct results (Tripathi and Siddiqi 2017, Tripathi, et al. 2020, Tripathi 2020).

3.3.1 Unmanned Aerial Vehicles

Indonesia's technical agriculture sector has expanded throughout time with the introduction of Unmanned Aerial Vehicles (UAV) as a substitute for satellite photography due to the higher quality of UAV photos in comparison to satellite photographs. Gonzalez-de-Soto et al. (2016) proposed a propose a low-cost method that utilises a $2,000 UAV that is easy for agricultural employees to operate. The implementation of their technology in a palm oil plantation led to the precise estimation of stand per hectare (SPH) and individual palm trees using Orthophotos. Even though their solution could count trees, they agreed that the accuracy of their system must be enhanced to 95% before their clients would accept it. Their answer was well-received at a sugar factory for estimating the taxation on their sugar cane assets with a precision of 15 centimetres.

The use of UAVs to collect data from farm vegetation has been studied and shown to be beneficial. It demonstrates that a low-cost UAV paired with a digital pocket camera may collect data used to count goods and help farms calculate annual expenses.

3.3.2 Unmanned Ground Vehicles

To reduce tractor emissions of greenhouse gases such as CO2 and NOx, as well as other pollutants that may cause respiratory illnesses and other health problems. Gonzalez-de-Soto et al. 2012) proposed a change to the power system of the commercial CNHi Boomer 3050 CVT small tractor that significantly reduces the production of these pollutants. In addition, they included a control system that enables the vehicle to operate autonomously, effectively making it into an excellent Unmanned Ground Vehicle solution. It was a successful application of technology based on a clean energy source, reducing pollutants by 50%.

3.3 Application of Internet of Things

The term Internet of Things (*IoT*) originates from the definition provided by recommendation, *ITU*, which states that IoT creates a network of physical things that can be controlled by software, such as gadgets, automobiles, and sensors. India is one of the first nations to employ cutting-edge solutions in this field, and farmers require the incorporation of technology on their farms in order to maintain a viable company. To increase the profitability of India's firms, scientists propose many methods. Some proposed a method with a *TI CC3200* launchpad interconnected with many sensors capable of measuring aspects of the land under study, hence reducing labour costs on those properties. Various researchers note that *GPRS*, *GPS*, *3G*, and *Wi-Fi* are technologies that are unreliable and present a barrier when providing solutions in real-time. To overcome this shortcoming, scientist develop a Bluetooth interaction that enables the system to function in a mode where internet is not required. These features modules that convey essential information to the owner of land in actual time, just like a Reminder module that sends SMS notifications to remind farmers about harvesting, fertiliser application, pesticide spraying, and irrigation schedules.

Monitoring Plant Growth is the process of measuring the altitude of a plant and assessing its development. Monitoring soil moisture levels and using information from IoT sensors, an irrigation manager calculates how much water each crop needs while keeping the water supply to the field at a steady rate. An online calculator for estimating agricultural profits, taking into account the specifics of the crop being evaluated. Use cloud computing to monitor for disasters like wildfires by retrieving relevant data from Yahoo's weather services. Detector of issues with the irrigation system and its peripherals (such as the power supply) that ensures everything is working as it should be. The *IoT* devices and Yahoo! Web services ascertain whether or not extra watering of the soil is necessary. In order to make sure the plant develops properly, one can use a light intensity sensor and an optical monitor to keep track of the amount of light hitting it. Both the Well Dry Check and the Field Dry Check measure the moisture content of the soil to determine if a given area is dry. Farmers have average knowledge of technology; therefore, new solutions cannot rely on their understanding to continue functioning. In addition, it is unfeasible to employ professionals to keep farming operating as intended.

4. CONCLUSION

This chapter provides a comprehensive review of multiple methodologies, including machine learning, image processing, deep learning, the internet of things, and machine vision. Since there are three major types of agricultural tasks, i.e., pre-harvest,

harvest, and post-harvest, as discussed earlier and also shown in *Fig. 3*, the *Table 1* presents the existing research conducted by various researchers discussing the different challenges related to agriculture. To overcome such challenges, the different existing approaches have been discussed in section *3*. The existing approaches aid in improving the agricultural yields by improving the various agricultural requirements like soil testing, weed identification, pest identification, early disease diagnosis, and seed detection. *Table* 2 presents the analysis of a variety of *ML* techniques that are utilised at various stages of agriculture. In today's society, technology is an essential component in agriculture. Utilizing a variety of technologies, farmers can earn a substantial profit with minimal effort. According to existing work, certain technologies, such as machine learning and image processing, produce exceptional results.

REFERENCES

Al-Saddik, H., Simon, J.-C., & Cointault, F. (2017). Development of spectral disease indices for 'Flavescence Doree' grapevine disease identification. *Sensors (Basel)*, *17*(12), 2772. doi:10.339017122772 PMID:29186057

Ali, M. M., Bachik, N. A., Muhadi, N. A., Yusof, T. N. T., & Gomes, C. (2019). Non-destructive techniques of detecting plant diseases: A review. *Physiological and Molecular Plant Pathology*, *108*, 101426–101437. doi:10.1016/j.pmpp.2019.101426

Barbedo, J. G. A. (2013). Digital image processing techniques for detecting, quantifying and classifying plant diseases. *SpringerPlus*, *2*(1), 660–671. doi:10.1186/2193-1801-2-660 PMID:24349961

Belasque, J. Jr, Gasparoto, M. C. G., & Marcassa, L. G. (2008). Detection of mechanical and disease stresses in citrus plants by fluorescence spectroscopy. *Applied Optics*, *47*(11), 1922–1926. doi:10.1364/AO.47.001922 PMID:18404192

Borges, E., Matos, A. P., Cardoso, J. M., Correia, C., Vasconcelos, T., & Gomes, N. (2012). *Early detection and monitoring of plant diseases by Bioelectric Impedance Spectroscopy*. In *Proceedings of the 2012 IEEE 2nd Portuguese Meeting in Bioengineering (ENBENG)*, Coimbra, Portugal. 10.1109/ENBENG.2012.6331377

Business Bliss Consultants. (2018). *Importance of Image Processing Technology in Agriculture*. UKDISS. https://ukdiss.com/examples/image-processing-for-agriculture-sector.php?vref=1

Cen, H., Weng, H., Yao, J., He, M., Lv, J., Hua, S., Li, H., & He, Y. (2017). Chlorophyll fluorescence imaging uncovers photosynthetic fingerprint of citrus Huanglongbing. *Frontiers in Plant Science*, *8*, 1509–1519. doi:10.3389/fpls.2017.01509 PMID:28900440

Choudhar, G. M., & Gulati, V. (2015). *Advances in Image Processing for Detection of Plant Diseases*, *2*(2), 135–141.

De Paul Obade, V., & Rattan, L. (2016). Towards a standard technique for soil quality assessment. *Geoderma*, *265*(8), 96–102. doi:10.1016/j.geoderma.2015.11.023

Dhingra, G., Kumar, V., & Joshi, H. D. (2017). Study of digital image processing techniques for leaf disease detection and classification. *Multimedia Tools and Applications*, *77*(15), 19951–20000. doi:10.100711042-017-5445-8

Elavarasan, Vincent, D. R., Sharma, V., Zomaya, A. Y., & Srinivasan, K. (2018). Forecasting yield by integrating agrarian factors and machine learning models: A survey. *Computers and Electronics in Agriculture*, *155*, 257–282. doi:10.1016/j.compag.2018.10.024

Fabio, V., Athina, K., Rene, B. & Martin, R. (2017). Assessing accuracy and geographical transferability of machine learning algorithms for wind speed modelling, 297-310. *Lecture Notes in Geoinformation and Cartography*. Springer. doi:10.1007/978-3-319-56759-4_17

Goel, D., Gulati, P., Jha, S. K., Kumar, N., Khan, A., & Kumari, P. (2022). *Evaluation of Machine Learning Techniques for Crop Yield Prediction, Artificial Intelligence Applications in Agriculture and Food Quality Improvement*. IGI Global. doi:10.4018/978-1-6684-5141-0.ch007

Gonzalez-de-Soto, M., Emmi, L., Benavides, C., Garcia, I., & Gonzalez-de-Santos, P. (2016). Reducing air pollution with hybrid-powered robotic tractors for precision agriculture. *Biosystems Engineering*, *143*, 79–94. doi:10.1016/j.biosystemseng.2016.01.008

Guilmartin, J. F. (2022). *Unmanned aerial vehicle*. Encyclopedia Britannica. https://clearpathrobotics.com/blog/2021/01/husky-ugv-drives-smart-precision-agriculture-for-steep-slope-vineyards/

Hill, M. G., Connolly, P. G., Reutemann, P., & Fletcher, D. (2014). The use of data mining to assist crop protection decisions on kiwifruit in New Zealand. *Computers and Electronics in Agriculture*, *108*, 250–257. https://cropforlife.com/list-of-agricultural-practices/ https://www.kdnuggets.com/2019/05/machine-learning-agriculture-applications-techniques.html. doi:10.1016/j.compag.2014.08.011

Jayalakshmi, R., & Balasubramanian, T. (2015). A brief survey of classification techniques applied to soil fertility prediction. *International Journal of Scientific Research in Computer Science Applications and Management Studies*. http://www.ijsrcsams.com/images/stories/Past_Issue_Docs/ijsrcsamsv8i2p121.pdf

Jia, H. Y. (2010). Soil fertility grading with Bayesian Network transfer learning. *IEEE International Conference on Machine Learning and Cybernetics*. IEEE. 10.1109/ICMLC.2010.5580915

Kapoor, A., Bhat, S. I., Shidnal, S., & Mehra, A. (2016). Implementation of IoT (Internet of Things) and Image processing in smart agriculture. *International Conference on Computation System and Information Technology for Sustainable Solutions*. IEEE. DOI:10.1109/CSITSS.2016.7779434

Khirade, S. D., & Patil, A. B. (2015). Plant Disease Detection Using Image Processing. *International Conference on Computing Communication Control and Automation*. IEEE. 10.1109/ICCUBEA.2015.153

Kumar, G., Kumar, S., & Kumar, N. (2015). Comparative study of wavelet and wavelet packet transform for denoising telephonic speech signal. *International Journal of Computer Applications*, *110*(15), 1–8. doi:10.5120/19389-9145

Kumar, N., Alam, K., & Siddiqi, A. H. (2016). Savitzky–Golay and Wavelet Transform Based Rama Spectroscopic data Denoising. *International Journal of Control Theory and Applications*, *9*(21), 297–305.

Kumar, N., Siddiqi, A. H., & Alam, K. (2014). Raman spectral data de-noising based on wavelet analysis. *International Journal of Computer Applications*, *102*(16), 20–22. doi:10.5120/17899-8864

Kumar, N., Siddiqi, A. H., & Alam, K. (2016). Raman spectra denoising using wavelet and wavelet packet transform. *Indian Journal of Industrial and Applied Mathematics*, *7*(01), 1–10. doi:10.5958/1945-919X.2016.00001.3

Kumar, N., Tripathi, P., & Alam, K. (2016). Non-Negative Matrix Factorization-Based EEG Signal Classification. *Indian Journal of Industrial and Applied Mathematics*, *7*(2), 212–219. doi:10.5958/1945-919X.2016.00019.0

Latha et al. (2014). Image Processing in Agriculture. *International journal of innovative research in electrical, electronics, instrumentation and control engineering*, *2*(6), 1562-1565.

López-López, M., Calderón, R., González-Dugo, V., Zarco-Tejada, P., & Fereres, E. (2016). Early detection and quantification of almond red leaf blotch using high-resolution hyperspectral and thermal imagery. *Remote Sensing (Basel)*, *8*(4), 276. doi:10.3390/rs8040276

Mahlein, A.-K., Hammersley, S., Oerke, E.-C., Dehne, H.-W., Goldbach, H., & Grieve, B. (2015). Supplemental blue LED lighting array to improve the signal quality in hyperspectral imaging of plants. *Sensors (Basel)*, *15*(6), 12834–12840. doi:10.3390150612834 PMID:26039423

Malarvizhi, K., Sowmithra, M., Gokula, P. D., & Kabila, B. (2021). Machine Learning for Plant Species Classification using Leaf Vein Morphometric. *International Journal of Engineering Research & Technology (Ahmedabad)*, *10*(04).

Mohanraj, I., Ashokumar, K., & Naren, J. (2016). Field Monitoring and Automation Using IOT in Agriculture Domain. *Procedia Computer Science*, *93*, 931–939. doi:10.1016/j.procs.2016.07.275

Mucherino, A., Petraq, P., & Panos, M. P. (2009). A survey of data mining techniques applied to agriculture. *Operations Research*, *9*(2), 121–140. doi:10.100712351-009-0054-6

Mustafa, N. B. A., Fuad, N. A., Ahmed, S. K., Abidin, A. A. Z., Ali, Z., Yit, W. B., & Sharrif, Z. A. M. (2008). Image processing of an agriculture produce: Determination of size and ripeness of a banana, *International Symposium on Information Technology*. DOI:10.1109/ITSIM.2008.4631636

Nahvi, B., Habibi., J, Mohammadi., K, Shamshirband, S., Saleh Al Razgan, O. (2016). Using self-adaptive evolutionary algorithm to improve the performance of an extreme learning machine for estimating soil temperature. *Computer Electron Agriculture*, *124*, 0168–1699. http://dx.doi.org/ doi:10.1016/j.compag.2016.03.025

Pantazi, X. E., Moshou, D., Alexandridis, T., Whetton, R. L., & Mouazen, A. M. (2016). Wheat yield prediction using machine learning and advanced sensing techniques. *Computers and Electronics in Agriculture*, *121*, 57–65. doi:10.1016/j.compag.2015.11.018

Pujari, J. D., Yakkundimath, R., & Byadgi, A. S. (2015). Image Processing Based Detection of Fungal Diseases in Plants. *Procedia Computer Science*, *46*, 1802–1808. doi:10.1016/j.procs.2015.02.137

Rashmi, P., Naik, S., & Marfatia, R. (2013). Image Processing and Machine Learning for Automated Fruit Grading System: A Technical Review. *International Journal of Computer Applications*, *81*(16), 29–39. doi:10.5120/14209-2455

Raza, S.-A., Prince, G., Clarkson, J. P., & Rajpoot, N. M. (2015). Automatic detection of diseased tomato plants using thermal and stereo visible light images. *PLoS One*, *10*(4), e0123262–e0123281. doi:10.1371/journal.pone.0123262 PMID:25861025

Ritz, C., Putku, E., & Astover, A. (2015). A practical two-step approach for mixed model-based kriging, with an application to the prediction of soil organic carbon concentration. *European Journal of Soil Science*, *66*(3), 548–554. doi:10.1111/ejss.12238

Rokhmana, C. A. (2015). The Potential of UAV-based Remote Sensing for Supporting Precision Agriculture in Indonesia. *Procedia Environmental Sciences*, *24*, 245–253. doi:10.1016/j.proenv.2015.03.032

Romero, J. R., Roncallo, P. F., Akkiraju, P. C., Ponzoni, I., Echenique, V. C., & Carballido, J. A. (2013). Using classification algorithms for predicting durum wheat yield in the province of Buenos Aires. *Computers and Electronics in Agriculture*, *96*, 173–179. doi:10.1016/j.compag.2013.05.006

Ruiz-Ruiz, S., Ambrós, S., Vives, M. D. C., Navarro, L., Moreno, P., & Guerri, J. (2009). Detection and quantitation of Citrus leaf blotch virus by TaqMan real-time RT-PCR. *Journal of Virological Methods*, *160*(1-2), 57–62. doi:10.1016/j.jviromet.2009.04.012 PMID:19406167

Sankaran, S., Maja, J. M., Buchanon, S., & Ehsani, R. (2013). Citrus Greening detection using visible, near infrared and thermal imaging techniques. *Sensors (Basel)*, *13*(2), 2117–2130. doi:10.3390130202117 PMID:23389343

Sankaran, S., Mishra, A., Ehsani, R., & Davis, C. (2010). A review of advanced techniques for detecting plant diseases. *Computers and Electronics in Agriculture*, *72*(1), 1–13. doi:10.1016/j.compag.2010.02.007

Schillaci, C., Acutis, M., Lombardo, L., Lipani, A., Fantappiè, M., Märker, M., & Saia, S. (2017). Spatio-temporal topsoil organic carbon mapping of a semi-arid Mediterranean region: The role of land use, soil texture, topographic indices and the influence of remote sensing data to modelling. *The Science of the Total Environment*, *601*, 821–832. doi:10.1016/j.scitotenv.2017.05.239 PMID:28578240

Searchinger, T., Waite, R., Hanson, C. & Ranganathan, J. (2019). *World Resources Report: Creating a Sustainable Food Future*. United Nations, World Resources Institute, and the World Bank.

Sivakumar, D. (2019). Computerized growth analysis of seeds using deep learning method, *International Journal of Recent Technology and Engineering (IJRTE)*, *7*(6S5), 1885-1892.

Söderström, M., Börjesson, T., Roland, B., & Stadig, H. (2014). Modelling within-field variations in deoxynivalenol (DON) content in oats using proximal and remote sensing. *Precision Agriculture, 16*(1), 1–14. doi:10.100711119-014-9373-6

Sofu, M. M., Er, O., Kayacan, M. C., & Cetisli, B. (2016). Design of an automatic apple sorting system using machine vision. *Computers and Electronics in Agriculture, 127*, 395–405. doi:10.1016/j.compag.2016.06.030

Suchithra, M. S., & Maya, P. (2020). Improving the prediction accuracy of soil nutrient classification by optimizing extreme learning machine parameters. *Information Processing in Agriculture, 7*(01), 72–82. https://www.sciencedirect.com/science/article/pii/S2214317318304906. doi:10.1016/j.inpa.2019.05.003

Suksawat, B., & Komkum, P. (2015). Pineapple quality grading using image processing and fuzzy logic based on Thai Agriculture Standards. *International Conference on Control, Automation and Robotics*. 10.1109/ICCAR.2015.7166035

Terhoeven-Urselmans, T., Vagen, T.-G., Spaargaren, O., & Shepherd, K. D. (2010). Prediction of soil fertility properties from a globally distributed soil mid-infrared spectral library. *Soil Science Society of America Journal, 74*(5), 1792–1799. doi:10.2136ssaj2009.0218

Tripathi, P. (2020). Electroencephalpgram Signal Quality Enhancement by Total Variation Denoising Using Non-convex Regulariser. *International Journal of Biomedical Engineering and Technology, 33*(2), 134–145. doi:10.1504/IJBET.2020.107709

Tripathi, P., Kumar, N., Rai, M., & Khan, A. (2022). Applications of Deep Learning in Agriculture. In *Artificial Intelligence Applications in Agriculture and Food Quality Improvement* (pp. 17–28). IGI Global., doi:10.4018/978-1-6684-5141-0.ch002

Tripathi, P., Kumar, N., & Siddiqi, A. H. (2020). De-noising Raman spectra using total variation de-noising with iterative clipping algorithm. In *Computational Science and its Applications, Taylor and Francis Group* (pp. 225–231). CRC Press. doi:10.1201/9780429288739-14

Tripathi, P., & Siddiqi, A. H. (2016). Solution of Inverse Problem for de-noising Raman Spectral Data with Total variation using Majorization-Minimization Algorithm. Int. J. *Computing Science and Mathematics, 7*(3), 274–282. doi:10.1504/IJCSM.2016.077855

Türkoğlu, M., & Hanbay, D. (2019). Plant disease and pest detection using deep learning-based features. *Turkish Journal of Electrical Engineering and Computer Sciences*, *27*(3), 1636–1651. doi:10.3906/elk-1809-181

Tyagi, N., Rai, M., Sahw, P., Tripathi, P., & Kumar, N. (2022). Methods for the Recognition of Human Emotions Based on Physiological Response: Facial Expressions. In *Smart Healthcare for Sustainable Urban Development* (pp. 183–202). IGI Global., doi:10.4018/978-1-6684-2508-4.ch013

Ucat, R. C., & Dela, C. J. C. (2019). Postharvest grading classification of Cavendish banana using deep learning and tensorflow, *International Symposium on Multimedia and Communication Technology (ISMAC)*. IEEE. 10.1109/ISMAC.2019.8836129

Uzal, L. C., Grinblat, G. L., Namías, R., Larese, M. G., Bianchi, J. S., Morandi, E. N., & Granitto, P. M. (2018). Seed-per-pod estimation for plant breeding using deep learning. *Computers and Electronics in Agriculture*, *150*, 196–204. doi:10.1016/j.compag.2018.04.024

Vasudevan, A., Kumar, D. A., & Bhuvaneswari, N. S. (2016). *Precision farming using unmanned aerial and ground vehicles. IEEE Technological Innovations in ICT for Agriculture and Rural Development*. TIAR.

Vibhute, A. K., Bodhe, S., Vibhute, A. K., & Bodhe, S. (2012). Applications of Image Processing in Agriculture: A Survey. *International Journal of Computer Applications*, *52*(2), 34–40. doi:10.5120/8176-1495

Wahabzada, M., Mahlein, A.-K., Bauckhage, C., Steiner, U., Oerke, E.-C., & Kersting, K. (2016). Plant phenotyping using probabilistic topic models: Uncovering the hyperspectral language of plants. *Scientific Reports*, *6*(1), 22482–22492. doi:10.1038rep22482 PMID:26957018

Wang, B., Waters, C., Orgill, S., Cowie, A., Clark, A., Li Liu, D., Simpson, M., McGowen, I., & Sides, T. (2018). Estimating soil organic carbon stocks using different modelling techniques in the semi-arid rangelands of eastern Australia. *Ecological Indicators*, *88*, 425–438. doi:10.1016/j.ecolind.2018.01.049

Wilf, P., Zhang, S., Chikkerur, S., Little, S. A., Wing, S. L., & Serre, T. (2016). Computer vision cracks the leaf code. *Proceedings of the National Academy of Sciences of the United States of America*, *113*(12), 3305–3310. doi:10.1073/pnas.1524473113 PMID:26951664

Yvon, M., Thébaud, G., Alary, R., & Labonne, G. (2009). Specific detection and quantification of the phytopathogenic agent 'Candidatus Phytoplasma prunorum'. *Molecular and Cellular Probes*, *23*(5), 227–234. doi:10.1016/j.mcp.2009.04.005 PMID:19401232

Zhang, N. (2022). *Plant Disease Detection*. Encyclopedia. https://encyclopedia.pub/entry/2567

Chapter 5
Role of Image Processing in Agricultural Practices

Agrima Saxena
Department of Computer Science Engineering, Galgotias University, Greater Noida, India

Ayush Bharadwaj
Department of Computer Science Engineering, Galgotias University, Greater Noida, India

Usha Chauhan
Department of Electrical, Electronics and Communications, Galgotias University, Greater Noida, India

Divya Sharma
Department of Electrical, Electronics and Communications, Galgotias University, India

Priti Kumari
National Institute of Technology, Patna, India

ABSTRACT

Agriculture is considered necessary to sustain life on this planet. With a growing population, we need agricultural productivity to increase significantly to meet the demands. Image processing has proven to be a powerful tool for analysis in many fields and applications. The agricultural sector, where parameters such as canopy cover, yield, and product quality were important for farmers. It is often the case that professional advice is not affordable, and that the availability of an expert and their services may take time. The situation could be greatly improved if image processing was combined with a reliable communication network, as this would eliminate the need to obtain expert advice within a limited time frame. This can be achieved because image processing is a powerful tool for parameter analysis.

DOI: 10.4018/978-1-6684-6418-2.ch005

INTRODUCTION

With continued population growth and increased global demand for food, combined with the unusual change in climate and instability of politics, the industry of agriculture must improve methods to increase field production. This resulted in research for new techniques and technologies that enable farming to get maximum output with minimal input. Image processing is one such technique that can help farmers to detect weeds on farms, control pesticides, observe water usage, and assess the colour of fruits and vegetables. In the late century, weed detection was done manually; special labour was used to detect weeds and plug them with hands from each and every part of the field. Later, herbicides was used to remove weeds by them, but manual labour was still used to detect weeds. After advancements in image processing techniques, they started using image processing for the detection of weeds. Here, we have stated and explained the various image processing and weed detection techniques, and the application of image processing.

Plant diseases that occur in different parts of the plant can be observed by changes in symptoms, colors, spots, etc. Observing these diseases with the naked eye is a typical method, but experts use image processing to identify them in their initial stages and stop them from spreading throughout the field. Color recognition of fruits and vegetables in the food and agriculture industry using image processing and with the help of machine learning has become a popular trend in recent years (Ayman & Ayman, n.d.; Feng & Sun, 2012; Zhang et al., 2014). Color is the main feature in identifying product quality. Color is probably the first parameter that consumers use to determine the quality of food products (Maldonado et al., 2012). Appearance is a factor that consumers use to reject or accept a product. This has a significant impact on product sales. Therefore, in recent years, significant efforts have been made in the field of machine learning to improve food quality. Building a machine that can detect colors like a human being has been a difficult task for the scientific community and industry in recent years (Wabali et al., 2017). An image processing system requires adequate capture conditions. Basic image processing components are:

- Image sensor (complementary metal oxide semiconductor (CMOS).
- Illuminate (fluorescent lamps).
- High contrast background for an interesting subject.
- Frame grabber to capture the real photograph.
- Start the usage of the device to begin taking photos.

Calorimetry is a scientific area that deals with measurement, quantification, and the representation of color. It's very useful in various fields because it has the ability to change color into an objective factor instead of a subjective factor (Collewet & Marchand, 2009). By supplementing this scientific field with the system of computer vision technology, it's possible to evaluate color in a non-invasive, non-contact way and to monitor color properties in a digital and standard way, making this technology suitable for use in food and the agriculture industry for quality assessment. Its versatility allows many other production systems to take advantage of the possibilities offered by colorimetry systems, image processing, and artificial vision.

IMAGE PROCESSING

Image processing is the strategy of changing an image into enhanced frames and performing some operation on it. For example, obtaining an enhanced picture or focusing important data on it. This is a type of label setting where the information is an image, similar to a video or photo edge, and the final result can be a picture or the qualities and attributes connected with that picture. These images are usually considered as two-dimensional signals, and a set of signal processing methods are included in image processing.

It is one of the fastest-growing innovations today that is applied in many business sectors. Image processing is a major area of exploration and research in computer applications and engineering. Processing of image consists of three important developments: importing images from a scanner or digital photography, image analysis and manipulation, including data compression, and image enhancement to find and show patterns, including those that are invisible to the human cyc, such as satellite imagery. The final step is where results may vary from image to image or report to report, depending on review or analysis. Some terms related to image processing are -

Grayscale Conversion

A grayscale image, also called monochrome, represents a lack of color variation (i.e., no color). Grayscale images often have an electromagnetic spectrum (ultraviolet, infrared, visible, etc.) and if only certain frequencies are recorded, they are monochromatic. The conversion of a color image to a grayscale image is not that unique. Color channels with different weights can effectively show the shooting effect of white and black videos with separate color photo filters of the camera. A common approach is to complement the brightness of the gray picture to that of the colored image. After image acquisition, image pre-processing includes grayscale conversion. Eerens et al (2014) and Jias et al (2000). Sun and Du (2004) emphasized the grayscale transformation as follows:

- In the middle stage of the food quality assessment model, they reported different programs.
- Evaluate foods such as fruits, seafood, grains, meat, and vegetables using pictures.
- Processing technology can be applied in various food quality assessments.
- Some more works of Wu et al. (2013) told about the operation of grey-level measurements of heterogeneous cotton fibres.
- Jayas et al. (2000) also showed an analysis of the picture methods to classify crops by utilizing conventional neural networks (Saxena & Armstrong, 2014).

Image Background Extraction

For applications with minimal background usage, pictures with areas of interest (solid-like objects) on different backgrounds can be easily extracted. This results in an uneven distribution of grey levels between objects. Following this interest and the picture backdrop, Eerens et al. (2014) and Jias et al. (2000) reported numerous operations where the backdrop isn't taken into consideration while checking the quality of foods, including burgers and germplasm corn, and so on.

Similarly, Wu et al. (2013) did an extraction of the background of textile pictures of the main products produced by cotton. This allowed a clear discernment of the fibers of cotton, which was previously a difficult task. Chase Advanced Technology Review by Sankaran et al. (2010) highlights fluorescence imaging, spectroscopy, infrared and also visible spectroscopy, hyper-spectroscopy picturing, and future advances in disease diagnosis of plants, including the anabolic activity of flora that emits compounds that are organic in nature and are volatile (Saxena & Armstrong, 2014).

Enhancement of Pictures

Digital enhancing of images is the enhancement and sprucing of image features, along with borders, edges, or contrast, to make a picture display beneficial for viewing and evaluation (Gonzalez, 1992). Color picture enhancement may be divided into classes consistent with the coloration space:

1. Enhancement of colored pictures in RGB color area, and
2. Enhancement of colored pictures primarily on the basis of transformed space (Anil, 1989).

Although it is possible to improve color images by applying the enhancement of the existing image to the grey level, we apply the algorithm to each of the blue, red, and green passages, and the final result is that images may not be optimally enhanced (Thomas & Strickland, 1997). It is an algorithm that applies regardless of the individual co-relation channel between the B, R, and G components of the picture. In a modified spatial algorithm, image transformation uses color spaces: RGB, HSV, YCbCr, YUV, Lab, etc. Thus, the colored picture must be converted to another space from RGB space (Gang & Qiao, 2008).

Color Image Segmentation

Color image segmentation is a term used in many computer programs such as Adobe Photoshop, MATLAB, etc. Photoshop uses various tools to segment digital images. It's a simple application that only uses images divided into parts like layers and magic stick tools. MATLAB is also a computer program for editing digital images. But this software uses algorithms to do many things. Researchers also use algorithms here to do segmentation. Many models are used in MATLAB software, such as the RGB model for dividing colors, the CMYK model, etc. Many authors also use these models and update them in their research articles. Research by Lee (2011) used photo statistics for citrus segmentation based on colored picture segmentation.

Although these procedures were effective, white noise in the picture data was not able to be suppressed. Observations made by Terasawa et al. (2009) monitoring plant growth by considering the skill of picturing system. Such monitoring has benefits like discerning crop growth rate and also its health rate, forecasting harvest quantity and time, diagnosing disease, and assessing quality. Another study by Zhenjiang and Yaju (2009) used images of growing leaves on young plants to develop a program to monitor the growth of plants with the help of computer vision techniques. Artizzu et al. (2009) developed a computer-based system for analyzing images to determine the proportion of soil, crops, and weeds in a picture. The system considered changes in light, the texture of soil background, and conditions for damaged crops such as stage of crop growth and size of weed as obstacles to process images.

Filtration of Image

When images are acquired by cameras or other image processing systems, they are often not directly usable by the intended vision system. Images can be damaged by intensity fluctuations which are random in nature, luminance fluctuations, or decreased contrast, which must be considered in the early stages of visual processing. Utilizing Gabor filters is a unique approach presented by Sansao et al (2012) to evaluate pictures of weeds. Here, additional green indexed pictures were used to filter crop areas and evaluate weed patches. The automatic picture segmentation algorithm created by Guijarro et al. (2011) was used to segregate weedy grain, maize, and barley from the soil, as shown in Figure 4 (Saxena & Armstrong, 2014).

Figure 1. Image processing by Artizzu (2009) image input
Source: Saxena and Armstrong (2014)

Figure 2. Image processing by Artizzu (2009) image segmentation
Source: Saxena and Armstrong (2014)

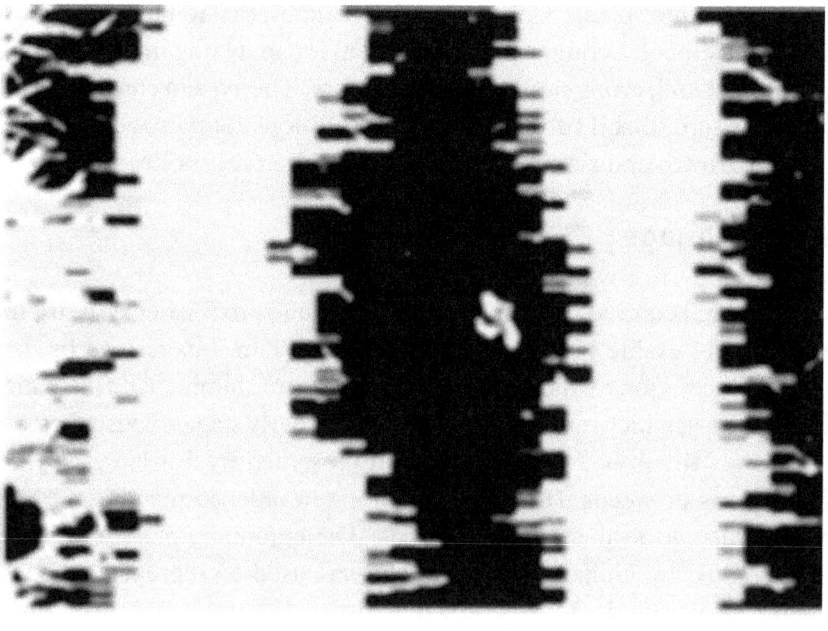

Figure 3. Image processing by Artizzu (2009) row elimination of crop image
Source: Saxena and Armstrong (2014)

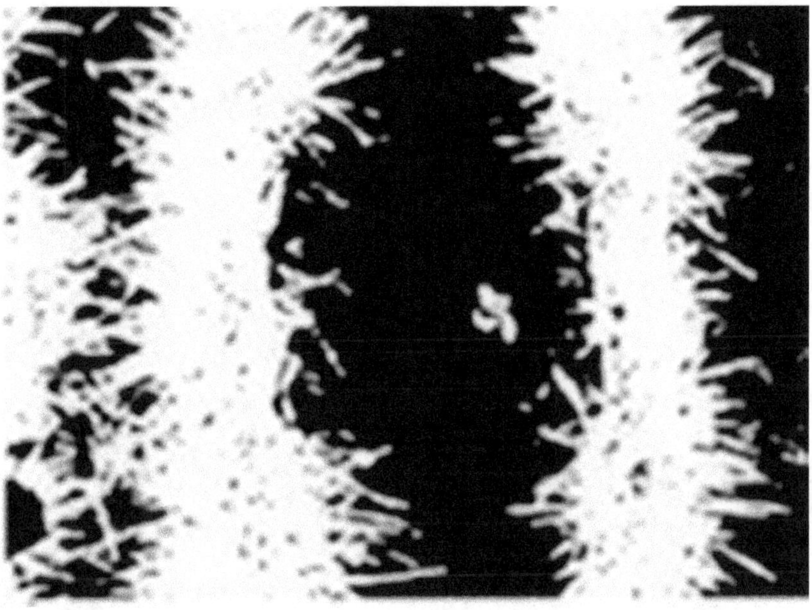

Figure 4. Image processing by Artizzu (2009) weed and filtration identification's final image
Source: Saxena and Armstrong (2014)

Figure 5. Results of classification of different kinds of classes of plants
Source: Guijarro et al. (2011)

Figure 6. Results of classification of different types of soil
Source: Guijarro et al. (2011)

Extraction of Features

An endeavor to identify weed from images using segmentation of pictures is reported by Gerhards and Weis (2009). Classification of weed was based on morphological attributes that grouped weeds completely into groups. Superior results have been reported from manual weed sampling. An attempt to identify weeds from images using image segmentation is reported by Gerhards and Weis (2009). Weed categorization was built on morphological attributes that grouped weed completely into weed categories. Superior results have been reported from manual weed sampling.

Kumar and Ehsanirad (2010) extricated these attributes for leaves detection for the classification of plants, reported that texture features, i.e., grey level co-occurrence matrix, require grey levels in the image. Another work by Jias et al. (2000) studies Fourier descriptors, morphology, transformation of wavelets, spatial moments, boundary chain codes, picture color and texture attributes, and techniques of extraction (Saxena & Armstrong, 2014).

Detection of Object Images

Artizo et al. (2010) made a system that uses computer vision-based methods to identify and extract regions of crop, weed, and soil as image objects. This system achieves 96%, 84%, and 84% co-relation for biomass images, seed pictures, and images of corn, respectively, and is planned to reduce the analytical complexity in futuristic development. Another study by Silva et al. (2012) gave an algorithm to detect vegetation and weed according to the color of the pictures.

Other studies by Cope et al. (2012) scrutinized different morphometric, computational, and photo processing ways for computing images of plants by measuring the contours of leaves, shapes of flowers, the structure of veins, and textures of leaves. These were trained in botany and proposed an automatic species identification system that could employ their expertise. This study suggested that system performance can be improved by using minimal number of categories and a limited feature set (Saxena & Armstrong, 2014).

Analysis of an Image Object

Colour and shape computations for ripened apple recognition in the images of trees proposed by Linker and Coleman (2014) were on the basis of 3-D convex computation. This method deals with analyzing of a three-dimensional convex object in the Delicious Golden orchard of apples under the influence of natural sunlight conditions. This method yields ninety-four percent accuracy in apple detection when using the Canny filter to detect edges. Rath and Pastrana (2013) developed a new way for the segmentation seedlings suffering from occlusion problems, which was tested on plants with a few numbers of leaves. This approach solved leaves complexities with elliptic approximations and used an active shape model to find leaf clusters (Saxena & Armstrong, 2014).

Figure 7. (a) and (b) 61% is the ratio of greenness determined by Yang et al. (2000)

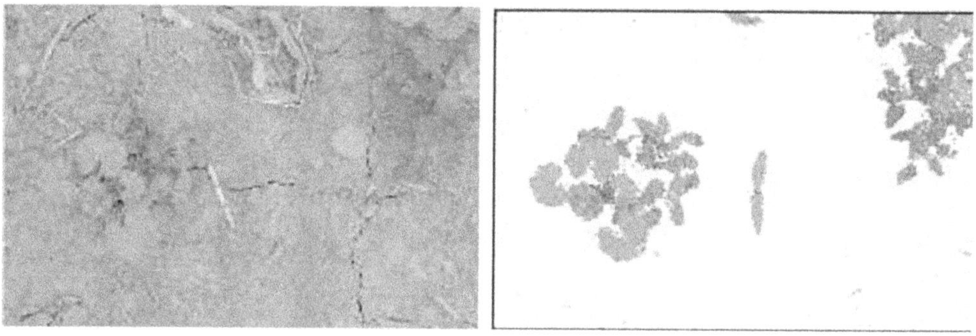

Transitioning in Images

In the case study written by Beers et al. (2010), he investigated Trans_Forum, an innovation program for the transformation of Dutch agriculture. Monitor and manage images using image visualization techniques to reduce migration complexity and increase future potential. The purpose of this research was focused mainly on managing of transfer and strategic niche, and also examining the change in the image of agriculture toward sustainable development (Saxena & Armstrong, 2014).

Image Histogram Analysis

As reported by Romeo et al. (2013) image histograms can be used to reflect the direct effects of illumination, where contrast is a distinguishing feature of vegetation. They designed a system with a decision module that determines whether the green color is sufficient based on the histogram analysis of the image. Another study on defect detection in yellow-skinned potatoes was published by Jin et al. (2009), who analyzed that many of the deficiency were in black or so-called dark spots having a very low percentage and no major peak in the grey-level graph of histogram. Figure 1 analyzed the histograms of the image and noticed an increase in the grey rate of the image. This helped to derive an advanced algorithm suitable for different fibers detected in products made by cotton (Saxena & Armstrong, 2014).

Figure 8. a and b. Results by Jin et al. (2009) with defects labeled and the histogram corresponding to it

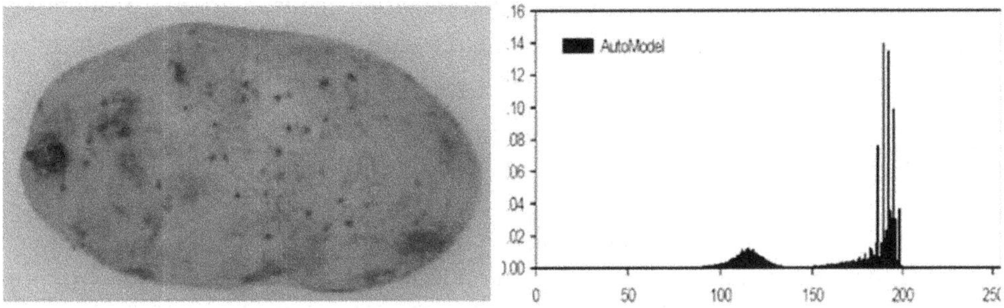

Binary Segmentation of Images

Sun and Du (2004) utilized picture segmentation to estimate several food quality valuation styles that divide food defects from other defects. They described the conclusiveness of diverse segmentation algorithms applied to apple defect discovery, pizza gravy division, and pistachio kernel connection discovery. In addition, another study by Lu and Mizushima (2013) suggested a picture segmentation system that uses Otsu's system to determine defects in apples and brace vector machines to grade and order apples.

Figure 9. Processed images of apples by Puchalski et al. (2008) examples

Using binary segmentation of image to split weeds from harvests, posed by Tellaeche et al. (2008), was exercised for planter decision-making for pesticide operation in weed-affected zones. A robotic expert system suggested by Montalvo et al. (2013) linked weeds and harvests, and Guerrero et al. (2013) reverie was suitable to distinguish crop rows in pictures from sludge fields utilizing a check of picture processing ways for farming picture segmentation. Downpour effects and fallen weeds are the recounted outcomes that may have caused counterproductive segmentation in weeds and harvests (Saxena & Armstrong, 2014).

WEED DETECTION

Weeds are flora that is viewed as undesirable. Weeds don't hold botanical category valuations. As a flora that's a weed in some contextures isn't a weed if it grows where it's required. This applies to all plants that diligently cultivate or breed or are outside their natural territory. The term is occasionally exercised to depict a common genus outside the flora kingdom that can reside in different surroundings and procreate fleetly. These contain seeds that stay in soil seedbanks for ages. They contend with desirable flora for resources that plants typically require soil, direct sunshine water, nutrients, and (to a lower extent) space to grow. Weed taxonomy is a grave problem in agrarian exploration. Weed grouping is mandatory to distinguish the weed species to be contained. There are two types of weeds based on the prevalence of edges present (Poojith et al., 2004):

1. Weeds that have low-frequency edges

Figure 10. Weeds that have high-frequency edges

Figure 11. Inter row weed detection

This approach allows you to see the weeds between the rows of shops. Here, real- time pictures are acquired and reprocessed to conserve weed zones. Take photos at 25 frames per second. Each frame should run for 0.04 seconds. Then we take the initial eight frames generated during 0.3 seconds and execute a logical 'AND' operation between them to obtain a reference picture referred to as the crop row photo. This reference crop row picture will be substituted after 0.3 seconds by the new crop row photo developed by applying the 'AND' operation of the frames developed in the ensuing 0.3 seconds (Poojith et al., 2004).

Once you have the image you are going to refer to, differentiate it with the next picture to get the weed output. Anything present afterwards the pre-processed image that is not seen in the reference image is commonly known as clipping. To do this step, XOR the reprocessed image with a reference image that we already have to get the output image with the weeds in-between the lines.

One of the major drawback of this process is the inability to categorize weeds between plants in a crop row. Crops are considered weeds even if they are not in a row. Therefore, this method cannot be relied upon to detect weeds.

Inter-Plant Detection of Weeds

Interrow detection of weeds no longer deals with weeds among flora in the rows. Interplant Weed Detection is hired to solve this problem. This area of the model makes a picture for the additional superior reprocessing and is composed of:

1. Obtaining of photo,
2. Segmentation of coloration, and
3. Detection of edges

Figure 12. Method used for interrow detection of weed

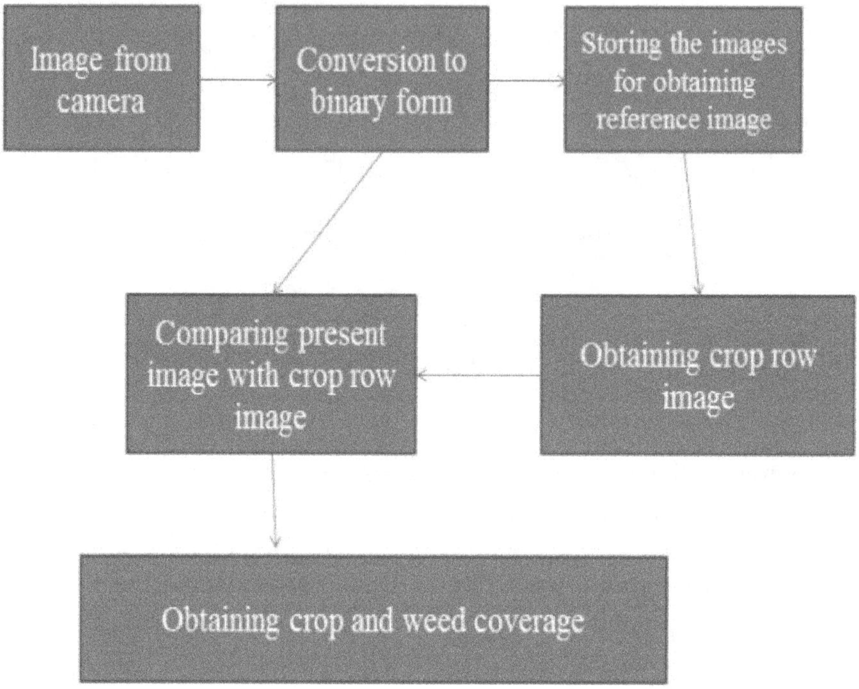

Segmentation of colours is a picture segmentation system employed to divide crops(involving weeds) from the backdrop. This is done by applying K- means clustering. This procedure helps to divide all visually distinguishable colors from each other. The L * a * b * color space(likewise called 'CIELAB' or CIE L * a * b *) allows us to evaluate these visual disparities. The L * a * b * place consists of a sheen layer ' L * ', a tier a chromaticity ' a * ' that indicates where the shade lies along the green-red axis, and a chromaticity subcaste ' a * ' that indicates where the color lies along the green-red axis. B * ". The colors are along the yellow-blue axis. All color data is in the " a * " and " b * " layers. The discrepancy between two colours can be gauged using the Euclidean-Distance-Metric. As we know, clustering can be used to cluster pixels of the same color into blocks of objects, smoothing segmentation. The final picture outcome consists of two colors only. The required picture after segmentation of colours contains green (for crops and weeds) and the rest of the picture, producing a qualified image for the edge detection step of the procedure (Poojith et al., 2004).

Figure 13. Method used for interplant detection of weed

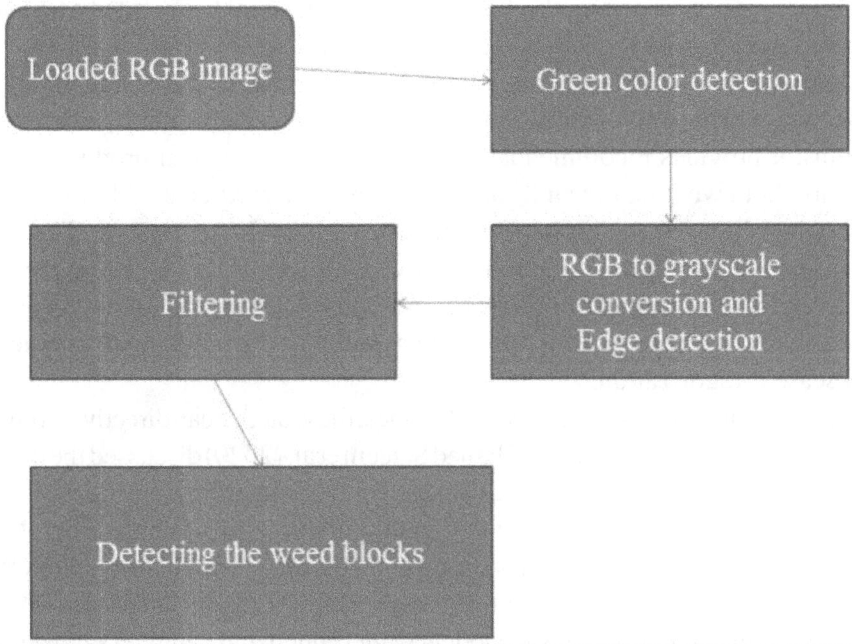

Detection of Diseases of Different Plants and the Pest That Affects Them

Kulkarni Omkar (2018) posed a deep CNN algorithm conditioned on photos of well-conditioned and damaged harvest leaves from open datasets. The programme achieves this objective by sorting leaf pictures into unhealthy or well-conditioned classes based on defect designs. For each category, the suggested procedure was sampled on five distinctive breeds of shops and three distinct forms of floral disease. The conclusions showed that the InceptionV3 model outperforms his Mobile Net model in terms of precision and validation misplacement. J. Kartika et al (2021) asserted that disease analysis is achievable and that the discovery of cotton leaf infirmities is achievable. As per J. Karthika et al., to ascertain cotton leaf diseases, K- means clustering is applied for attribute extraction and a support vector motor is stationed to discover them. R. Mounika et al. (2020) portrayed how to determine and categorize numerous forms of flora diseases by applying picture processing strategies. One such approach for automatically detecting and classing diseases is segmentation of pictures segmentation. It's applied to execute strategies to recognise and separate leaf diseases. K-means are used to resolve disease segments. The face highlights of the GLCM are also segregated, and the assembly is finalized with his

SVM. This methodology has been sampled to localise infections in citrus leaves. By operating image processing methods, et al. Md. Awal Abdul has created a technique to give a system that was made with Android Studio.

Pictures of infested crops are compared to the CDDASS database by the Crops-Diseases-Detection-And-Solution-System. When CDDASS detects ailment symptoms, it provides recommendations to planters so they can produce the best judgments about whether to handle affected crops with medicines. Multiple tactics of separating diseased region of flora are depicted by Khirade et al. This study likewise analyzed attribute extraction and division strategies for extracting infected leaf traits and distinguishing flora diseases. ANN approaches like as tone- organizing attribute charts, backpropagation algorithms, and SVM are frequently applied for flora disease categorization.

By applying picture processing methodologies, these tactics can directly distinguish and classify multiple plant diseases. Hamad Saleem et al. (2020) discussed the difficult job of finding and grading flora leaf diseases. The Tensor-Flow identification frame for objects is used to appertain three DL meta- infrastructures Single. Shot. Multi-Box Sensor (SSD), Faster Region - Based Convolution-Neural-Network(RCNN), and Region-Based Fully Convolution Networks (RFCN). To ascertain flora species' diseases, all deep learning models were trained and sampled on restrained environmental datasets.

Aitor Gutierrez et al. (2019) deep learning solutions contrast with solutions that combine computer vision and machine learning. Generate regional proposals using a Region Proposal Network (RPN) and identify diseases in selected regions using a Disease Detection Network (DDN). RPN suggests a collection of zones of varying sizes that appear to be diseased. Inspect the proposed locations for repressors and classifiers to identify areas containing insects and eggs. An automated extraction and discovery system was invented by Miranda et al. used numerous picture processing ways to determine and extract pests in captured pictures. The system presented was simple but efficient. The process for extraction of detected objects from a photo is direct. We examine the picture vertically and horizontally to decide each of the coordinates and deliver the object's photo. Praise for Fina et al. provides a malware spotting system that combines the K- means algorithm for clustering with a parallel filter or purifier. The dataset then, is procured by separating the data space into Voronoi cells. Voronoi cells tend to detect packages of parallel geographic boundaries, separating objects(pests) from the backdrop (habitat of pest). Detection is completed by re-collecting several salient features of the habitat of pests (leaves, stems) by applying correspondence purifiers to identify plants pests, and generating correlation peaks for diverse data sets. The study also established that the probability of discovery from pictures of pests was proportionate to the degree of the output signal and inversely proportionate to the viewing phase. This verified that crop pest discovery was a function of pest position and angle by which it is viewed. It's emboldening to witness that the filter corresponding can achieve invariance rotation of pest up to three-sixty degrees, depicting the usefulness of the algorithm used in crop pest discovery.

APPLICATIONS OF IMAGE PROCESSING

The importance and impact of image processing of photos may be judged by their applications in numerous regions like satellite imaging, medical and aerial imaging, inspection of various industries, law enforcement, applications in sectors like defense, and the most common agriculture sector. Every field in the private or government sector has a relationship with image processing techniques. One image processing method may produce superior performance in one field while producing poor

Figure 14. Processing of images in the field of agriculture's applications
Source: Surya and Satheesh,(2014)

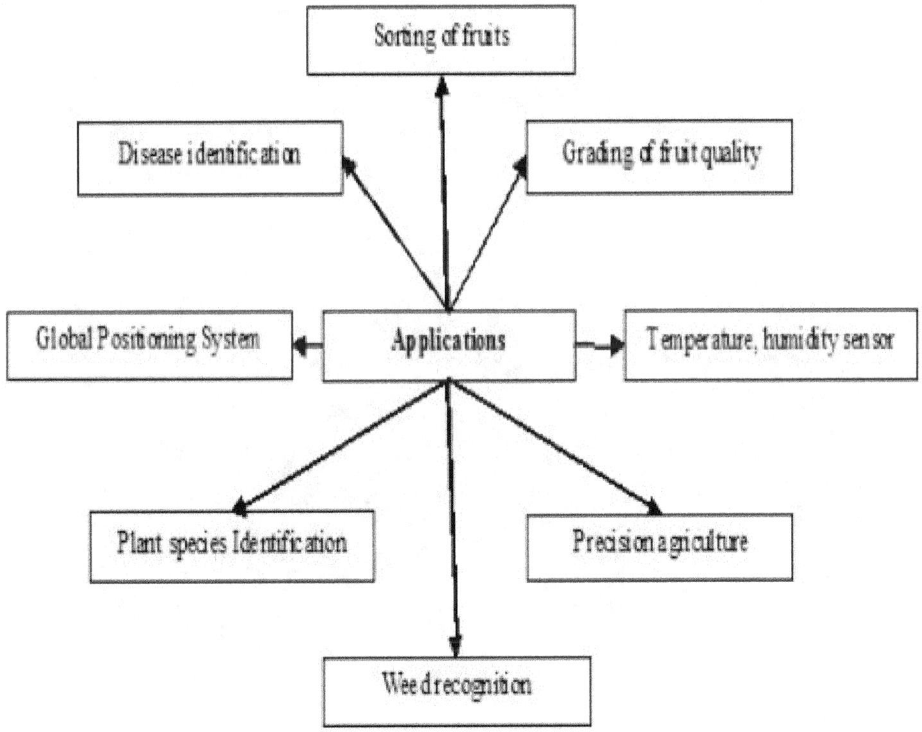

performance in another. Therefore, it is essential to choose the appropriate modules related to the application (Surya & Satheesh, 2014).

Plant Disease Identification

Picture reprocessing is an important operation in the identification of ailments in multiple agriculture harvests. Diseases, and pests, in the harvests pose a substantial hazard to agriculturists. Bacteria, nematodes, fungi, and viruses are some of the common causes of disease infections. The reasons that cause the emergence of new threats are unpredictable climatic conditions and societies' increasing focus on nature.

The severity and spread of diseases are determined by the resistance capability of the crop (Prabha & Satheesh, 2012). If the crop has a high level of resistance, then the susceptibility of the crop will increase gradually. If the crop has a low level of resistance, then the susceptibility of the crop will increase faster. Symptoms of the disease begin to appear in the crop with moderate variations in size, color, shape, and growth from the earlier stage. Other commonly noted symptoms include abnormal bulging in the root region, falling of fruit in the immature stage, and the

Figure 15. Leaf input image

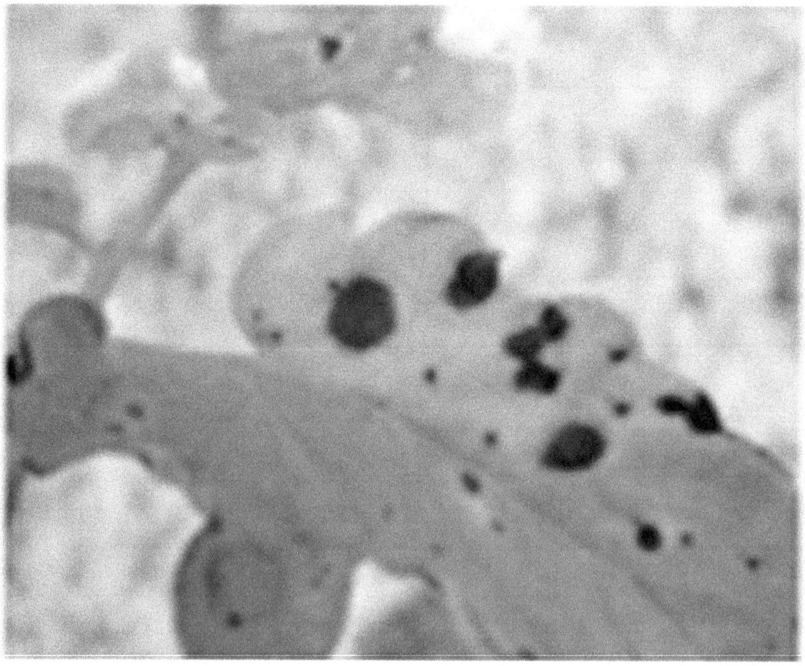

Figure 16. Infected region highlighted image's output

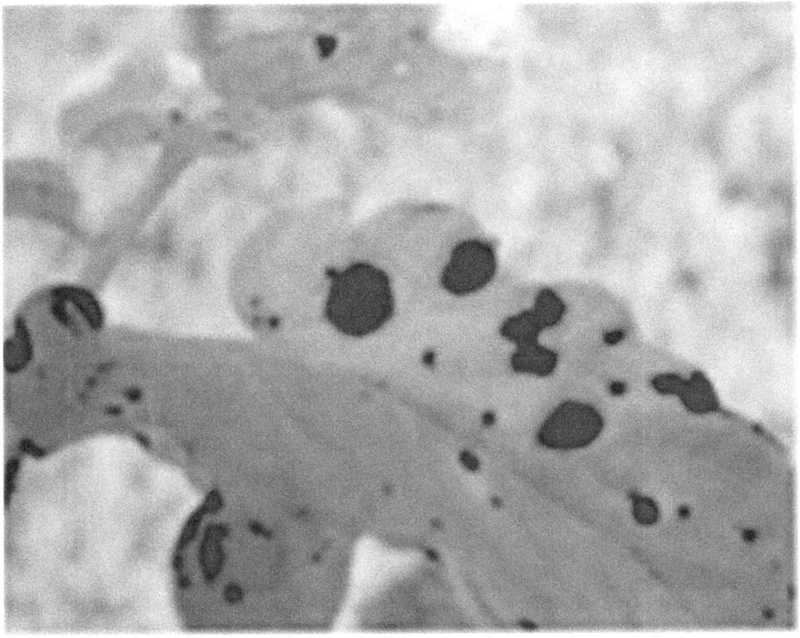

Figure 17. Input leaf image

Figure 18. Extraction of infected region

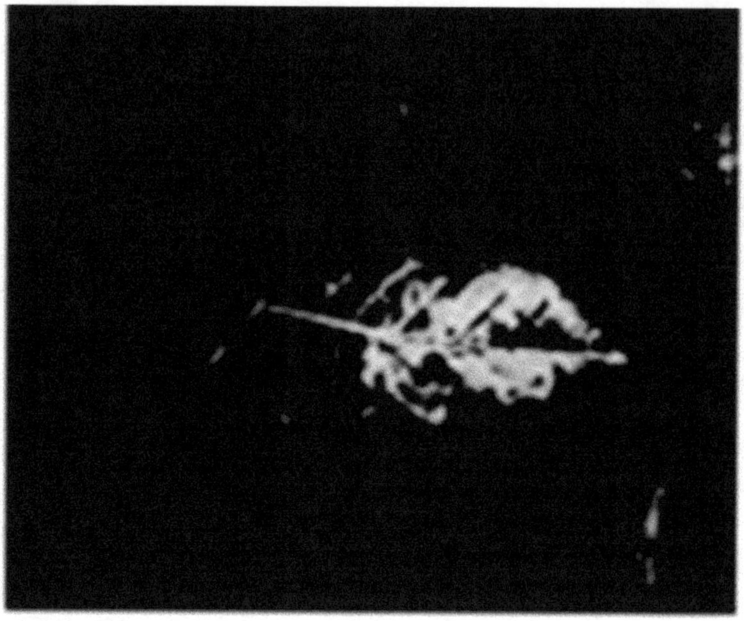

appearance of dark spots or dots in fruits and leaves of the crop (Camargo & Smith, 2009). Hence, early diagnosis of disease infections can protect the crops from entire damage (Figure 5).

Classification of Fruits and Their Sorting

Fruit breeds are accessible in the market for daily consumption as a regular activity due to raised demand from customers. The fruits entered from the markets are categorized manually, which is quite a time – taking and tiresome procedure. Huge amounts of fruits have to be categorized in a short of time (Butz et al., 2005). This type of job has prompted the need for an automated tool for fruit sorting, which can classify tons of fruits in a shorter span of time. This tool is both cost-effective and time-saving, with higher accuracy when compared to other classification methods. Post-harvest sorting and classification of agricultural produce is a technology where processing of photos is produced for auto-mation in order to automatically sort and classify the produce. This technique is more beneficial than the usual manual labor, with higher reliability, accuracy, consistency, and speed.

The industrial region is profited by the automated sorting of vegetables and fruits by employing the non-destructive methodology of categorizations (Costa et al., 2011). Different breeds of fruits, with varying sizes, colors, shapes, and textures, can be

Figure 19. Mango input image
Source: Surya and Satheesh (2014)

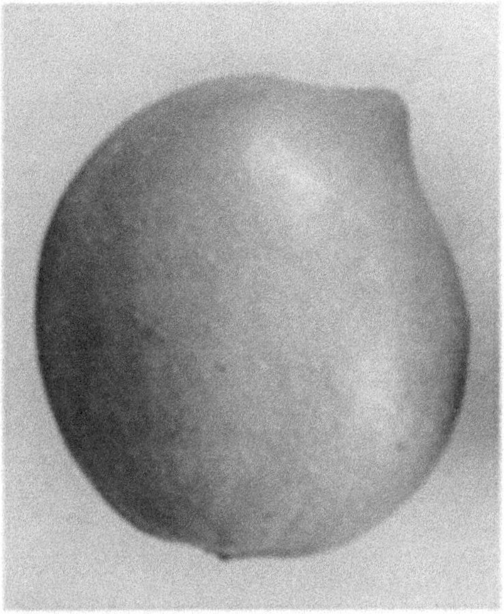

Figure 20. Corresponding histogram
Source: Surya and Satheesh (2014)

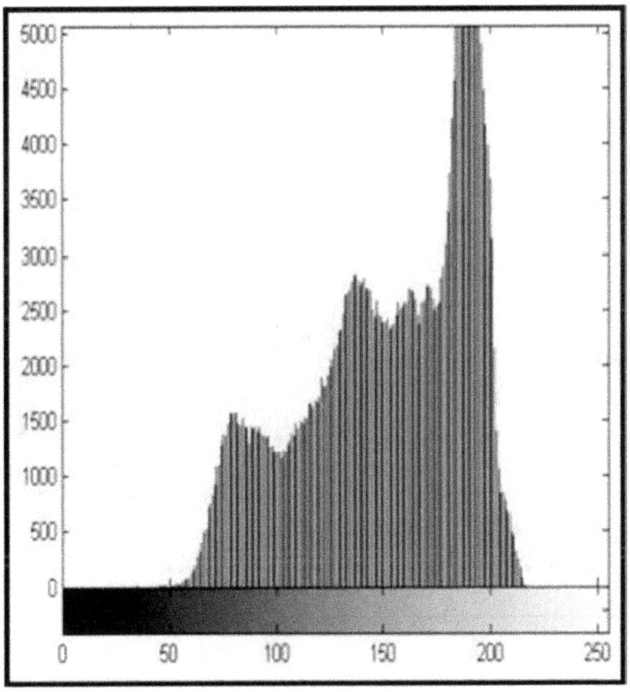

Figure 21. Chikoo image's input
Source: Surya and Satheesh (2014)

Figure 22. Corresponding histogram
Source: Surya and Satheesh (2014)

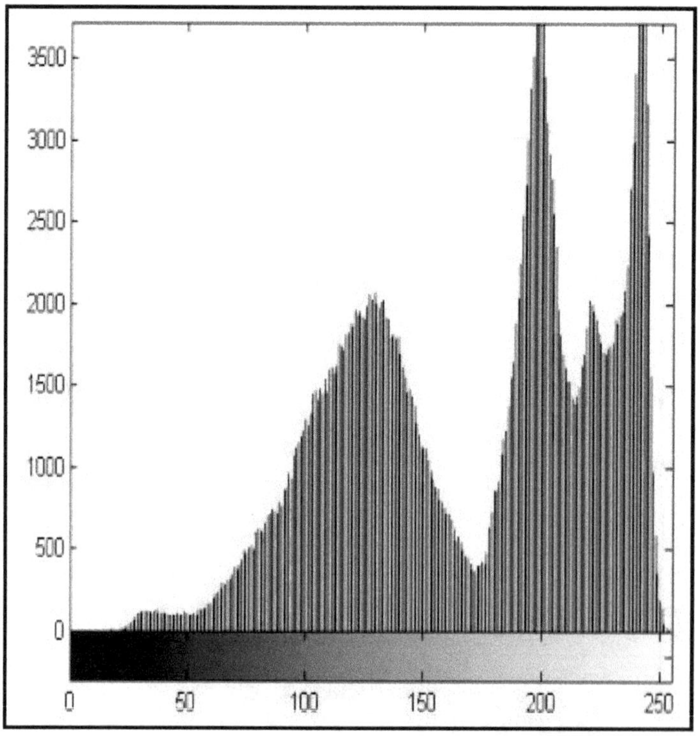

freely identified and separated. Damaged vegetables and fruits can likewise be freely identified and discarded. This automation helps companies, supermarkets, and other commercial fruit stores to classify fruits with high precision within a short time.

Identification of Species of Plants

Flora identification is also a significant task for botanists, experimenters, and also common people. Content based picture reclamation strategies are applied for species identification from a group of flora pictures. Floras are identified on the basis of on their morphological characteristics, for example their as texture, size, shape, and color of the flowers and their leaves. There's a requirement for trained and proficient botanists.

Figure 23. Image's input
Source: Surya and Satheesh,(2014)

Figure 24. Region-based output
Source: Surya and Satheesh (2014)

Figure 25. Boundary output
Source: Surya and Satheesh (2014)

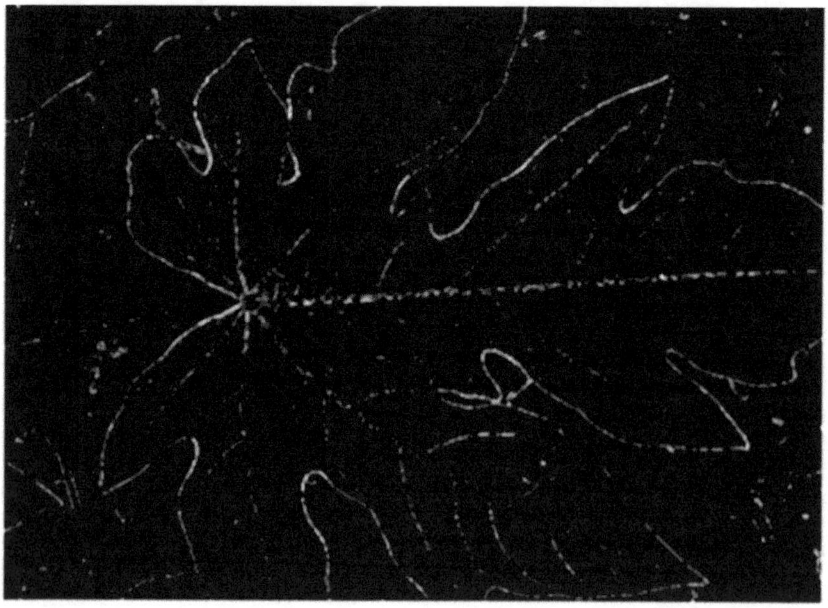

Figure 26. Signature for the output
Source: Surya and Satheesh (2014)

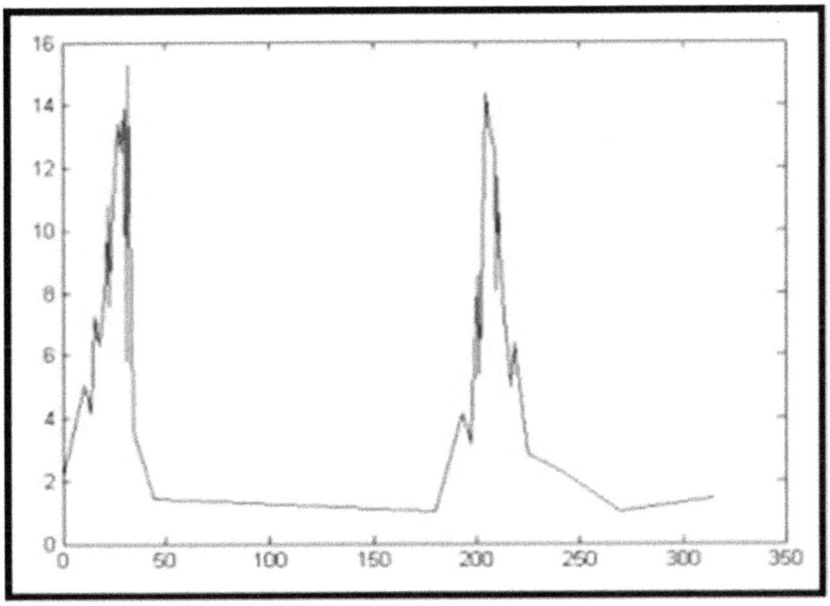

Technologies such as smartphones with an Android application and real-time image capturing devices with laptops and palmtops could help overcome the difficulty of species recognition for the common man. Species of various plants' identification with a collection of photos of species requires feature extraction and image analysis.

Precision Farming

The advancements in information tech and science of agriculture have made it feasible to connect these two regions, leading to the advancement of precision agriculture. It's an info and tech – based. farm operation engine developed to distinguish, examine, and handle variability within fields in order to maximize not only profitability, but also sustainability, and the production of resources of land (Prabha & Satheesh, 2012). It is a better alternative to traditional farming with regard to profit and production. Precision farming techniques can help farmers make better decisions regarding optimal crop production. It requires an understanding of and skill in using the natural resources found within a given field. It provides maximum profit and production with minimum input and optimal use of resources.

Pre-acquired knowledge about technology is important for farmers. Formal education is a must for agriculturalists to procure knowledge about precision farming. There are various technological tools available to collect and implement information

Figure 27. Various technological tools for implementing precision farming
Source: Surya and Satheesh (2014)

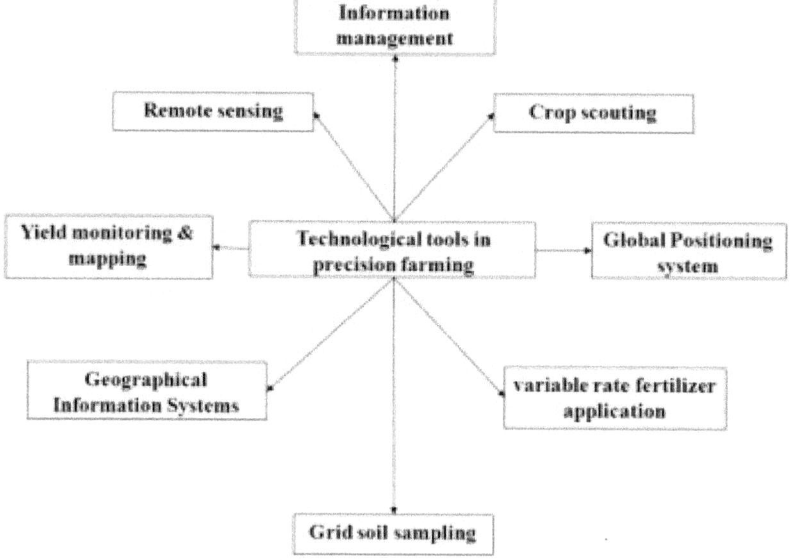

effectively, as shown in Figure 8. The success of these technologies depends on the farmer's understanding (Surya and Satheesh, 2014).

Fruit Quality Analysis

The request for quality products in shops has urged the evolution of an automatic system for grade assessment. Thus, ensuring the examination of fruit grade in order to produce quality fruits in the shops. The grade of fruit is estimated by characteristics like size, color, shapes, textures, and flavors (Freixenet et al., 2002). Manual inspection can be time taking and very prejudiced. Therefore, it's quite mandatory to develop a non-destructive quality assessment method to determine grade of fruits with high speed and accuracy, in contrast to certain destructive assessment methods. Now, many companies use computer-vision-based (CV-based) automatic quality or grade assessment equipment. These are thus replacing the traditional way of inspection i.e., by hand (Gao et al., 2010).

Assessment of Land and Crops Using 'Remote Sensing'

Remote sensing is a significant data source utilized in GIS for retrieving data collected with the help of satellites. It's a procedure of carrying data about an object, region,

Figure 28. (a) Image's input, (b) background removal, (c) extracted fruit region
Source: Surya and Satheesh (2014)

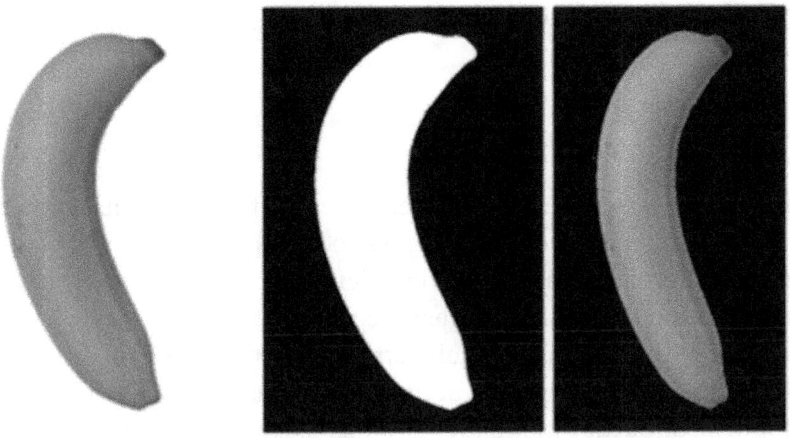

or incident without bodily contact. Remote sensing data detectors are handheld gadgets arranged to put on satellites, aircrafts, or directly on farming machineries.

The passive system is the most commonly used remote sensing type to sense electromagnetic energy that is reflected from the plant. The factors which are necessary for remote sensing are the reflecting behavior of visible fluorescence energy from a foreign source. Passive systems depend on foreign sources of energy, for example sunlight. Satellite imagery had been upgraded through the application of picture detectors. Images retrieved through satellites can be examined and assessed in their digitalized form.

The advancements in photo processing strategies, like as image refinement, restoration, and analysis, have created remote sensing advancements independent of GIS. The use of image processing software packages combined with remote sensing provides better analysis and interpretation of gathered data.

The major purpose for using remote sensing is to watch the Earth's surface and hence gauge variables including biological, physical, and geographical to distinguish materials on the land for additional investigation (Surya & Satheesh, 2014).

Recognition of Weeds

Weed infestation is a significant risk to agriculturists, which reduces harvest yield and quality. Thus, more attention is warranted in the monitoring of weeds. A common strategy for restraining the growth of weeds is the use of herbicides. With the help of the latest inventions, the recognition of weeds is automatic such that the machines distinguish crops from weeds automatically. An automatic machine monitors the growth of weed periodically and determines when weeds need to be regulated. Classifiers, alongside picture processing techniques, makes it effortless to distinguish and destroy the weed in an earlier stage of growth (Lamb & Brown, 2001).

CONCLUSION

The image processing strategy has proven to be an efficient machine-vision strategy for farming. Different spectrum imaging methods, like hyperspectral, infrared, imaging, and X-rays were applicable in judging flora index, canopy measures, and irrigation mapping of land with higher precision.

Weed division can be appropriately carried out with photos processing algorithm, given that the yield is taken into account. The precision of division depends on limitation of picture acquisition and algorithms used and ranges from 85%-9%6. With such good precision in bracket, agriculturists will be capable to apply insecticides in the right way. This approach will also help to minimize the environmental impact and also expenditure.

REFERENCES

Anil, K. (1989). *Fundamentals of Digital Image Processing*. Prentice-Hall.

Ayman, H. A. E., & Ayman, A. A. K. Understanding color image processing by machine vision for biological materials. In *Structure and function of food engineering* (pp. 227–274). INTEHC.

Butz, P., Hofmann, C., & Tauscher, B. (2005). Recent developments in NoninvasiveTechniques for freshfruit and vegetable internalquality analysis. *Journal of Food Science*, *70*(9), 131–141. doi:10.1111/j.1365-2621.2005.tb08328.x

Camargo, A., & Smith, J. S. (2009). An imageprocessing based algorithm to automatically.identify plant disease visual. symptoms. *Biosystems Engineering*, *102*(1), 9–21. doi:10.1016/j.biosystemseng.2008.09.030

Chaudhary, P., & Chaudhari, A. (2012). Color Transform Based Approach for Diseases SpotDetection on Plant Leaf. *International Journal of Computer Science and Telecommunications, 3*(6).

Collewet, C., & Marchand, E. (2009). Colorimetry-based visual servoing. In *2009 IEEE/RSJ Interna-tional Conference on Intelligent Robots and Systems.* IEEE. .10.1109/IROS.2009.5354416

Costa, C., Antonucci, F., Pallattino, F., Aguzzi, J., Sun, D. W., & Menesatti, P. (2011). Shapeanalysis of agriculturalproducts: A review of recentresearch advances andpotential application to computer vision. *Food and Bioprocess Technology*, *4*(5), 673–692. doi:10.100711947-011-0556-0

Feng, Y.-Z., & Sun, D.-W. (2012, November). Application of Hyperspectral imaging in food safety inspection and control: A review. *Critical Reviews in Food Science and Nutrition*, *52*(11), 1039–1058. doi:10.1080/10408398.2011.651542 PMID:22823350

Freixenet, J., Munoz, X., Raba, D., Marti, J., & Cuff, X. (2002). Yet another survey on imagesegmentation: region andboundary information integration. In A. Heyden (Ed.) ECCV, LNCS 2352. Springer.

Gang, S., & Qiao, X.-L. (2008). AdaptiveColor Image Enhancement basedon Human Visual Properties. *Proceedings of International Congress on Image and Signal Processing*.

Gao, H., Zhu, F., & Cai, J. (2010). A review of Non-destructiveDetection forfruit Quality" *IFIP Advances in Information and Communication Technology*, *317*, 133–140. doi:10.1007/978-3-642-12220-0_21

Gonzalez, R. C. (1992). *DigitalImage Processin Reading*. Wesley Addison.

Gutierrez, A., Ansuategi, A., Susperregi, L., Tub-o, C., T, I., & Da, L. (2019). A Benchmarking of Learning Strategies for Pest Detection and Identification on Tomato Plants for Autonomous Scouting Robots Using Internal Databases. Journal of Sensors.

Johnny, M., Gerardo, B., & Tanguilig, B. (2020). Pest Detection & Extraction UsingI mage Processing Techniques. *International Journal of Computer and Communication Engineering, 3*.

Karthika, J., Mathankumar, K., Santhose, M., Sharan, T., & Shariharan, S. (2021). Disease Detection in Cotton-Leaf Spot Using. Image-Processing. *Journal of Physics*.

Khirade Sachin, D., & Patil, A. B. (2015). *Plant Disease Detection Using Image Processing. In International Conference on Computing Communication Control and Automation*. IEEE.

Lamb, D.W. & Brown, R.B. (2001). Remote-Sensing& Mapping of Weeds in Crops. *Journal of Agricultural Engineering Research*, *78*(2), 117-125.

Maldonado, A., Vigneault, C., & Nakaji, K. (2012). Postharvest Technologies of Fresh Horticulture Produce. In Horticulture (pp. 164-172). InTech.

Mounika, R., & Bharathi, P. (2020). Detection of Plant Leaf Diseases Using Image Processing. *Journal of Critical Reviews, 7*(6).

Oimkar, K. (2018). Crop Disease Detection Using Deep Learning. In *Fourth International Conference on Computing Communication Control and Automation (ICCUBEA)*. IEEE.

Poojith, L., Reddy, B., Kumar G., & Vittal, K. (2004). *International journal of innovative research in electrical, electronics, instrumentation, and control engineering*. Academic Press.

Prabha, D. & Satheesh, K. (2012a). *Crop Disease Identification using Image processing Methods*. In The National Conference on Green Computing, ST Hindu College, Nagercoil, TamilNadu, India.

Prabha, D. & Satheesh, K. (2012b). *A study on Image Processing Methods for Fruit Classification*. In Proceedings of IPC & IPEeL_ACT & CIIT_CENT & CSPE, Chennai, TamilNadu, India.

Saleem, M., Sapna, K., Potgieter, J., & Arif, K. (2020). Image-Based Plant Disease Identification by Deep Learning Meta Architectures. MDP Journals, 9.

Saxena, L., & Armstrong, L. (2014). A survey of image processing techniques for agriculture. In Proceedings of Asian Federation for Information Technology in Agriculture. (pp. 401-413). Australian Society of Information and Communication Technologies in Agriculture.

Surya, DSatheesh, K. (2014). Image processing methods and its Role in agricultural sector – A study'. *International Journal of Business Intelligents*, *03*(01), 366–373.

Thomas, B. A. Strickland, H., & J, R. (1997). Color Image Enhancement using Spatially Adaptive Saturation Feedback. *IEEE International Conference on Image Processing*, *3*, 30-33.

Wabali, V. C., Esiri, A., & Zitte, L. (2017). A sensory assessment of color and textural quality of refrigerated tomatoes preserved with different concentrations of potassium permanganate. *Food Science & Nutrition*, *5*(3), 434–438. doi:10.1002/fsn3.410 PMID:28572927

Zhang, Huang, Li, Zhao, Fan, Wu, & Liu. (2014). *Application & developments and of computer vision for external quality inspection of vegetables and fruits: A review.* Food Research International.

Chapter 6
Applications of Machine Learning in Agriculture

Padmesh Tripathi
https://orcid.org/0000-0001-9455-1652
KCC Institute of Technology and Management, India

Nitendra Kumar
https://orcid.org/0000-0001-7834-7926
Amity Business School, Amity University, Noida, India

Mritunjay Rai
https://orcid.org/0000-0002-8911-4826
Shri Ramswaroop Memorial University, India

Pushpendra Kumar Shukla
Vindhya Institute of Technology and Science, India

Kailash Nath Verma
IIMT College of Engineering, India

ABSTRACT

Today's epoch christened as the modern epoch, is the epoch of information and technologies. No field is untouched with the emergence of technologies. In different segments of agriculture, numerous technologies have been employed extensively. Machine learning has extensively been employed in every field today. It is composed of high-performance computing and big data technologies. Machine learning (ML) and its variants have been employed in different domains of agriculture like yield prediction, weed detection, disease detection, water management, soil management, livestock production, etc. Machine learning techniques have emerged as a boon to agriculture and have increased the quality and quantity of agricultural products. In this chapter, some applications of machine learning technologies in agriculture have been explored.

DOI: 10.4018/978-1-6684-6418-2.ch006

INTRODUCTION

Food is an elementary necessity for human existence that is fulfilled through agriculture. Agriculture is an indispensable segment for economy of any country (Meshram et al., 2021). Inopportunely, its market is unpredictable as several factors affect agriculture. For example: drought or flood can easily influence future commodity prices and have grave consequences on all food prices. The only approach to come across the growing food demand is to keep track of the environment, crops and the market. This is the time when machine learning plays role in agriculture. Agriculture decision making is empowered by machine learning applications through the analysis of real-time sensor data and past trends. This supports the manufacturers for improving crop yields, better predicting demand and reducing the food production costs. Thus, ML applications in agriculture can optimize the food production and revolutionize the economy of any country.

Smart Farming and Precision Agriculture

Last seven decades have witnessed the use of modern techniques and tools in agriculture and new concepts like smart farming and precision agriculture have emerged. Though, a large number of farmers are far far away from these, but the future is calling to all such farmers in order to improve quality and quantity in agriculture. Smart farming and precision agriculture are very often used interchangeably, but these have slight difference in sense. In Smart farming, data and information techniques are used for optimizing complicated farming systems. Emphasis is on the smart way of applications of collected information instead of on access to data and its application. In precision agriculture, for monitoring and optimizing the production processes, recent farming management ideas together with digital techniques are employed. Here, optimization plays vital role. Precision agriculture consents the determination of actions on the basis of per plant or per square meter. For example, if one has to use fertilizers in his field, the precision agriculture suggests him not to use equal amount of fertilizers over whole field, but to measure the soil variations within the field and use the fertilizers accordingly.

In smart farming, the technologies like deep learning, machine learning, artificial intelligence, sensors, robots, internet of things (IoT), etc. (Tripathi et al., 2022, Goel et al., 2022) are used extensively. Farmers have great opportunity to upsurge both the quantity and quality of crops (Tyagi, 2016).

Some technologies being used in smart farming are:

- Location systems: This includes satellites, GPS, etc.

- Sensors: These are used for monitoring temperature, moisture, light soil, water, etc.
- Robots: These are used for seeding, spaying the herbicides, pesticides, etc.
- Drone cameras: These are used to monitor the crops status.
- Precision irrigation and plant nutrition
- Different software platforms
- Platforms for analysis and optimization.

Figure 1. Smart farming in action (Thales/Nesta)

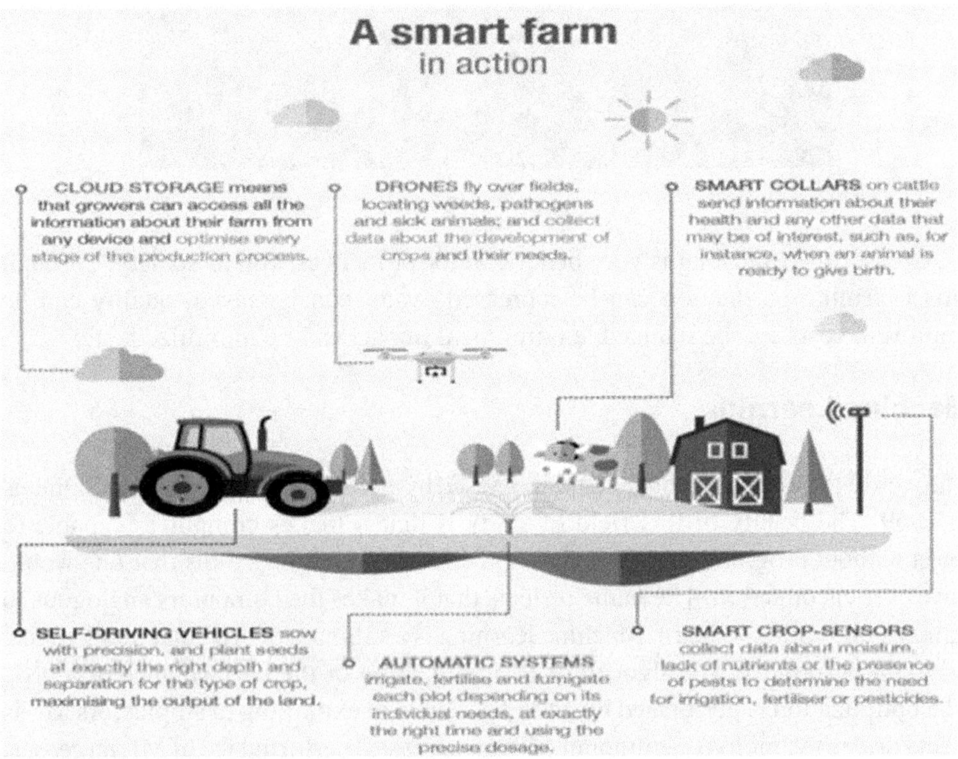

A UN report says that by 2050, the world population will be 9.7 billions that is 1.73 billions more people will require food in comparison to September 2022. As per the estimation of United Nations, world population in September 2022 is 7.97 billions. Unless and until, more and more farmers use the smart farming, it will not be possible to feed all in 2050.

Figure 2. Use of drones in smart farming

Use of Smart farming is very beneficial for farmers as well as society. Through smart farming production can be increased, water can be saved, quality can be improved, costs can be reduced, environment impact can be minimized, etc.

Machine Learning

The word Machine Learning (ML) was devised by an IBM employee, Arthur Samuels in 1959. The beauty of this field of study is that it makes computers capable to learn without programming. It is one of the most electrifying tools that one would have ever encountered. The name reflects that it makes the computers analogous to humans: the ability to learn. Machine learning is a sub-area of artificial intelligence.

Main purpose of ML algorithms is optimization of the performance of a task. The optimization is performed by past experience or exploiting examples. As ML is a data driven technology, as amount of data increases, performance of ML increases, just as it happens in case of humans: more experience better performance (Vieira et al., 2020). More precisely, data encompass a collection of examples, labeled by a collection of characteristics, normally called as features. In Broad sense, working of ML systems consists of two processes, the first one is learning (used for training) and second one is testing. For learning, a feature vector is formed by these features. This feature vector can be any of nominal, ordinal, binary or numeric (Lopez-Arevalo et al., 2020) and this vector is employed as an input. Learning phase allows machines to learn to do the assignment from experience by depending on the training data. As soon as, the learning performance attains a satisfactory level, it ends. Consequently, the model established in training process may be utilized for clustering, classification or prediction.

Figure 3. Machine learning as sub-area of Artificial Intelligence

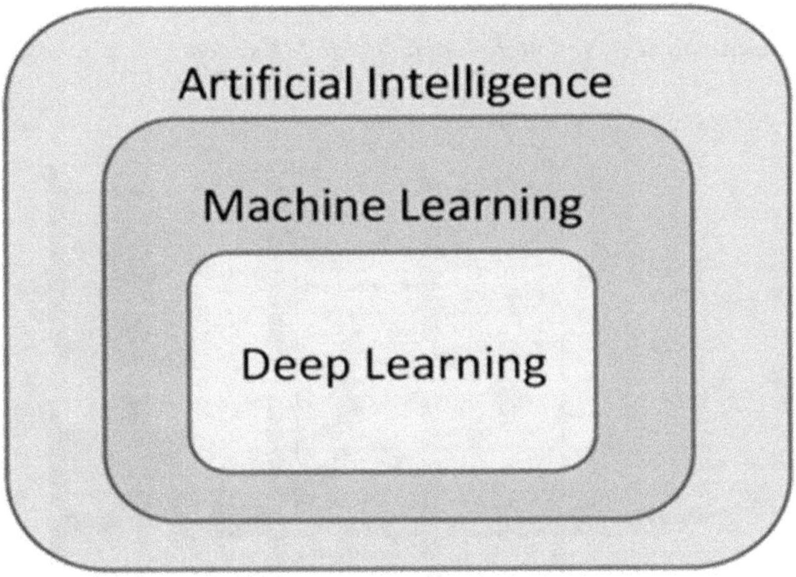

A classic ML system is summarized in figure (Benos et al., 2021). First of all, the complex raw data is pre-processed to transform it into a suitable state. This pre-processing has the three components (Anagnostis et al., 2020):

1. **Data Cleaning:** Inconsistent, irrelevant or missing items are removed and de-noising of data is performed (Tripathi & Siddiqi, 2016; Tripathi, 2020; Tripathi et al., 2020; Kumar et al., 2014, 2015, 2016).
2. **Data Integration:** Data is integrated when more than one data sources are available.
3. **Data Transformation:** Discretization, normalization, etc. are data transformations.

Next, the extraction/selection feature targets at creation or/and identification of the significant informative subclass of features where, learning model is accepted to be employed during the entire training phase (Zheng & Casari, 2018).

As far as the feedback loop is concerned, it helps for adjustments related to the feature extraction/selection unit and the pre-processing one that further increases the overall performance of the learning model. In the duration of the phase of testing, earlier unnoticed samples are imported to the trained model generally characterized as feature vectors. At last, a suitable decision is taken by the model (e.g., regression or classification) in dependence of features present in each sample.

Figure 4. A graphical representation of a classic ML system

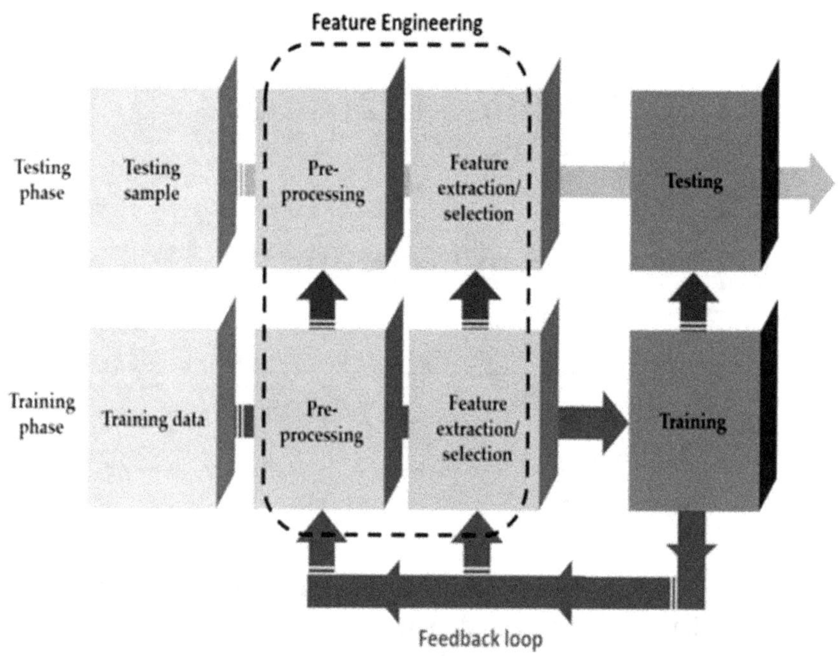

ML may be classified into following four categories depending on the learning type (Simeone, 2018; Choi. 2020):

1. **Supervised Learning:** Forms a mathematical model of a collection of data having inputs as well as outputs. Machine attempts to get the optimal way to reach a desired output when input is given. Classification, regression and active learning are types of supervised-learning algorithms (Alpaydin, 2010).
2. **Unsupervised Learning:** Uses a collection of data having only inputs, gets the structure in data, like clustering and grouping. Algorithms learn from unclassified, unlabeled or non-categorized test data.

3. **Semi-Supervised Learning:** Appears in between unsupervised and supervised learning. Input data create a combination of non-labeled and labeled data.
4. **Reinforcement Learning:** In reinforcement learning, decisions are taken towards exploring the actions that can move towards more positive outcome. Trial and error method is employed to determine it. Dynamic programming techniques are extensively used in reinforcement learning algorithms.

As of today, ML is widely being used in many fields like speech recognition (Anvarjon et al., 2020), image recognition (Cheng et al., 2020), autonomous driving (Fujiyoshi et al., 2019), stock market forecasting, credit card fraud detection (Rai & Dwivedi, 2020), medical diagnosis (Kokkotis et al., 2020), fluid mechanics (Sofos & Karakasidis, 2021), email, spam and malware filtering (Gangavarapu et al., 2020), activity recognition (Anagnostis et al, 2021), agriculture, etc.

MATHEMATICS BEHIND MACHINE LEARNING

Mathematics plays a key character in machine learning. Mathematics in machine learning is to explore about what is going on, why it's going on, and how one can use different things to acquire the required results. Machine learning turns around learning and implementing algorithms, Mathematical skill is essential in applying ML algorithms. It is difficult task to deal with the complex data features and parameters. In such situations, Mathematical skills are very helpful in eradicating uncertainty and forecasting data values precisely.

The most important parts of Mathematics that are used extensively in machine learning are Linear Algebra, Multivariate Calculus, Probability & Statistics and Optimization (Deisenroth et al., 2020) based on which a machine learning model can construct algorithms to extemporize as per human intelligence.

Linear Algebra, which is also called the mathematics of 21st century, is the main pillar of machine learning. It acts as a platform for machine learning algorithms where all the activities are performed to get desired results. It is a systematic representation of the information that a computer can recognize and all the operations performed in linear algebra are systematic rules. It is used to build Linear algorithms, Decision trees, Logistic regressions, Support Vector Machines, etc.

Multivariate calculus concerns with the facet of numerical optimization, which drives the machine learning algorithms. Without familiarity with its concepts, it would be very difficult to forecast the outcomes using an available data set. Differentials, Integrations, limits, etc. are helpful in training deep neural networks. Partial differentials are used for optimization of a given function.

Figure 5. Contribution of different concepts of mathematics in ML

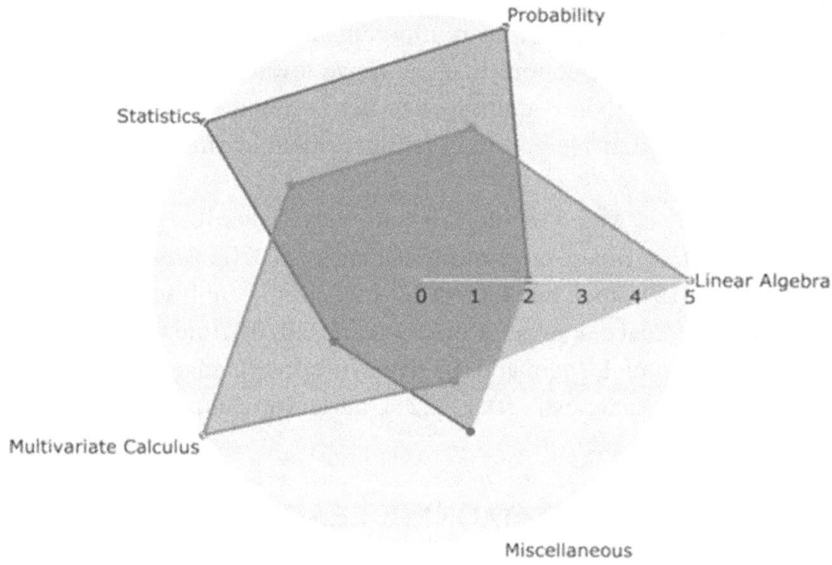

Probability in machine learning forecasts the set of outcomes. It can be described into two ways: Joint probability and conditional probability. In machine learning the basic concepts of probability like measure of central tendency, distributions (Gaussian, Bernoulli, etc.), probability and cumulative density functions, hypothesis testing are used widely.

Statistics emphasizes the qualitative and quantitative facets of the algorithm. It benefits in identification of goals by transforming the collected data into accurate observations. Statistics in machine learning emphases on Descriptive as well as Inferential statistics. Descriptive statistics deals with describing and summarizing the small dataset of a model while Inferential statistics focuses on extracting insights from a given sample of a large dataset.

Optimization is crucial in machine learning as our goal always happens to find optimal. Several learning problems are articulated as minimization of a loss function over a training set of examples. Loss functions represent the difference between the actual problem instances and the predictions of the training model. Mostly, the convex optimization is used in machine learning.

APPLICATIONS OF MACHINE LEARNING IN AGRICULTURE

There is abundance of applications of machine learning in agriculture. These applications can be divided into four categories namely: soil, water, crop and livestock management (Liakos et al., 2018).

Figure 6. Categories of machine learning applications in agriculture

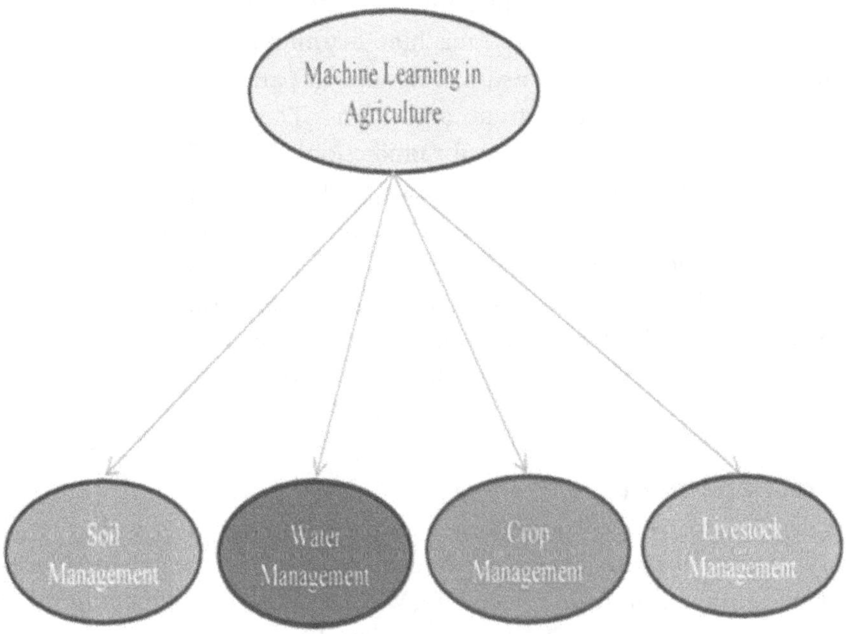

Soil Management

In order to have maximum yield of crops, the health of soil must be assessed. For this purpose, the properties of soil namely moisture content, soil drying, soil temperature etc. must be considered before seeding. A survey on soil management have been presented for identification/prediction of these properties using ML algorithms by Liakos et al. (2018) and Sharma et al. (2021).

Applications of ML algorithms for prediction and classification of soil attributes helps the farmers in many ways like minimum use of fertilizers, improvement in soil health, reduction in soil analysts demand, etc. and thus allowing increase in profits.

Suchithra and Pai (2020) used the Extreme Learning Machine together with diverse activation functions, e.g., sine-squared, Gaussian radial basis, etc. for classification of soil features like available Potassium (P), available Phosphorus (P), Boron (B), Organic Carbon (OC) and parameter soil reaction (pH).

Yang et al. (2019) employed four ML techniques namely Cubist regression model, partial least squares regression, ELM and LSSVM for predicting pH and Soil Organic Matter of paddy soil. The prediction accuracy was compared based on vis-NIR full bands and genetic algorithm was used to reduce bands. They concluded that extreme learning machine with genetic algorithm for band reduction produced best result and is very helpful in prevention of paddy soil degradation.

Morellos et al. (2016) employed machine learning algorithms to predict total nitrogen, organic carbon and moisture content soil parameters and compared the performance with linear multivariate algorithms. They concluded that machine learning algorithms performed better for prediction of MC, TN and OC.

Johann et al. (2016) employed an stochastic model Auto-regressive error function in combination of computational models to estimate soil moisture and compared the performance with multiple linear regression models, one model having two independent variables and other having six independent variables. To measure the performance of methods, RMSE, determination coefficients and average percentage error were evaluated. They concluded that their model had superior performance than others.

Water Management

Water is vital for agriculture. Therefore, water management is essential for maximizing the production in agriculture. In agriculture, Water management affects agronomical, climatological, and hydrological balance. Water quality can be improved and pollution as well as health risks can be reduced by proper water management (Bilali et al., 2021). Numerous ML techniques are employed in water management.

Mohapatra et al. (2021) have employed SVM, DNN and ANFIS for forecast of seasonal groundwater levels. The usefulness of these techniques was assessed using suitable graphical and statistical indicators. The conclusion was that DNN technique worked most proficiently in forecasting seasonal groundwater levels.

Singha et al. (2021) proposed a model built on deep learning (DL) for forecasting the quality of ground water and compared it with three ML based models: XGBoost, random forest and ANN. Results revealed that DL model has best prediction performance in term of performance metrics R^2.

Yu et al. (2022) estimated quality index of irrigation water employing four machine learning techniques namely: random forests, ELM, ANN and SVM as inputs, only physical groundwater parameters were used. Physical parameters included electrical conductivity, temperature, dissolved oxygen and pH. Results showed that these models worked efficiently in estimation of quality index of irrigation water.

Crop Management

One of the essential needs for farmers in order to achieve maximum production with quality, is crop management. Several production situations like climate conditions, soil types, social and economic contexts, etc. are faced by small farmers. Farmers may have differences in their purpose with respect to quantity, quality and products (Therond et al., 2017). These differences in purposes bring about differences of management of their farm by each farmer. Differences in management practices are classified broadly in conservation or conventional agriculture and organic farming system (Seufert et al., 2012; Muneret et al., 2018). Crop management includes farming techniques for management of physical, chemical and biological crop environment having the objective of achievement of both the qualitative and quantitative targets (Yvoz, et al., 2020).

Applications of cutting-edge methodologies like crop quality, crop recognition, weed detection, disease detection, yield prediction, etc. to manage crops help in increasing the production and subsequently, increasing the financial income.

Crop Quality

Crop quality is a major factor which highly affects the price of crop in market. This depends on several factors like crop characteristics, cultivation practices, climate and soil conditions, etc. Better quality of products guarantees better prices of products. This offers more earnings to farmers. Papageorgiou et al. (2018) observed skin colours, flesh firmness and soluble solids contents as maturity indices applied for harvesting to maintain the fruit quality. Timing of harvesting affects the quality features of harvested products very much. Hence, development of a decision support system may certainly support farmers in proper decisions to increase product quality. For this purpose, auto harvesting robots, ML algorithms with imaging technologies are achieving cheering results. Hua et al. (2019) surveyed on automated fruit harvesting techniques for tomato, kiwifruits, sweet pepper and apple to establish the modern developments in automatic harvesting robots in horticulture.

Crop Recognition

Automatic crop recognition has extensively been used in numerous agriculture sectors like new species discovery, plant taxonomy and botanical gardens. Recognition and classification of plant species can be done by analyzing various organs like fruits, seeds, leaves, flowers, stems, and roots (Bonnet et al., 2016; Seeland et al., 2017). Using leaf-based Plant recognition based on leaf is very popular and it exams characteristic of specific leaf like texture, shape and colour (Zhang et al., 2020). Now a days, remote sensing has got very popularity in crop classification. ML algorithms together with image processing has been employed in automatic classification and recognition of crops. Tian et al. (2021) proposed a technique in which machine learning has been used as a classifier and aerial imagery as a data source for identification of crops.

Weed Detection

Weeds are unwanted vegetation that are very prominent and harmful for crops. They consume the resources available for crops like sunlight, water, space and nutrients faster than crops and usually spread and grow very fast. Thus, they affect badly the growth of crops. Hence, weed control is vital for preventing the crop yield reduction. This can be done by applying herbicides or mechanical treatment. Mechanical treatment is little bit difficult. Therefore, herbicides are used mostly. But, herbicides have several drawbacks like they are harmful for environment and products itself. They contaminate the food crops. Besides, if used uniformly over entire agricultural area, they are very costly. To overcome these drawbacks, the help of Smart and precision agriculture is inevitable. The distinction between the crop and weed can be done using proximal or remote sensing attaching the sensors on satellites as well as ground and aerial vehicles. For localization and differentiation of target weeds, ML algorithms in combination with imaging technologies are used and allow the specific area for applying the herbicides (Slaughter et al., 2008).

Disease Detection

Presence of diseases in crops is a normal phenomenon in agriculture. Diseases cause major loss in production. Therefore, in order to have maximum production, timely detection of diseases is an essential task. Two types of tasks are performed generally by farmers: either leaves or branches are analyzed regularly during growing to detect the diseases or pesticides are applied on entire crop to avoid diseases. Both are risky and prone to error. Type of pesticides, time of application of pesticides and area of its applications is reliant on kind of disease, its stage and affected area. Use of pesticides on entire area may be harmful for the farmer's health as well as the crops. To overcome such problems, Smart and precision agriculture play important role. Using precision agriculture, farmers decide precise pesticide at precise time and precise area. Numerous techniques based on ML are proposed for detecting diseases.

Karadağ et al. (2018) used three ML algorithms namely Naive Bayes, K-nearest neighbor and ANN for classification to detect fusarium diseased peppers from the leaf reflections and concluded that the method was useful in detection of disease using leaf reflections. Alagumariappan et al. (2020) projected a decision support system for identification of plant diseases in which the performance of three ML algorithms: SVM, ELM, linear and polynomial kernels were evaluated. Extreme learning machine performed better than the rest two algorithms. Sujatha et al. (2021) paralleled the ML algorithms and DL algorithms in terms of citrus plant disease detection. They concluded that in comparison to ML algorithms, the performance of DL algorithms were much better.

Yield Prediction

Yield prediction is a necessity for decision-makers at regional and national levels for fast decision-making. A precise yield prediction model may be helpful for farmers in deciding on what to grow and when to grow. There are several methodologies based on machine learning like Neural Networks, Linear Regression, SVM, Random Forest, etc. which has been extensively used in yield prediction (Klompenburg et al., 2020). Crane-Droesch (2018) proposed a new technique in which parametric statistical models were augmented with DNN (deep neural networks) and termed it as semiparametric neural networks (SNN). This model was used in yield prediction and it out performed. Reddy and Kumar (2021) have applied several artificial techniques and got good results for yield prediction. Chlingaryan et al. (2018) presented a survey on application of machine learning in yield prediction. They concluded that the fast advancements in ML techniques and sensing technologies generate a profitable solution and provide better crop estimation. Elavarasan and Vincent (2020) proposed a Deep Recurrent Q-Network model to predict the yield and concluded that the model outperformed for yield prediction.

Livestock Management

Animal welfare and livestock production are the sub-categories of livestock. Wellbeing and health of animals are included in animal welfare while livestock production includes the matters related to the production system. Numerous ML techniques have been employed in livestock production and animal welfare.

Animal Welfare

For classification of cattle behavior, Dutta et al. (2015) employed ensemble classification using bagging, AdaBoost and random subspace techniques accompanied by the supervised classification methods like Naive Bayes classifier, Binary Tree,, ANFIS, k-Nearest Neighbour classifier and LDAC classifiers and compared them. The performance of the classifier bagging ensemble with Tree learner was superior to others with accuracy of 96%. Pegorini et al. (2015) proposed a ML based system for the automatic classification and identification of chewing patterns.

Livestock Production

Livestock production deals with the accurate estimation and prediction of farming parameters to have maximum economic productivity of the production system. Several ML based techniques like support machine vector, support machine regression, etc. have been employed for this (Alonso et al., 2013, 2015).

DISCUSSION

Numerous machine learning techniques are devised for the management of different data arising from agriculture. These techniques are classified in different categories of ML models. Artificial neural networks have been proved to be most efficient models. Ensemble Learning is other technique which got popularity. Support vector machine which works at high performance on image data, completes the group. Sensors have been proved as a boon for agriculture as input data applied in ML models correspond to different sensors. Remote sensing as well as imaging from UGVs, UAVs and satellites are the common means of obtaining measurements used in ML techniques. In animal welfare, activity recognition has been performed using devices like accelerometers while in livestock production, the growth characteristics of animals are utilized.

CONCLUSION

Machine learning in agriculture is one of the rapid growing areas. It is extensively used to solve complicated problems in agriculture and thus, helping the formers in their economic growth. It has been observed that machine learning based models have gained significant outcomes in solving the problems related to agriculture. Automated machine learning (AutoML) is recent technique which can be utilized in building ML models of high quality, highly efficient and more accurate without consuming much time. The need of time is to develop customized and precise ML models which are able to automatically analyze bigger, perform fast with complex data and be helpful in optimization of processes like predictions and classifications in agriculture.

REFERENCES

Alagumariappan, P., Dewan, N. J., Muthukrishnan, G. N., Raju, B. K., Bilal, R. A. A., & Sankaran, V. (2020). Intelligent plant disease identification system using machine learning. *Engineering Proceedings*, *2*, 49. doi:10.3390/ecsa-7-08160

Alonso, J., Castañón, Á. R., & Bahamonde, A. (2013). Support Vector Regression to predict carcass weight in beef cattle in advance of the slaughter. *Computers and Electronics in Agriculture*, *91*, 116–120. doi:10.1016/j.compag.2012.08.009

Alonso, J., Villa, A., & Bahamonde, A. (2015). Improved estimation of bovine weight trajectories using Support Vector Machine Classification. *Computers and Electronics in Agriculture*, *110*, 36–41. doi:10.1016/j.compag.2014.10.001

Alpaydin, E. (2010). *Introduction to Machine Learning*. MIT Press.

Anagnostis, A., Benos, L., Tsaopoulos, D., Tagarakis, A., Tsolakis, N., & Bochtis, D. (2021). Human activity recognition through recurrent neural networks for human-robot interaction in agriculture. *Sensors (Basel)*, *11*, 2188.

Anagnostis, A., Papageorgiou, E., & Bochtis, D. (2020). Application of Artificial Neural Networks for Natural Gas Consumption Forecasting. *Sustainability (Basel)*, *12*(16), 6409. doi:10.3390u12166409

Anvarjon, T., Mustaqeem, & Kwon, S. (2020). Deep-Net: A Lightweight CNN-Based Speech Emotion Recognition System Using Deep Frequency Features. *Sensors (Basel)*, *20*(18), 5212. doi:10.339020185212 PMID:32932723

Benos, L., Tagarakis, A. C., Dolias, G., Berruto, R., Kateris, D., & Bochtis, D. (2021). Machine Learning in Agriculture: A Comprehensive Updated Review. *Sensors (Basel)*, *21*(11), 3758. doi:10.339021113758 PMID:34071553

Bonnet, P., Joly, A., Goëau, H., Champ, J., Vignau, C., Molino, J.-F., Barthélémy, D., & Boujemaa, N. (2016). Plant identification: Man vs. machine. *Multimedia Tools and Applications*, *75*(3), 1647–1665. doi:10.100711042-015-2607-4

Cheng, Q., Zhang, S., Bo, S., Chen, D., & Zhang, H. (2020). Augmented Reality Dynamic Image Recognition Technology Based on Deep Learning Algorithm. *IEEE Access : Practical Innovations, Open Solutions*, *8*, 137370–137384. doi:10.1109/ACCESS.2020.3012130

Chlingaryan, A., Sukkarieh, S., & Whelan, B. (2018). Machine learning approaches for crop yield prediction and nitrogen status estimation in precision agriculture: A review. *Computers and Electronics in Agriculture*, *151*, 61–69. doi:10.1016/j.compag.2018.05.012

Crane-Droesch, A. (2018). Machine learning methods for crop yield prediction and climate change impact assessment in agriculture. *Environmental Research Letters*, *13*(11), 114003. doi:10.1088/1748-9326/aae159

Deisenroth, M. P., Faisal, A. A., & Ong, C. S. (2020). *Mathematics for Machine Learning*. Cambridge University Press. doi:10.1017/9781108679930

Dutta, R., Smith, D., Rawnsley, R., Bishop-Hurley, G., Hills, J., Timms, G., & Henry, D. (2015). Dynamic cattle behavioural classification using supervised ensemble classifiers. *Computers and Electronics in Agriculture*, *111*, 18–28. doi:10.1016/j.compag.2014.12.002

Elavarasan, D., & Vincent, P. D. (2020). Crop yield prediction using deep reinforcement learning model for sustainable agrarian applications. *IEEE Access : Practical Innovations, Open Solutions*, *8*, 86886–86901. doi:10.1109/ACCESS.2020.2992480

Fujiyoshi, H., Hirakawa, T., & Yamashita, T. (2019). Deep learning-based image recognition for autonomous driving. *IATSS Research*, *43*(4), 244–252. doi:10.1016/j.iatssr.2019.11.008

Gangavarapu, T., Jaidhar, C. D., & Chanduka, B. (2020). Applicability of machine learning in spam and phishing email filtering: Review and approaches. *Artificial Intelligence Review*, *53*(7), 5019–5081. doi:10.100710462-020-09814-9

Goel, D., Gulati, P., Jha, S. K., Kumar, N., Khan, A. & Kumari, P. (2022). Evaluation of Machine Learning Techniques for Crop Yield Prediction. *Artificial Intelligence Applications in Agriculture and Food Quality Improvement*, 112-128. doi:10.4018/978-1-6684-5141-0.ch007

Hua, Y., Zhang, N., Yuan, X., Quan, L., Yang, J., Nagasaka, K., & Zhou, X. (2019). Recent Advances in Intelligent Automated Fruit Harvesting Robots. *The Open Agriculture Journal*, *13*(1), 101–106. doi:10.2174/1874331501913010101

Johann, A. L., de Araújo, A. G., Delalibera, H. C., & Hirakawa, A. R. (2016). Soil moisture modeling based on stochastic behavior of forces on a no-till chisel opener. *Computers and Electronics in Agriculture*, *121*, 420–428. doi:10.1016/j.compag.2015.12.020

Karadağ, K., Tenekeci, M. E., Tasaltın, R., & Bilgilic, A. (2018). Detection of pepper fusarium disease using machine learning algorithms based on spectral reflectance. *Sustain Comput, 8*. Advance online publication. doi:10.1016/j.suscom.2019.01.001

Klompenburg, T. V., Kassahun, A., & Catal, C. (2020). Crop yield prediction using machine learning: A systematic literature review. *Computers and Electronics in Agriculture, 177*, 105709. doi:10.1016/j.compag.2020.105709

Kokkotis, C., Moustakidis, S., Papageorgiou, E., Giakas, G., & Tsaopoulos, D. E. (2020). Machine learning in knee osteoarthritis: A review. *Osteoarthritis and Cartilage Open, 2*(3), 100069. doi:10.1016/j.ocarto.2020.100069 PMID:36474688

Kumar, G., Kumar, S., & Kumar, N. (2015). Comparative study of wavelet and wavelet packet transform for denoising telephonic speech signal. *International Journal of Computer Applications, 110*(15), 1–8. doi:10.5120/19389-9145

Kumar, N., Alam, K., & Siddiqi, A. H. (2016). Savitzky–Golay and Wavelet Transform Based Rama Spectroscopic data Denoising. *International Journal of Control Theory and Applications, 9*(21), 297–305.

Kumar, N., Siddiqi, A. H., & Alam, K. (2014). Raman spectral data de-noising based on wavelet analysis. *International Journal of Computer Applications, 102*(16), 20–22. doi:10.5120/17899-8864

Kumar, N., Siddiqi, A. H., & Alam, K. (2016). Raman spectra denoising using wavelet and wavelet packet transform. *Indian Journal of Industrial and Applied Mathematics, 7*(01), 1–10. doi:10.5958/1945-919X.2016.00001.3

Kumar, N., Tripathi, P., & Alam, K. (2016). Non-Negative Matrix Factorization-Based EEG Signal Classification. *Indian Journal of Industrial and Applied Mathematics, 7*(2), 212–219. doi:10.5958/1945-919X.2016.00019.0

Liakos, K., Busato, P., Moshou, D., Pearson, S., & Bochtis, D. (2018). Machine Learning in Agriculture: A Review. *Sensors (Basel), 18*(8), 2674. doi:10.339018082674 PMID:30110960

Lopez-Arevalo, I., Aldana-Bobadilla, E., Molina-Villegas, A., Galeana-Zapién, H., Muñiz-Sanchez, V., & Gausin-Valle, S. (2020). A Memory- Efficient Encoding Method for Processing Mixed-Type Data on Machine Learning. *Entropy (Basel, Switzerland), 22*(12), 1391. doi:10.3390/e22121391 PMID:33316972

Meshram, V., Patil, K., Meshram, V., Hanchate, D., & Ramkteke, S. D. (2021). Machine learning in agriculture domain: A state-of-art survey. *Artificial Intelligence in the Life Sciences, 1*, 100010. doi:10.1016/j.ailsci.2021.100010

Mohapatra, J. B., Jha, P., Jha, M. K., & Biswal, S. (2021). Efficacy of machine learning techniques in predicting groundwater fluctuations in agro-ecological zones of India. *The Science of the Total Environment*, *785*, 147319. doi:10.1016/j.scitotenv.2021.147319 PMID:33957597

Muneret, L., Mitchell, M., Seufert, V., Aviron, S., Pétillon, J., Plantegenest, M., Thiéry, D., & Rusch, A. (2018). Evidence that organic farming promotes pest control. *Nature Sustainability*, *1*(7), 361–368. doi:10.103841893-018-0102-4

Papageorgiou, E. I., Aggelopoulou, K., Gemtos, T. A., & Nanos, G. D. (2018). Development and Evaluation of a Fuzzy Inference System and a Neuro-Fuzzy Inference System for Grading Apple Quality. *Applied Artificial Intelligence*, *32*(3), 253–280. doi:10.1080/08839514.2018.1448072

Pegorini, V., Karam, L. Z., Pitta, C. S. R., Cardoso, R., da Silva, J. C. C., Kalinowski, H. J., Ribeiro, R., Bertotti, F. L., & Assmann, T. S. (2015). In vivo pattern classification of ingestive behavior in ruminants using FBG sensors and machine learning. *Sensors (Basel)*, *15*(11), 28456–28471. doi:10.3390151128456 PMID:26569250

Rai, A. K., & Dwivedi, R. K. (2020). Fraud Detection in Credit Card Data using Unsupervised Machine Learning Based Scheme. In *Proceedings of the International Conference on Electronics and Sustainable Communication Systems (ICESC)*, Coimbatore, India, 2–4 July 2020. Institute of Electrical and Electronics Engineers Inc. 10.1109/ICESC48915.2020.9155615

Reddy, D. J., & Kumar, M. R. (2021). Crop Yield Prediction using Machine Learning Algorithm. *2021 5th International Conference on Intelligent Computing and Control Systems (ICICCS)*, 1466-1470. 10.1109/ICICCS51141.2021.9432236

Seeland, M., Rzanny, M., Alaqraa, N., Wäldchen, J., & Mäder, P. (2017). Plant species classification using flower images—A comparative study of local feature representations. *PLoS One*, *2017*(12), e0170629. doi:10.1371/journal.pone.0170629 PMID:28234999

Seufert, V., Ramankutty, N., & Foley, J. A. (2012). Comparing the yields of organic and conventional agriculture. *Nature*, *485*(7397), 229–232. doi:10.1038/nature11069 PMID:22535250

Singha, S., Pasupulet, S., Singha, S. S., Singh, S., & Kumar, S. (2021). Prediction of groundwater quality using efficient machine learning technique. *Chemosphere*, *276*, 130265. doi:10.1016/j.chemosphere.2021.130265 PMID:34088106

Slaughter, D. C., Giles, D. K., & Downey, D. (2008). Autonomous robotic weed control systems: A review. *Computers and Electronics in Agriculture*, *61*(1), 63–78. doi:10.1016/j.compag.2007.05.008

Sofos, F., & Karakasidis, T. E. (2021). Machine Learning Techniques for Fluid Flows at the Nanoscale. *Fluids (Basel, Switzerland)*, *6*(3), 96. doi:10.3390/fluids6030096

Sujatha, R., Chatterjee, J. M., Jhanjhi, N. Z., & Brohi, S. N. (2021). Performance of deep learning vs machine learning in plant leaf disease detection. *Microprocessors and Microsystems*, *80*, 103615. doi:10.1016/j.micpro.2020.103615

Therond, O., Duru, M., Roger-Estrade, J., & Richard, G. (2017). A new analytical framework of farming system and agriculture model diversities. A review. *Agronomy for Sustainable Development*, *37*(3), 21. doi:10.100713593-017-0429-7

Tian, Y., Yang, C., Huang, W., Tang, J., Li, X., & Zhang, Q. (2021). Machine learning-based crop recognition from aerial remote sensing imagery. *Frontiers of Earth Science*, *15*(1), 54–69. doi:10.100711707-020-0861-x

Tripathi, P. (2020). Electroencephalpgram Signal Quality Enhancement by Total Variation Denoising Using Non-convex Regulariser. *International Journal of Biomedical Engineering and Technology*, *33*(2), 134–145. doi:10.1504/IJBET.2020.107709

Tripathi, P., Kumar, N., Rai, M., & Khan, A. (2022). Applications of Deep Learning in Agriculture. In *Artificial Intelligence Applications in Agriculture and Food Quality Improvement* (pp. 17–28). IGI Global. doi:10.4018/978-1-6684-5141-0.ch002

Tripathi, P., Kumar, N., & Siddiqi, A. H. (2020). De-noising Raman spectra using total variation de-noising with iterative clipping algorithm. In *Computational Science and its Applications, Taylor and Francis Group* (pp. 225–231). CRC Press. doi:10.1201/9780429288739-14

Tripathi, P., & Siddiqi, A. H. (2016). Solution of Inverse Problem for de-noising Raman Spectral Data with Total variation using Majorization-Minimization Algorithm. *Int. J. Computing Science and Mathematics*, *7*(3), 274–282. doi:10.1504/IJCSM.2016.077855

Tyagi, A. C. (2016). Towards a second green revolution. *Irrigation and Drainage*, *65*(4), 388–389. doi:10.1002/ird.2076

Vieira, S., Lopez Pinaya, W. H., & Mechelli, A. (2020). Introduction to Machine Learning. Academic Press. doi:10.1016/B978-0-12-815739-8.00001-8

Yu, H., Wen, X., Wu, M., Sheng, D., Wu, J., & Zhao, Y. (2022). Data-based groundwater quality estimation and uncertainty analysis for irrigation agriculture. *Agricultural Water Management*, *262*, 107423. doi:10.1016/j.agwat.2021.107423

Yvoz, S., Petit, S., Biju-Duval, L., & Cordeau, S. A. (2020). framework to type crop management strategies within a production situation to improve the comprehension of weed communities. *European Journal of Agronomy*, *115*, 126009. doi:10.1016/j.eja.2020.126009

Zhang, S., Huang, W., Huang, Y., & Zhang, C. (2020). Plant species recognition methods using leaf image: Overview. *Neurocomputing*, *408*, 246–272. doi:10.1016/j.neucom.2019.09.113

Zheng, A., & Casari, A. (2018). *Feature Engineering for Machine Learning: Principles and Techniques for Data Scientists*. O'Reilly Media, Inc.

Chapter 7
Achieving Sustainable Agriculture:
An AI-Based Approach

Syed Asif Ali
SMI University, Karachi, Pakistan

ABSTRACT

Sustainable agriculture means an integrated agricultural system (IAS), which is a prolonged or long-term support system that fulfils the requirements of agricultural products, improves the environment (by controlling, improving, and monitoring natural biological cycles), and boosts the economy. Crop product quality and quantity have increased significantly over the last half-century, owing in large part to organic chemistry. Organic chemistry, however, remains a key component of crop protection and helps to create techniques to boost crop output. In the modern era of technology, artificial intelligence approaches are used for the analysis of soil nutrients and their optimal management on the basis of supervised learning, depending on behaviour analysis and prior experience. Soil nutrients are an essential component of soil fertility and environmental sustainability. This chapter will cover soil nutrient measurements, water irrigation, and predator control through artificial intelligence techniques such as decision making, artificial neural networks, and machine learning.

INTRODUCTION

The Latin words "agri," which means "field," and "cultura," meaning "piece of land," are indeed the basis of the English term agriculture. Agriculture is the art and science that instructs or educates the farmer on how to cultivate the land in order to

DOI: 10.4018/978-1-6684-6418-2.ch007

grow crops or plants and raise livestock or products from livestock (including food, wool, and other products) for human use. The contribution made by agriculture to the economy is an important aspect in the development of the economies of emerging nations.

Although the history of agriculture is as old as human history, the Egyptians are considered at the beginning of civilization to have started the practice of agriculture on a very large scale (between 10,000 BC and 4,000 BC). American agricultural scientist Norman Ernest Borlaug, on the other hand,

Types of Agriculture

Agriculture usually comes in two distinct types.

1. Commercial Agriculture
2. Subsistence Agriculture

Agriculture performed on a small scale for individual or family subsistence is known as subsistence agriculture.

Commercial agriculture is the process of growing crops for the purpose of trading and earning profit on the national or international market.

Table 1. Subsistence and commercial agriculture

S.No	Subsistence Agriculture	Commercial Agriculture
1	Referred to as small-scale agriculture	Referred to as large-scale agriculture
2	To fulfill the needs of farmers and their families are the main goal.	To make a profit is the main goal.
3	It is a common practice in tropical areas of the world and is dependent on animal input.	It is a common practice in tropical areas of the world and is dependent on animal input.
4	Usually seen in less developed nations,	Usually appear in wealthy nations,
5	Food or items in excess may be put up for sale.	The goal of Nain, sometimes known as agribusiness, is to produce food for sale in the global market.

BACKGROUND

Sustainable Agriculture

The Latin term "sus," which means "up" or "hold," is the origin of the English word "sustain." An Integrated Agricultural System (IAS), generally referred to as sustainable agriculture, is a protracted or long-term support system that meets the requirements of agricultural products, improves the environment (by controlling, developing, and tracking natural biological cycles), and enhances the economy.

The quantity and quality of crop products have improved significantly in the last half-century in large part due to organic chemistry. Organic chemistry, however, remains an important component of agricultural production and aids in developing techniques to improve crop output.

Proper water use, soil quality (nutrients from the soil), crop protection against predators (mainly Insect pests), as well as sunlight, of course, are essential for improving crop products. The world's Population is growing gradually, and as an outcome, there really are parallel challenges related to food. The role of organic chemistry is crucial in crop protection, predator control, and improving the provision of food in order to fulfill or increase food provision.

Chemicals to boost crop yields and kill pests are provided in order to raise the demand for increased crop yields for future organic chemistry. Moreover, the chemicals kill the biological processes taking place in the soil, harm the environment and have negative effects on human health, which reduces the overall advantages of organic chemistry. By controlling the ratio or composition of chemicals, artificial intelligence-based technology and efficient algorithms for agriculture can help safeguard against by the negative effects of chemicals. It is also the best way to deal with difficulties we are currently facing as well as prevent challenges down the road (Bhagat et al., 2022; Alreshidi, 2019; Neményi, 2018). The overall process is stated in Figure 1 Sustainable Agriculture process.

MAIN FOCUS OF THE CHAPTER

Sustainable agriculture means an integrated agricultural system (IAS), which is a prolonged or long-term support system that fulfils the requirements of agricultural products. It is a burning research topic for researchers working on artificial intelligence. For this purpose, there is a need to collaborate with people related to agriculture for better results. A lot of work has been done on the smart agriculture system and its monitoring and controlling strategies. To improve and increase the yield of crops, the very important thing is the analysis of the nutrients in the soil and its composition. A digital instrument for measuring the ratio of npk meters is used to identify the npk in soil. A pH meter is used for checking the pH of soil. Sensors are available to calculate the level of moisture and temperature. The data picked from these digital instruments helps in identifying the deficiency of soil if it rains. In this chapter, an artificial intelligence (AI) based solution is proposed to improve the crop yield by monitoring soil composition, water irrigation and protecting the crops from predators by applying the control conditions.

Figure 1. Sustainable agriculture process

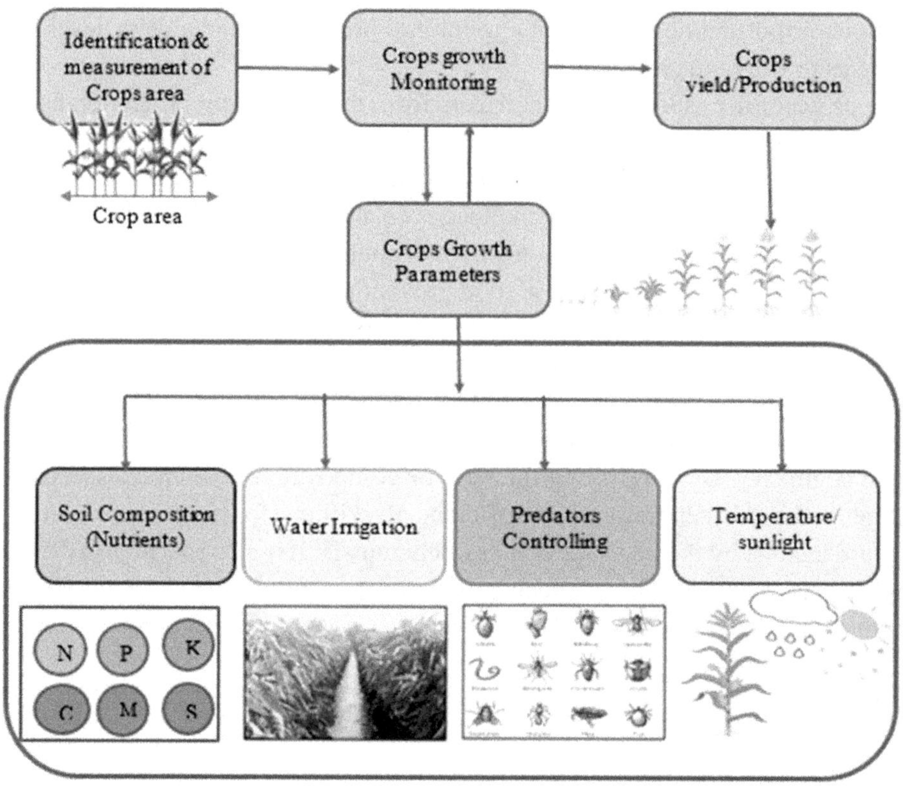

SOLUTIONS AND RECOMMENDATIONS

Crop Yield Enhancement Through Artificial Intelligence-Based Technologies

The yield of crops is always impacted negatively by fungi, weeds, pests, water, heat (sunlight), and nutrients. Across the globe, those who work in agriculture make a variety of decisions to preserve their crops and achieve the maximum yield. They utilized a range of equipment and agricultural supplies for that goal. To do it, they used a variety of tools and agricultural materials. Artificial intelligence-based solutions are suggested to enhance crop yields. The method of machine learning for soil assessment and nutrient management is presented in this chapter. The topic of increasing water irrigation and predator control through pattern matching and artificial neural networks and machine learning approaches is discussed (Sharma, 2021; Khaki & Wang, 2019).

Machine Learning-Based Soil Assessment and Nutrient Management

The act of controlling how much, where, when, and how fertiliser is applied in order to maximise farm yield while minimising nutrient losses that might harm the environment. Crop quality and growth are strongly influenced by the type and nutrients of the soil. Traditional methods for assessing soil nutrients are fairly difficult to utilise, which makes them quite challenging for practical applications. Agriculture is primarily a non-technical industry; however artificial intelligence technology is utilized to enhance agriculture (Agrahari et al., 2021; Hengl et al., 2017; Motia et al., 2021).

This helps in analysing the soil's chemical, physical, biological, and biological (N&C microbes, natural manure, and soil texture) factors. First, datasets of soil samples will be collected in next step databases of soil samples will be analyses. The training phase is an important part of the entire system; machine learning algorithms (Sharma et al., 2020; John et al., 2020) are used to train the system for taking decisions without human involvement (see Figure 2). Data science transforms traditional practices into an integrated system focused on sustainable use (Li, 2021).

Artificial Intelligence techniques are used for assessment of soil nutrient and its proper management on the basis of behavior analysis and previous experience, Neural Networks are used for the assessment of soil nutrients and their proper management on the basis of supervised learning. Soil nutrients are an important aspect that contributes to soil fertility and environmental effects.

Real-time data drives the precise application of resources, and state-of-the-art software allows for quick, easy analysis of crop management and protection. There are five methods for measuring soil nu

- Soil npk meter
- pH meter for soil
- Sensor for soil moisture
- Soil temperature sensor

Figure 2. Machine learning based soil assessment and nutrient management

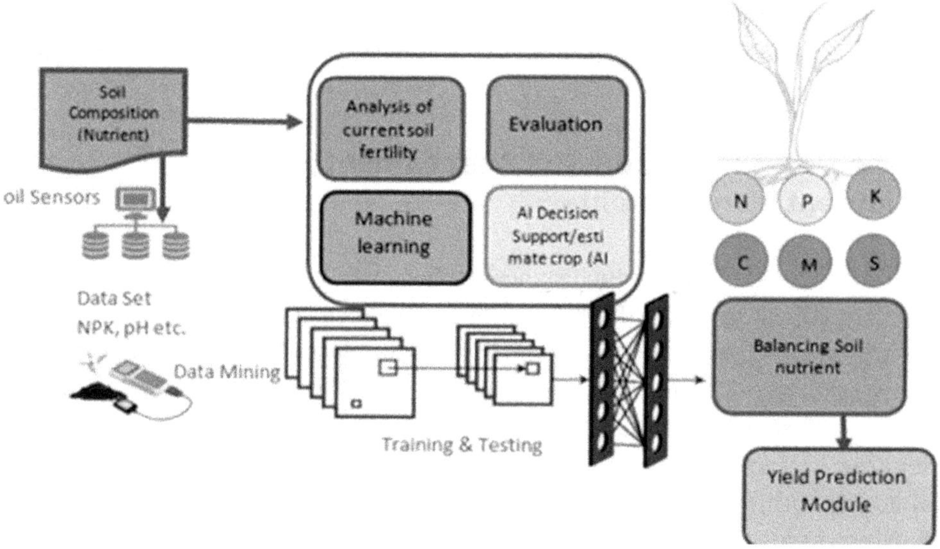

Soil NPK Meter

There are multiple methods for measuring the soil nutrients (N, K, and P) using optical sensors or spectrometers. The soil NPK meter (shown in Figure 3) is used to determine the amount of nitrogen (N), phosphorus (P), and potassium (K), which are essential and most consumed elements for crop development and are basically organic compounds that are returned to the soil as leftovers. This modern artificial intelligence tool and techniques assist farmers in detecting fertility deficiencies. Soil NPK sensor is a comparatively low-cost, quick-responsive, high-precision, and portable sensor. Then there are traditional chemistry methods.

Figure 3. Soil NPK meter use for measuring fertility of soil

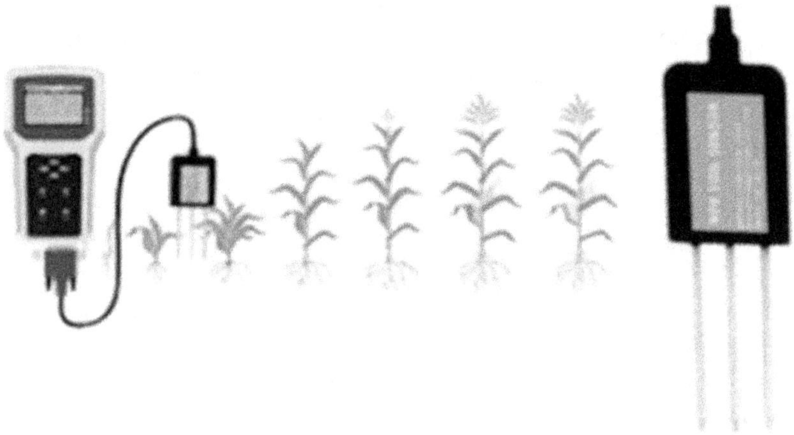

pH Meter for Soil

Indicative of soil pH is soil pH. Its concentration has a big influence on the life processes of soil microorganisms in addition to how readily available and in what ionic form major and trace elements are in the soil. Absorption of nutrients will be hampered by an improper pH balance.

Figure 4. pH meter for soil

Sensor for Soil Moisture

A soil moisture sensor is used to measure soil moisture, which is essential for the health of the soil and the growth of crops. Monitoring soil moisture is important for applying water irrigation, and digital network using moisture sensors and artificial intelligence is used to analyze and regulate the information.

Figure 5. Sensor use for soil moisture

Soil Temperature Sensor

In order to keep the ecosystem and environment in balance, soil temperature is a crucial element and soil parameter. Crop growth is impacted by it on a biochemical, physical, and chemical level. In artificial intelligence, the temperature sensor is essential for modelling and forecasting crop yield.

Artificial Neural Networks and Machine Learning Approaches to Water Irrigation

Water is essential for the optimal growth of crops. Different crops need different amounts of water during the growth process. In fact, agriculture is the area that consumes a large amount of water, and the condition of water resources all over the world is very critical. The water irrigation process can easily be managed by using artificial intelligence to reduce the volume of water by providing the concept of a smart water irrigation system. Soil moisture and temperature sensors facilitate controlling and saving massive amounts of water by embedding moisture and temperature parameters in an intelligent or smart irrigation system (Machine Learning and Internet of Things) (Klyushin et al., 2021; Nagappan et al., 2020; Jain et al., 2021).

A smart system also helps in ensuring the even accessibility of water for every plant in the field. Smart systems also observed diseases in crops due to improper irrigation systems. The texture or layout of land and types of crops may require little modification in smart irrigation. The availability of the exact amount of water on time is a bigger challenge and a major concern in gaining a smart amount of crop yield. In the technological world using sensors and artificial intelligence, now it is possible to sense the water requirements of crops using smart or intelligent systems. Due to a lack of technological awareness, farmers are still using traditional methods for irrigation, which makes it difficult to manage the desired amount of water, which causes low yield. After managing the irrigation system via artificial intelligence, basically, decision support algorithms and machine learning training are applied to collect irrigation requirements data and, after examining the system, easily predict the yield of crops. ANN (Kamyshova et al., 2022; Abdel-Fattah et al., 2021; Karasekreter et al., 2013). and machine learning algorithms can easily estimate the time and need for water in marking or identifying areas of crops, which, of course, brings disaster change in the field of agriculture.

Figure 6. Water Irrigation system using artificial neural networks (ANN) and machine learning (ML) approaches

Predators Control Analyzing Crop Health by Drones

Every year, crop pests and diseases cause the farmer to lose revenue. Crop yields will increase if crop security is enhanced. One can minimize the impacts of pests on crops using technology, which allows us to use drones and other sensors. The use of drones for data collection is facilitated by technology (Kumar et al., 2022; Li, 2021; Crop Protection Innovations, 2022).

The use of artificial intelligence in domestic pest management, however, might potentially be quite beneficial. Imagine the drones immediately scanning a meadow to determine that no treatment is required. Artificial intelligence cameras swivel in the subterranean farm area. Based on the identification and classification of plant leaf diseases, image processing, and pattern recognition Future application options are essentially limitless. Although pest management can be entirely automated, technology may undoubtedly be useful.

Domestic pest management may benefit from the application of artificial intelligence in agriculture or become safer, faster, and more affordable for business owners and clients (Pudasainee et al., 2020).

Increase productivity, food security, and farmer incomes with the use of big data in agriculture.

It stated that "making farmers the owners of their data and providing them options to manage the flow of their data to stakeholders should help create confidence with farmers in exchanging data and reaping the benefits of big data analysis."

Figure 7. Predators are controlled by image processing and pattern matching

FUTURE RESEARCH

The use of technology, including artificial intelligence algorithms, tools, and procedures, has greatly improved the agricultural industry. Artificial intelligence technology has recently been employed by farmers in wealthy nations to effortlessly control and manage crop quality. Sensor technologies make it relatively simple to test the soil's composition, including NPK, pH, moisture, and temperature. The water irrigation system and the predator control system are likewise under the authority and supervision of the farmers. Agriculture intelligence also uses Global Positioning System (GPS) technologies. By enhancing the composition and quality of the soil and controlling pest attacks on crops, artificial intelligence technology not only aids in the production of healthy crops but also controls other tasks related to the crop supply chain process.

Figure 8. Artificial intelligence based predators are controlled system

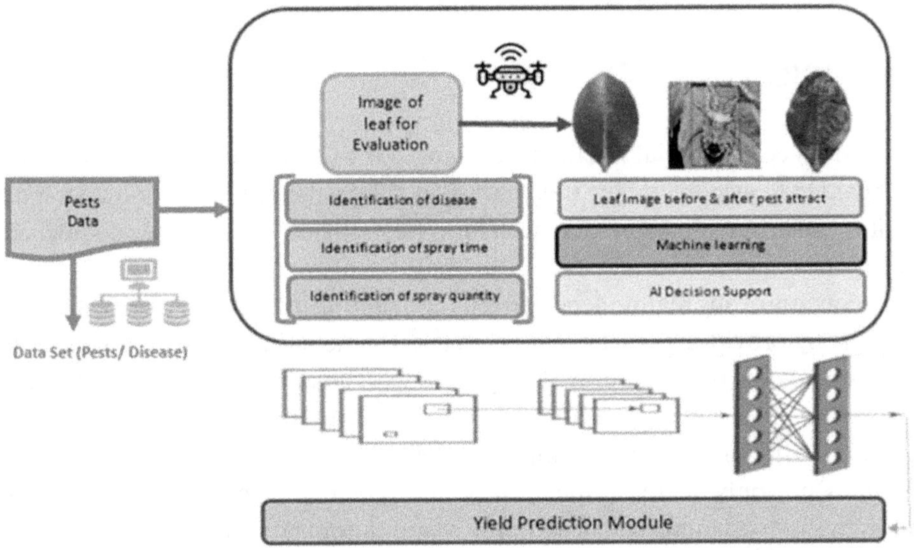

Farmers may decide if crops need irrigation, fertilizer, or pesticide application using real-time AI data. Also, cutting down on resource usage while increasing food production is possible with current farming techniques like vertical agriculture. AI can scan images from satellites and drones to help farmers manage their livestock and crops. Since they will be notified as soon as something seems unexpected, they won't need to constantly monitor the fields. Moreover, aerial photography helps improve the accuracy and efficacy of pesticide application. Future farming will be done without the help of farmers since robots will monitor and manage every stage of the crop's growth more efficiently and, of course, within the allocated budget. Some people and people related to the farming industry think how robots can be dealt with all the issues related to the agriculture? The answer is that if robots can play a role in healthcare why not they deal with agriculture field.

In my nation, the chilly temperature makes it impossible for wheat to grow in the northern region. While the climate may be warmer in some sections of the world, artificial intelligence systems now offer the required control to enable wheat cultivation in the northern regions. In other places of the world that experience excessive heat, the summers are typically quite hot and dry while the winters are pleasant. Rainfall is in low supply, and the scarcity of water is a hindrance to agriculture's ability to produce. The difficulties facing agriculture will increase as global climate change continues. With artificial intelligence tools and techniques, we can simply address all the issues facing the agriculture industries.

The future of the agriculture sector lies in the full integration of robotics with other AI tools and approaches.

CONCLUSION

In the modern era Artificial intelligence changes the lifestyle of human, concept of smart healthcare system, smart transportation system, smart education and smart agriculture make unbelievable change in lifestyle. Artificial intelligence could make life more convenient.

The economy of any regime depends on its agricultural industries. Agriculture helps any regime by providing a clean and green environment, increasing their financial support, social networks and providing people with an opportunity to earn for their families other than getting food products and other by-products from agriculture industries. Artificial intelligence improves the overall yield quality with accuracy.

AI does not only increase the quality and quantity of corpus yield but also improve the fundamental business needs using automated systems and decision support systems that apply to existing data and predict profitability on the basis of previous data patterns.

The people associated with the agriculture business can get more work done with fewer people with an automated water irrigation system, a pesticide spraying system, a fertilizing system, driverless tractors or other agriculture vehicles, and artificial intelligence-based robots for the harvesting process. On the other hand, agriculture businesses can be improved by intelligent customer interaction, supply chain management, etc.

In summary, both the farmers themselves and the government's economy benefit from the artificially based smart agricultural system. Farmers now recognize the value of agriculture automation and are using innovative, intelligent harvesting techniques to obtain a greater and more precise yield with higher quality while using less resources. Besides crop productivity, other IoT applications in agriculture that assist agrarian automation include management of clean and green environment, controlling of farm animals and agriculture vehicle and machineries, drones etc. It also necessitates providing a contribution to the long-term viability of agricultural food production.

REFERENCES

Abdel-Fattah, M. K., Mokhtar, A., & Abdo, A. I. (2021). Application of neural network and time series modeling to study the suitability of drain water quality for irrigation: A case study from Egypt. *Environmental Science and Pollution Research International*, *28*(1), 898–914. doi:10.100711356-020-10543-3 PMID:32822008

Agrahari, R. K., Kobayashi, Y., Tanaka, T. S. T., Panda, S. K., & Koyama, H. (2021). Smart fertilizer management: The progress of imaging technologies and possible implementation of plant biomarkers in agriculture. *Soil Science and Plant Nutrition*, *67*(3), 248–258. doi:10.1080/00380768.2021.1897479

Alreshidi, E. (2019). Smart sustainable agriculture (SSA) solution underpinned by internet of things (IoT) and artificial intelligence (AI). arXiv preprint arXiv:1906.03106. doi:10.1080/00380768.2021.1897479

Bhagat, P. R., Naz, F., & Magda, R. (2022). Artificial intelligence solutions enabling sustainable agriculture: A bibliometric analysis. *PLoS One*, *17*(6), e0268989. doi:10.1371/journal.pone.0268989 PMID:35679287

Crop Protection Innovations. (2022, June 30). Retrieved October 10, 2022, from https://www.bayer.com/en/agriculture/crop-

Hengl, T., Leenaars, J. G., Shepherd, K. D., Walsh, M. G., Heuvelink, G., Mamo, T., Tilahun, H., Berkhout, E., Cooper, M., Fegraus, E., Wheeler, I., & Kwabena, N. A. (2017). Soil nutrient maps of Sub-Saharan Africa: Assessment of soil nutrient content at 250 m spatial resolution using machine learning. *Nutrient Cycling in Agroecosystems*, *109*(1), 77–102. doi:10.100710705-017-9870-x PMID:33456317

Jain, T., Garg, P., Tiwari, P. K., Kuncham, V. K., Sharma, M., & Verma, V. K. (2021). Performance prediction for crop irrigation using different machine learning approaches. In Examining the Impact of Deep Learning and IoT on Multi-Industry Applications (pp. 61-79). IGI Global. doi:10.4018/978-1-7998-7511-6.ch005

John, K., Abraham Isong, I., Michael Kebonye, N., Okon Ayito, E., Chapman Agyeman, P., & Marcus Afu, S. (2020). Using machine learning algorithms to estimate soil organic carbon variability with environmental variables and soil nutrient indicators in an alluvial soil. *Land (Basel)*, *9*(12), 487. doi:10.3390/land9120487

Kamyshova, G., Osipov, A., Gataullin, S., Korchagin, S., Ignar, S., Gataullin, T., Terekhova, N., & Suvorov, S. (2022). Artificial neural networks and computer vision's-based phytoindication systems for variable rate irrigation improving. *IEEE Access : Practical Innovations, Open Solutions*, *10*, 8577–8589. doi:10.1109/ACCESS.2022.3143524

Karasekreter, N., Başçiftçi, F., & Fidan, U. (2013). A new suggestion for an irrigation schedule with an artificial neural network. *Journal of Experimental & Theoretical Artificial Intelligence*, *25*(1), 93–104. doi:10.1080/0952813X.2012.680071

Khaki, S., & Wang, L. (2019). Crop yield prediction using deep neural networks. *Frontiers in Plant Science*, *10*, 621. doi:10.3389/fpls.2019.00621 PMID:31191564

Klyushin, D., & Tymoshenko, A. (2021). Optimization of drip irrigation systems using artificial intelligence methods for sustainable agriculture and the environment. In *Artificial intelligence for sustainable development: Theory, practice and future applications* (pp. 3–17). Springer. doi:10.1007/978-3-030-51920-9_1

Kumar, P. P., Akshay, R., Achha, N., Sagar, K., Thirupathi, V., & Reddy, G. R. (2022, May). An affordable multitasking drone for smart farming with artificial intelligence features. In. AIP Conference Proceedings: Vol. 2418. *No. 1* (p. 020002). AIP Publishing LLC. doi:10.1063/5.0081755

Li, D. (2021). Application of artificial intelligence and machine learning based on big data analysis in sustainable agriculture. *Acta Agriculturæ Scandinavica. Section B, Soil and Plant Science*, *71*(9), 956–969. doi:10.1080/09064710.2021.1965650

Motia, S., & Reddy, S. R. N. (2021, August). Exploration of machine learning methods for prediction and assessment of soil properties for agricultural soil management: A quantitative evaluation. *Journal of Physics: Conference Series*, *1950*(1), 012037. doi:10.1088/1742-6596/1950/1/012037

Nagappan, M., Gopalakrishnan, V., & Alagappan, M. (2020). Prediction of reference evapotranspiration for irrigation scheduling using machine learning. *Hydrological Sciences Journal*, *65*(16), 2669–2677. doi:10.1080/02626667.2020.1830996

Neményi, M. (2018). Precision crop production and artificial intelligence–the future of sustainable agriculture. *Acta Agraria Debreceniensis*, 47-58.

Pudasainee, A., Pudasaini, U., & Chaulagain, B. P. (2020). Prospects of ICT based agricultural technology in the context of Nepal. *Nepalese Journal of Agricultural Sciences*, *19*, 211.

Sharma, R. (2021, May). Artificial Intelligence in Agriculture: A Review. In *2021 5th International Conference on Intelligent Computing and Control Systems (ICICCS)* (pp. 937-942). IEEE. 10.1109/ICICCS51141.2021.9432187

Sharma, R., Kamble, S. S., Gunasekaran, A., Kumar, V., & Kumar, A. (2020). A systematic literature review on machine learning applications for sustainable agriculture supply chain performance. *Computers & Operations Research, 119*, 104926. doi:10.1016/j.cor.2020.104926

Chapter 8
An Outlook Over Smart Irrigation System for Sustainable Rural Development

R. Muthuminal
https://orcid.org/0000-0001-9931-2204
SNS College of Technology, India

R. Mohana Priya
SNS College of Technology, India

ABSTRACT

The process of applying water to the soil artificially by means of systems such as pumps, tubes, sprays, pipes is called irrigation. As irrigation plays a vital role in the growth of plants or cultivating species to satisfy the basic needs of human beings, it is necessary to manage it in a proper way. Usually, irrigation is done at the regions which are subjected to irregular rain or dry or drought prone areas in order to manage the crop yield along with the consideration of quality. Since the mode of applying the water in a uniform manner is the main task in irrigation, several irrigation systems are adopted. The growth of the agricultural crops, maintaining landscapes, reducing weeds, and also to regenerate the disturbed soils in dry areas or also during the season of less rainfall are dependent on irrigation for their prosperity. Hence, in this chapter, the authors detail the background of irrigation, how it impacts the crop yield and demand, smart technologies available, growth analysis, and intelligent and smart irrigation systems with its benefits and necessities.

DOI: 10.4018/978-1-6684-6418-2.ch008

INTRODUCTION

Irrigation is a process that exists for several eras. This was proved by the archaeology department by collecting records where irrigation system was used in places where rainfall was very less and this would greatly support the crops in times of less rainfall. For example, this irrigation was used in the alluvial plains in Indus valley during 4500 BC since the water availability during that was very low this irrigation application helps in residents to manage their prosperity by maintaining the agricultural production application of water at the alluvial plains of Indus valley has been done during in 4500 BC alluvial plains were supported by water, as a result, it helps the resident to increase their prosperity by increasing the agriculture production. During the Indus valley civilization, the growth of refined irrigation and nature reservoir systems along with the art of man-made reservoirs for directing the water flow around 3000 BCE, and a permanent canal irrigation system from c. 2600 BCE were successful.

Irrigation can be regulated with the help of these structures which implement large-scale agriculture practices. In the third millennium, farmers in the Mesopotamian plain implemented agriculture in their agriculture practices. This helped them in creating perennial irrigation where crops were watered through small channels which were developed in their field. This progression had been evident in the year 2000, the total fertile land was found to be 2,788,000 km2 (689 million acres) and it was equipped with an irrigation structure globally. Around 68% of the area is in Asia, 17% in the Americas, 9% in Europe, 5% in Africa, and 1% in Oceania. The major flanking areas of high irrigation density are found. They are northern and eastern India and Pakistan along the Ganges and Indus rivers.

Apart from these areas, all populated areas had smaller irrigated systems all around the world. In the year 2012 the irrigated land was found to be 3,242,917 km which was due to the drastic evolution that happened in the agriculture sector with the help of an irrigation system, the estimated irrigated area in 2012 was practically the size of India. It was also found that 40% of food production which was accounted for 20% of the irrigated world. To increase food production, it is necessary to enhance the crop yield which could be possibly done by implementing proper irrigation methods. Crop monitoring cannot be done manually throughout the season so to make it possibly done smart irrigation system was adopted where moisture data, soil data, and weather data are used which could help determine the required amount of water for a particular plant. This technique is also helpful in reducing the wastage of water in fields. This technology includes a product that helps in maximize irrigation efficiencies by reducing water waste the it is also helpful in monitoring crop health and quality along with maximum efficiency of irrigation to plants. It also has the ability to irrigate water to fields automatically by adjusting the watering schedule to the actual condition of the soil.

SMART CITY

The smart cities mission was launched on June 25, 2015 to increase the growth economic growth of the country. This can be done by implementing the sustainable development model. There are a few strategies to achieve smart cities. They are the development of cities, based on a three-fold model: 1. Retrofitting 2. Redevelopment 3. Greenfield. A smart city is a project where different types of technologies are implemented to make the city more sustainable, and habitual and also helpful in reducing the climatic footprints through innovation. United Nations estimates that in the upcoming years that is between 2015-2050 world population is going to increase by 32% which means from 7.2-9.7 billion inhabitants, the urban population will increase by 63% from 3.9 to 6.3 billion. The current estimations are suggesting that by 2030 over 60%of the world's population will live in cities and the significant growth will be in Asia, Africa, and Latin America. Smart city leads to urbanization. Urbanization becomes inevitable due to the migration of a huge number of populations from rural areas to urban areas hoping for a better life i.e for jobs, education, medical care, access to culture etc. in 2050, India will reach 1.7 billion inhabitants with its mega-cities Mumbai, and New Delhi. The increases in urbanization will result in evolving smart cities. The characterization of a smart city is shown in Figure 1

Figure 1. Smart city

WHAT IS A SMART IRRIGATION SYSTEM?

Smart irrigation systems are a combination of an advanced technology where sprinklers along with different types of nozzles are used which will increase the efficiency of irrigation and also provide maximum efficiency in irrigating the field. The watering of plants also could be done automatically by implementing different types of controllers.

There are two general types of smart irrigation technology: weather-based and soil moisture based which is shown in Figure 2.

Figure 2. Types

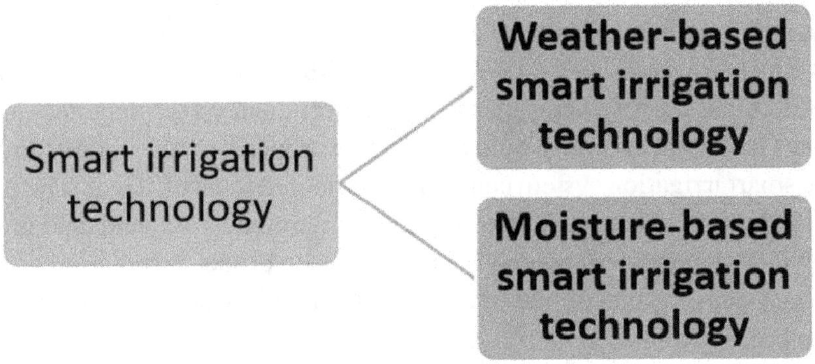

Weather-Based Smart Irrigation Technology

Weather-based smart irrigation is a technology where weather monitoring systems are implemented on-site. This type could also be enhanced by using weather sensors where temperature, rainfall, and solar radiation can be monitored.

Moisture-Based Smart Irrigation Technology

The other type of smart irrigation system is a moisture-based system. In this method moisture content of the soil is measured. It then adjusts the time of irrigation water based on this data.

Be aware that there are two distinct types of moisture-based systems available. The first, referred to as suspended-cycle irrigation, uses a traditional timed and automated watering schedule. The only difference is that it doesn't allow watering to happen if the soil moisture threshold is too high. However, if the soil is too dry it won't adjust the pre-programmed schedule to allow for additional watering. Many of these sensors can be added to an existing traditional controller.

The second type of moisture-based smart irrigation system is a water-on-demand system (National Association of Landscape Professionals, Inc., n.d.). This one maintains two soil moisture thresholds – high and low – and automatically calls for water as needed to maintain soil moisture between the two. This type requires that the existing controller be replaced with a controller with its sensors. When properly calibrated this can be an extremely efficient setup because it allows more site data to be programmed (National Association of Landscape Professionals, Inc., n.d.).

Other Components of a Smart Irrigation System

Sensors and controllers are considered to be one of the parts of any irrigation system. Apart from them, there are other various factors available such as water-efficient emitters which are rotors and spray heads. As with any irrigation system, zoning is also an important aspect to consider for overall efficiency and plant health.

A smart irrigation system should be installed properly to save money and water, if not the smart irrigation system cannot be considered to be efficient. The properly installed irrigation system will also help monitor crop health, yield, and growth.

Many factors must be taken into account for the proper installation of a smart irrigation system. Some of these include (National Association of Landscape Professionals, Inc., n.d.):

- Sprinkler pressure
- Sprinkler placement
- Sprinkler nozzle selection
- Condition of the components
- Soil type, condition, and slope
- Plant requirements

The instrument must be properly installed, programmed, and calibrated. It must also be monitored over time and may need periodic re-calibration as conditions change. Most homeowners choose to consult a landscape professional as the easiest, most effective way to get the most value from the technology. Switching to a smart irrigation system is a truly "smart" way to enjoy the benefits of a beautiful lawn and landscape (National Association of Landscape Professionals, Inc., n.d.)

WHY DO WE NEED SMART IRRIGATION?

At the being of Indian civilization, there was an abundant amount of water available for agriculture which in turn maintained the prosperity of the residents and farmers. As the population increases the availability of water was reduced which leads to the use of irrigation systems. There is an immediate necessity to conserve water which is possibly done with the help of smart irrigation whereas irrigation is not efficient as smart irrigation since the use of traditional irrigation does not monitor the requirements of water for the plants.

SMART IRRIGATION FOR SMART CITIES

As said earlier increased urbanization leads to diminishes greenery, to protect the greenery, one can adopt the smart irrigation concept. Smart irrigation's objective is to maintain rural production though agriculture is dependent on monsoons where sufficient water availability is not present. Nowadays not only in agriculture but smart irrigation is also used in cities to prevent the vanishing of greenery due to rapid urbanization

A smart city means everything is done in a smarter way this smart irrigation is also done in a smart way using the sensors of the required type which can water the plant at the required time resulting in healthy growth.

WHAT ARE THE ADVANTAGES OF A SMART IRRIGATION SYSTEM?

One of the greatest advantages of a smart irrigation system is its ability to save water (Chandler, 2019).

Smart irrigation system uses sensor of real-time or uses weather data to inform the time to water the plants which can reduce the wastage of water it is considered to be more efficient than traditional irrigation where only a 50% amount of water can be saved.

There are two important aspects of smart irrigation: control types -- the way the irrigation is controlled—and delivery types—the type of water delivery systems used (Chandler, 2019).

There are also two basic types of control for smart irrigation systems, weather-based and soil-based, each varying in its technical method of sensing and supplying information (Chandler, 2019).

Weather-based smart irrigation systems are called evapotranspiration which is based on local weather systems which was obtained through sensor or weather data about water scheduling. Water schedules can be calculated from analytical assessment which is a combination of temperature, humidity, isolation, and wind.

Soil-based smart irrigation systems are based on soil moisture content which was obtained from sensors. This will support the use of local soil moisture data drawn from sensors in the ground to support informed decisions about watering schedules (Chandler, 2019). Users can configure these systems to manage irrigation on demand, for example, when a particular land area is too dry and start an irrigation routine or stop irrigation when a particular saturation point is met because a soil moisture level has been reached (Chandler, 2019). Controlling these two set points reduces the amount of water used by linking it to the moisture level needed in the soil for a particular crop (Chandler, 2019).

The major advantage of smart irrigation is that precision watering and also deals with the efficiency of smart irrigation in delivering water to plants. There are generally four types of delivery: surface, sprinklers, trickle, and subsurface methods.

- The oldest technique is surface irrigation, which uses irrigation channels to transport water while allowing gravity to do the work.
- Sprinklers are means of water distribution like rain that can be fixed or mobile.
- Trickle irrigation is a method where water spreads to the ground surface.
- Subsurface methods are also like the trickle method where water is applied to plants by burring next to plants root zone

The loss of evaporation can be reduced by the use of trickle systems and subsurface methods.

When considering the smart irrigation concept two important technologies are involved which are the new controller and add-on sensor as shown in Figure 3 along with the kinds of controllers.

NEW CONTROLLERS

The smart irrigation system can be utilized efficiently by incorporating new controllers which help achieve potential water savings. Irrigation controllers can be separated into two main categories: Climate based controllers and soil moisture-based controllers (Oklahoma State University, 2017).

Figure 3. Kinds of irrigation technologies

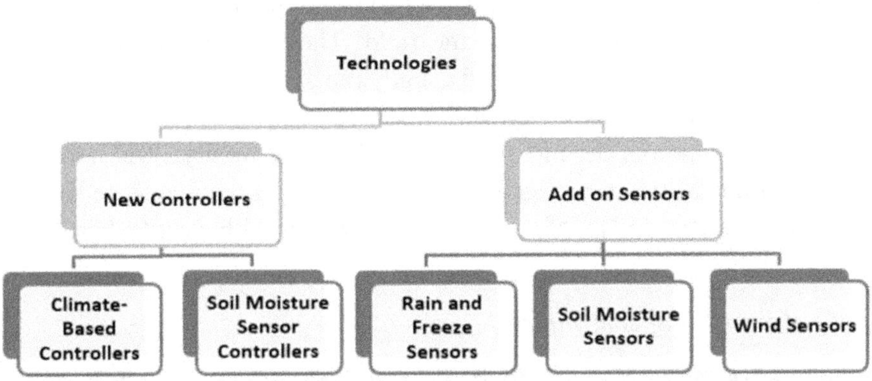

Climate-Based Controllers

The climate-based controllers use local weather data to adjust irrigation schedules, also called evapotranspiration. Evapotranspiration is the combination of evaporation from the soil surface and transpiration by plant materials (Saswell, n.d.). These climate-based controllers gather local weather information and make irrigation run-time adjustments so the landscape only receives the appropriate amount of water (Oklahoma State University, 2017).

There are three basic types of ET controllers:

- Signal-based controllers use climatological data from a publicly available source and the ET value is calculated for a grass surface at the site (Saswell, n.d.). The ET data is then sent to the controller by a wireless connection.
- Historic ET controllers use a pre-programmed water use curve, based on historic water use in different regions (Saswell, n.d.). The curve can be adjusted for temperature and solar radiation (Saswell, n.d.).
- On-site weather measurement controllers use weather data collected on-site to calculate continuous ET measurements and water accordingly (Saswell, n.d.). Figure 4 shows the illustration of the controllers.

Soil Moisture Sensor Controllers

The second type of smart irrigation controller includes soil moisture sensor controllers which measure the soil moisture (Oklahoma State University, 2017) of the field to determine the water requirements of the plants. The soil moisture sensor estimates the soil volumetric water content (Oklahoma State University, 2017). Volumetric water content is the ratio of the volume of water to the unit volume of soil. Once the volumetric content reaches the user-defined threshold, the controller will automatically adjust the valve to open and irrigate the field. The appropriate threshold value depends on soil and vegetation type and usually ranges from about 10 percent to 40 percent (Oklahoma State University, 2017). Soil moisture sensors must be installed in a representative area of the turf (Oklahoma State University, 2017); far enough from sprinkler heads, tree roots, sidewalks, and walls (Oklahoma State University, 2017). Figure 5 shows a view of a soil moisture sensor with a controller.

Figure 4. Illustration of controllers

Figure 5. Soil moisture sensor with controller

SMART IRRIGATION TECHNOLOGY: ADD-ON SENSORS

In many scenarios, a scheduling irrigation controller is already implemented and used and the upgradation of smart agriculture becomes impractical. The already existing systems can be enhanced by the implementation of sensors like soil moisture, rain, wind or freeze sensor can be added to upgrade the existing system. Some manufacturers produce devices capable of measuring multiple environmental elements using one apparatus (Oklahoma State University, 2017). The installment of this sensor can produce similar results as of smart irrigation. These add-on sensors are also more affordable than smart irrigation controllers. The installment is also easy compared to smart agriculture technology.

Soil Moisture Sensors

Soil moisture can be implemented into the existing system itself. The soil sensor measures the moisture content of the root zone and schedules the irrigation timings and maintains the soil moisture above the threshold period. There are different varieties of soil sensor are available, user have to check the compatibility of the sensor for the intended purpose. Soil sensor will also include a soil freeze sensor which interrupts the irrigation cycle when the temperature falls below 32 F.

Rain and Freeze Sensors

The rain and freeze sensor are very much useful in preventing water runoff in the fields. This sensor will interrupt the irrigation cycle in the rainy season or freezes the irrigation cycle when irrigation is unnecessary. This sensor use could efficient when used along with a soil moisture sensor. Irrigating the crops during the rainy season will lead to a waste of water, and money and causes unnecessary runoff. Three different types of rain sensors are available and each function is based on separate concepts (Oklahoma State University, 2017).

- The original type of rain sensor still in use today works with a small cup or basin that collects water, once a pre-determined amount is collected, the weight of the cup interrupts the irrigation cycle (Oklahoma State University, 2017). Debris in the cup can also interrupt the irrigation cycle and should be checked and cleared of litter periodically (Oklahoma State University, 2017).
- The second type of rain sensor uses a dish with two electrodes that are a specific distance from the bottom of the cup (Oklahoma State University, 2017). The distance can be adjusted to allow for small rain events and similar to the first type of rain sensor, debris can reduce accuracy by displacing water

in the cup (Oklahoma State University, 2017). When the water reaches the electrodes, the irrigation cycle is interrupted (Oklahoma State University, 2017).

- The third type of rain sensor does not have a rain catch cup, which makes it low maintenance and reliable (Oklahoma State University, 2017). Instead, the sensor uses several disks that expand as they get wet. The expanded disks trigger the switch and interrupt the cycle (Oklahoma State University, 2017). The system will resume the scheduled cycles once the disks dry out. The disks should be checked at least once a year to determine if they need to be replaced (Oklahoma State University, 2017). All of the devices should be mounted in an open area where they will receive rainfall (Oklahoma State University, 2017). A view of the rain and freeze sensor is shown in Figure 6

Figure 6. Rain and freeze sensor

Wind Sensors

Irrigating during windy conditions reduces the distribution of irrigation uniformity across the field and also affects the water infiltration into the soil profile. Wind sensors interrupt the irrigation cycle if wind speed exceeds a specific threshold (Oklahoma State University, 2017).

The Environmental Protection Agency (EPA) has created performance criteria for irrigation technology manufacturers under the Water Sense program (Oklahoma State University, 2017). Often, it depends on consumer preference when deciding which irrigation controller or add-on sensor is appropriate for the end user. Many local irrigation distributors have smart irrigation technology available for customers (Oklahoma State University, 2017).

THE BENEFITS OF A SMART IRRIGATION SYSTEM

Growing at a CAGR of 15.3%, the smart irrigation market is expected to reach $2.1 billion by 2025 as agribusinesses and farmers actively embrace smart irrigation technology to improve their day-to-day operations (Oklahoma State University, 2017). The advantages of smart irrigation are far-reaching. By monitoring soil moisture levels, a smart water irrigation system allows farmers to automate their irrigation processes and reduce water use. In addition to more efficient consumption of resources, other benefits include (Oklahoma State University, 2017):

- Cost savings due to minimized water waste
- Reduced human efforts
- A unified view of soil characteristics, including moisture and nutrient contents
- Smart notifications in case of abnormalities
- Better long-term landscape health
- IoT ecosystem for smart irrigation

The advantage can be enhanced by implementing the excessive use of IoT. These sensors are real-time sensors that could be placed in fields placed in the field, and send real-time data to a central gateway (Lenniy, 2023) that then automatically switches on a water pump whenever moisture or temperature values are outside the predetermined range.

Wireless low-power networks like LoRa are used to empower IoT sensors and make it possible for the information to flow in real-time to and from the central gateway (Lenniy, 2023). The entire smart irrigation system can be managed by an end user through a custom cloud-based platform or mobile application (Lenniy, 2023).

SOLUTION TO SMART IRRIGATION WITH IoT

The irrigation system based on the Internet of Things (IoT) could be enhanced in consideration of the functionality which uses advanced software. This will help us to create our solution to the smart water irrigation problem by customizing them and also by integrating third-party APIs to get valuable additional data, to help in taking a decision over-irrigation of farmlands.

Monitoring Weather

Weather monitoring usually incorporates remote sensing technology, where the required data is collected from the satellite along with the weather reports necessary from weather stations. This will involve incorporating the data in predicting the occurrence of rain, and conditions of water requirement in other weathers so that the system will recalculate the amount automatically based on actual precipitation received.

Location Technology

Location technology like precision farming falls under the agribusiness which has an upper hand in increasing smart farming by making the cost economical and collecting deep insights. The mapping methodology involved in precision farming is said to be as smart mapping which also helps in enhancing agribusinesses. The maps that are collected previously could be used again to bring together data from sensors, and images from satellites so that it will help the farmers to utilize the drone technology to keep an eye over critical lad characteristics.

Artificial Intelligence

The foundation of farming automation is evolving to include AI. Aside from the innovative yet infrequently utilized agricultural robot systems, AI aids in automating routine chores like data labeling, report creation, and notification sending. Such modest moves toward automation through AI are getting simpler to deploy, but their influence on agribusinesses is significant and obvious in a matter of months.

HOW TECHNOLOGY IMPACTS ON TYPES OF IRRIGATION SYSTEMS?

You can select from a variety of irrigation system types that can be improved with intelligent irrigation software depending on how water is dispersed around the field. Flood, sprinkler, center pivot, drip, and micro-irrigation systems are the most popular. Let's examine how each method of irrigation might be more effective thanks to technological advancements.

Sprinkler Irrigation

A sprinkler irrigation system uses high-pressure overhead sprinklers to scatter water after it has been pumped through pipes. These sprinklers can be placed on a movable platform or in the middle of the field as shown in Figure 7 (Muthuminal & Mohana Priya, 2022).

Figure 7. Sprinkler irrigation

Role of Software: At the sprinkler irrigation system, the usage of thermal and acoustic rain sensors will help in identifying the rainfall along with measuring its intensity to program the next irrigation after the rain stops. The water budget for the following month is calculated by this system after data analysis. To avoid excessive water consumption and overwatering due to rain, sprinklers receive automated notifications.

Center Pivot Irrigation

This type of sprinkler irrigation sometimes referred to as water-wheel and circular irrigation is the most common. In a conventional center pivot system, plants are irrigated by sprayers as a long irrigation conduit tied to a central tower gently circles the field as shown in Figure 8.

Figure 8. Center pivot irrigation

Role of Software: In-field sensors provide data insights that the system using circular irrigation sprinklers uses to modify the stream of water or angle of flow. By doing so, you can irrigate plants that are far from the source of water and prevent overwatering of nearby plants. The system schedules irrigation and determines possible yield and crop periods by examining weather reports and soil moisture.

Drip Irrigation

In this method of irrigation, utilizing pipelines containing tiny perforations termed drippers, water is delivered precisely to the root system. This enables farmers to dramatically lower runoff and evaporation as shown in Figure 9.

Role of Software: The biggest difficulty with this kind of irrigation is making the watering process visible. The technology tells the user when watering begins and ends via an app. Additionally, it assesses soil characteristics both before and after irrigation.

Figure 9. Drip irrigation

Micro-Irrigation

The system which works under the low-pressure, low-volume system is said to be micro irrigation. This system offers specific control over watering. In this, it applies water directly to the plant's roots, enhancing irrigation effectiveness and ensuring standardized distribution as shown in Figure 10.

Figure 10. Micro irrigation

Role of Software: As the water is carefully managed, the system can predict the appropriate dosage for each plant. Plant recognition and irrigation adjustments can be made using AI algorithms. Whatever sort of irrigation system you decide on, putting sophisticated IoT sensors in it will enable you to acquire data-driven, useful insights and keep track of your irrigation requirements.

ANALYZING DATA IN REAL-TIME FOR DELICATE AND INTELLIGENT AGRICULTURE

The majority of the underground water extracted by India can be used for farming (approximately 90%). The rear settlements do not acquire enough water regarding the efficacy of canal systems. Groundwater dependence has increased substantially due to factors such as several crops grown each year, rising demand from other users, and industrialization. Groundwater has been overused, the water's quality has deteriorated and salinity has entered.

To optimize crop growth and yield, create the best use of water and other inputs, and preserve soil health, smart agriculture tries to assist farmers in making decisions. This is crucial for the livelihood of small-scale farmers, ranchers, and other related professions, especially in light of the uncertainty surrounding climate change. It requires regular data monitoring of weather, water, soil, and crop growth as well as a deep awareness of the local agroecological zone, geomorphology, subsurface conditions, and local socio-economic conditions. Decisions are then made utilizing this integrated data.

Smart Agriculture: Data Monitoring System

The increase in quality and yield of the crop is the main objective of smart agriculture, through the efficient usage of inputs with a data-based decision at the farm level. The data collection can be done by:

1. Topography, aquifer, hydrogeology, geomorphology
2. Water sources in the vicinity, in the village, in the taluka,
3. Soil type

Regularly and frequently collected data are composed through the cropping season, from pre-sowing to harvest, for:

1. Local weather conditions and forecast,
2. Soil quality and moisture,
3. Water availability from all sources, and
4. Crop growth status

The information may come from publicly available sources as well as information gathered through automated or manual measurements. For monitoring and evaluation by experts as well as farmers, local agriculture extension workers, or others, the data is transferred to an integrated system and archived. This makes it possible for farmers to receive professional advice at the farm and village levels, aiding them in decision-making throughout the crop cycle, maximizing the use of inputs like water and fertilizers, and preserving the health of the soil. The strategy includes teaching farmers and other skilled locals how to use technology and integrate data-driven judgments into their conventional agricultural practices.

Over time, the system will gather information from many farms to build a data library that will aid in making important choices. A localized size of data collection will aid in providing a clear picture of the field's conditions, and such collection carried out at numerous field locations will allow for the gathering of the regional scenario (Vadapalli & Pervalli, 2020). As a result, the system will require data collecting locations at different scales, starting with clusters of farms and expanding to villages, clusters of villages, talukas, districts, and so forth. The development of AI/ML (Artificial Intelligence / Machine Learning)-based decision-making will also benefit from the accumulation of this massive data over time as shown in Figure 11.

Figure 11. Smart agriculture

The technology will also support long-term choices such as altering irrigation patterns, budgeting for village water use, and creating a community water security policy.

PER DROP MORE CROP: A CASE STUDY

Optimize Irrigation Water Use

The center pivot irrigation system is Georgia's main agricultural irrigation method. The majority of the state's 23,000 center pivot systems are situated below the Fall Line. Statewide, drip irrigation is also used to grow orchards and vegetables. The less popular solid set sprinklers and moving guns are other types of irrigation. Any sort of irrigation system can use some of the techniques listed below, including remote control and irrigation scheduling. Retrofits and variable rate irrigation are two other techniques that are unique to center pivot irrigation.

Enhance On-Farm Efficiency

The tools for smart irrigation range from simple and cost-free to complex and expensive. Farmers in Georgia have access to the greatest management techniques listed below. A brief explanation, instances, an estimated cost, and an assessment of the benefits of water usage efficiency or water savings are provided for each activity. Water savings refers to actual water preserved through a decrease in groundwater withdrawal, whereas irrigation efficiency refers to the extent of pumping water that is consumed by the crop (more crop per drop). Since each efficiency and savings situation depends on specific field trials and administrative choices, the calculations below are not always cumulative.

USAGE OF IRRIGATION RETROFIT

- Evolution of efficiency

In Georgia, center-pivot irrigation systems were first used by farmers in the 1970s. These systems used impact sprinklers that sprayed water from the apex of the pivoting core and functioned at increased pressure. With spray-type sprinklers on drop hoses, these systems can be converted from high to low pressure, which results in significant energy and water savings. The low-pressure method dramatically lowers transpiration and wind drift losses by supplying water at a reduced pressure more precisely to the surface of the ground.

Modern low-pressure spray-type sprinklers are made to distribute water droplets across the field with a high degree of uniformity. By spraying water nearer to the crop in fine droplets in conjunction with drop hoses, these sprinklers assist in lowering losses due to wind drift and evaporation. The water is sprayed far higher in the air with tiny droplets from high-intensity impact sprinklers, in contrast. Additionally, low-pressure spray-type emitters are distinctive in that they use less energy because they function at pressures between 10 and 30 pounds per square inch (psi).

- Efficiency

Due to water losses from evaporation, wind, and drift, the effectiveness of the high-pressure system is just 60 percent respectively. Renovating a pivot may boost water consumption efficiency to 80 to 90% when integrated with the implementation of end-gun controls to keep irrigation inside of the field border and the correction of defects in the system.

- Estimated cost

Costs for changing a current high-pressure to a low-pressure system range between $2,500 and $5,000. Low-pressure drop hose sprinklers are a common feature of modern center pivot systems.

USAGE OF REMOTE IRRIGATION CONTROL

- Monitor irrigation from phone or web

When checking irrigation systems in fields, farmers have logged endless hours and petrol tanks traveling back and forth at all hours of the day and night. With the help of this remote-control technology, a farmer can now automatically track and monitor irrigation systems from their phone or computer, saving time and fuel.

This feature enables the remote control of many different components in the system, including the ability to turn on, turn off, change its direction, set the timer for the system to begin, precisely angle-control the end-gun sprinkler, verify the device's position, and sometimes even speed-control adjustable - rate irrigation on some systems. All systems made possible by this technology may be found on a single dashboard, and some of them are even linked to soil moisture sensor systems that send data in real-time from the field. Both center pivot and drip irrigation systems can use this technique.

- Estimated cost

Depending on the device, remote pivot control systems can cost anywhere between $2,000 and $5,000, plus a yearly membership fee of around $300.

- Efficiency

The effectiveness of remotely controlled technologies for irrigation systems has not been officially studied. Any efficiency improvement is very subjective by supplier and field. Restrictions on timing are removed from irrigation scheduling by the ability to avoid the system at specific angles, regulate the end-gun by degree, and create automatic timers to start and finish the system (e.g. if the farmer is unable to arrive at a system at the precise moment it needs to be stopped).

HOW TO BUILD AN IOT-BASED IRRIGATION SYSTEM FOR SMART AGRICULTURE

Many worldwide agricultural problems can be solved by smart agriculture, including enhancing team production, evaluating the outcome, interacting with other groups and sub-units, and regulating market trends. Farms across the world are quickly considering integrating IoT technology into their operational processes. The IoT agriculture industry is anticipated to exceed $30 billion globally by 2023.

In their daily job, farmers have been actively experimenting with high-tech solutions. For instance, they use sensors to map out their farms in great detail and to examine the peculiarities of the soil. Smartphones are used to track output, and crop development, and learn about market shifts. All of these technologies can be connected thanks to IoT (Digiteum Team, 2022)

HOW TO BUILD AN IRRIGATION SYSTEM FOR FARMS

Smart irrigation is where IoT in agriculture first appears for the bulk of farms. We can conserve resources to achieve that crops receive the optimum care by adjusting the watering schedule and quantity. IoT solutions based on sensors gather information about the soil, keep track of a crop's status, and send this data from the sensors to agricultural irrigation systems. A platform responds to this alarm and the water spray activates as quickly as there is insufficient water in the soil (IoT Solutions World Congress, 2019).

Smart irrigation is based on flow pressure and temperature updates; it will cease once the moisture level reaches a certain level. An end-user can monitor the pump's condition to confirm that a pipe is operational and that the water is flowing in the appropriate quantities. Users of automated irrigation systems can monitor their irrigation status, start and stop the operation with the push of a button, and get reports on crop health, moisture levels, and weather conditions. A system must have six key elements to fulfill these various goals:

- Remote control

A smartphone or laptop is commonly used to monitor agricultural irrigation systems. An individual downloads a mobile app, enters their profile information in it and is then taken to the administration page. Some farmers opt to have a second physical irrigation control device due to the ease with which devices can be compromised.

The controller, whether it be a real object or a smartphone, is linked to sensors that gather and transmit information about the soil quality, temperature, and relative humidity.

- Precision agriculture

Depending on their biological requirements, different crops require different irrigation regimens. Based on the prevailing soil and crop conditions, precision agriculture employing IoT and advanced analytics technology enables on-demand planning and implementation of farming activities like irrigation or pesticide control. The irrigation system should recognize multiple zones based on the type of crop, soil, and weather conditions to ensure that each type of plant obtains the optimum treatment. According to the real-time sensor readings, farmers should be able to install a different irrigation valve for each of these locations and set up a different timetable for each of them.

- Smart schedules

The intricacy of smart irrigation systems varies, particularly when it comes to irrigation modification. With the ability to pause or end irrigation at any time, some tools enable basic scheduling for a specific irrigation zone. Other methods can accommodate elaborate weekly plans with planned breaks and varying levels of intensity. While this is going on, a water supply system using IoT sensors can change the frequency and intensity of watering based on readings from moisture, humidity, and other sensors.

- Notification systems

The system should alert the farmer when watering begins and ends, as well as any errors. The system ought to be able to recognize irregularities in the irrigation flow and alert the farmer right away in the event of a technical issue.

- Weather and environment analytics

Modern methods also include the field's location and particulars, the present weather, and the forecast in addition to the soil moisture levels. The technology can produce a unique long-term watering flow that can be sequential on the forecasts and computational modeling predictions.

- Interoperability

Field irrigation systems ought to be compatible with other laptops and cell phones. With this interconnectivity, a farmer should be able to get insights from any device or platform.

STEPS TO BUILD YOUR DRIP IRRIGATION SYSTEM

Customizable agricultural watering schedules should be possible, reducing water waste and allowing for the assessment of moisture levels. A microcontroller serves as a medium for information transmission to do. The microcontroller is linked to sensors that send data on temperature, moisture, and other variables. The simple irrigation system construction process can be broken down into the five parts listed below:

Step 1. Prepare Hardware

For an irrigation system, you need to have the following farm irrigation equipment:

- Soil moisture sensors are necessary for assessing the dielectric data of soils as well as water levels and volumes.
- SMS controllers are necessary for scheduling on-demand irrigation
- Temperature sensors, outfitted with RTDs—Resistance Temperature Detector Components—for on-demand watering scheduling

Step 2. Set Up LED Light

The LED light can be set to switch on constantly to warn both the watering system and the operator of changes in temperature and moisture. Water sprinklers in various field zones are kept under check using LED lamps.

Step 3. Substituting Static Sprinklers With Rotating Nozzles

Water waste can be reduced by using rotating nozzles on sprinklers since they deliver targeted water currents to numerous zones. Sprinklers can reach all regions of the land because of the variety of angular positions.

Step 4. Internet Connection

IoT systems communicate to the controller as well as other devices over wireless connections to transmit data. First, low-power networks like Sigfox and LoRaWAN are used to send sensor data to the data processing center. The following stage involves using a laptop or mobile device as a microcontroller to convey data insights to farmers' interfaces.

Step 5. Locating Sensors

The condition of the surface of the soil should indeed be monitored by sensors, which should send real-time data to the cloud (or other data storing and processing facilities). They should be placed in areas that accurately reflect the overall state of the soil, which is typically close to the plant's base. To keep the sensor out of direct touch with water, it should be covered by soil.

Drones can be used as extra support to monitor vast fields from a distance and identify the areas that require watering immediately.

Step 6. Monitoring Technical Issues

Many irrigation systems commonly experience issues with cracks and tank leaks. Building a drip irrigation system with permeability control is therefore essential. The sensors should identify anomalous water flow and give notifications to the overall network. The algorithms for notifications alert the farmer right away. Additionally, even though the farmer doesn't manually stop the irrigation, it should stop immediately as the moisture levels exceed the maximum permitted threshold. Automation methods are therefore necessary for the effective mitigation of agricultural risks.

PROMINENT PROVIDERS OF SMART IRRIGATION SERVICE

Over the past five to ten years, the efficient irrigation market has grown quickly, and leading companies have introduced innovative irrigation solutions for farms of various sizes and crops. Let's peek at the most well-known companies that offer smart irrigation services.

- Hunter industries

For small and large farms, the company provides three different irrigation systems, each with a different type and number of controllers.

- IMMS is a Windows program for central water management that can support more than 10,000 controllers. The network's core features include irrigation statistics and warnings, flow reporting and control using ACC controllers, and compatibility with wireless Ethernet and GPRS protocols. Solar Sync sensors for analyzing temperature and sunlight are compatible with the system.

- The Hydrawise Cloud platform is an iOS, Android, and Windows Phone app for managing water usage on the go. The range of controllers supported by the solution is limitless. The software enables simultaneous access to several control systems for farmers and anticipates the required water costs based on weather predictions using pattern recognition.

- Calsense

Using CS3000 controllers and real-time humidity sensors, Calsense created a web and mobile platform for IoT irrigation management. The technology alerts the farmer to mainline outages and current weather conditions. Additionally, the online program creates real-time irrigation interfaces and links farmers to encrypted cloud storage, where all records are kept.

- Galcon

With the aid of this irrigation, farmers who are utilizing the Internet of Things can transmit data both offline and online from sensors to the DC controller. A farmer may plan new tasks, change flow intensity, and keep track of water usage data even without an Internet connection. They can also track the status of their irrigation system.

Unique passwords are used to protect the data, and each user has an admin panel. Without the need for hard drive storage, cloud storage allows integrated information exchange from mobile and desktop devices.

- IoT irrigation system solution with Digiteum

Smart irrigation is a cutting-edge method of watering made possible by remotely controlled equipment, connected sensors, and other gadgets. Farmers and their staff can automate their workflow while getting a real-time view of the irrigation operation.

The farm manager can ultimately foresee potential hazards, base irrigation plans on weather updates and moisture levels, and utilize statistics from earlier irrigation sessions. Farming is a less wasteful and more predictable process because farmers can adapt to changes and plan for the future by drawing on their past knowledge.

REFERENCES

Chandler, S. (2019, February 7). *What are the advantages of a smart irrigation system? TechTarget*. IoT Agenda. Retrieved April 26, 2023, from https://www.techtarget.com/iotagenda/answer/What-are-the-advantages-of-a-smart-irrigation-system

Digiteum Team. (2022, November 25). *How to build an IOT solution for an agricultural irrigation system*. Digiteum. Retrieved April 26, 2023, from https://www.digiteum.com/iot-solutions-agricultural-irrigation-system/

IoT Solutions World Congress. (2019, July 10). *IoT transforming the future of Agriculture: IoT Solutions World Congress: May 21 – 23 Barcelona*. IOT Solutions World Congress. Retrieved April 26, 2023, from https://www.iotsworldcongress.com/iot-transforming-the-future-of-agriculture/

Lenniy, D. (2023, March 10). *Smart irrigation in agriculture*. Intellias. Retrieved April 26, 2023, from https://intellias.com/smart-irrigation-in-agriculture/

Muthuminal, R., & Mohana Priya, R. (2022). *An Overview on Techniques Involved in Recharging Ground Water and Its Impacts*. Intechopen, doi:10.5772/intechopen.105448

National Association of Landscape Professionals, Inc. (n.d.). *Shopping for a smart irrigation system?* Love Your Landscape.org. Retrieved April 26, 2023, from https://www.loveyourlandscape.org/expert-advice/water-smart-landscaping/smart-irrigation/shopping-for-a-smart-irrigation-system/

Oklahoma State University. (2017, February 1). *Smart Irrigation Technology: Controllers and sensors*. Smart Irrigation Technology: Controllers and Sensors. Retrieved April 26, 2023, from https://extension.okstate.edu/fact-sheets/smart-irrigation-technology-controllers-and-sensors.html

Saswell. (n.d.). *Smart irrigation controllers have difference between weather-based and soil moisture sensors*. Shenzhen Saswell Technology Inc. Retrieved April 26, 2023, from https://www.saswell.com/smart-irrigation-controllers-have-difference-between-weather-based-and-soil-moisture-sensors_n100

Vadapalli, A., & Pervalli, S. (2020). Smart Agriculture System using IoT Technology. *International Journal of Advance Research in Science and Engineering*, *9*(9).

Chapter 9
Artificial Intelligence Technology in Different Fields

Sakshi Rajput
Maharaja Surajmal Institute of Technology, India

Deepak Sarangi
Maharaja Surajmal Institute of Technology, India

Preeti Sehrawat
Maharaja Surajmal Institute of Technology, India

ABSTRACT

The chapter consists of information regarding different usage of artificial intelligence systems in the health sector and the transportation sector in different types of ways which will help to maintain a good experience for the users. The chapter includes the data that shows the sections where AI is used in healthcare and transportation sectors like smart devices used in hospitals such as chatbots for patients, virtual nurse assistants, robotic surgeries, smart traffic control system which will help the peoples to avoid unnecessary long trips and save their time, drone systems, etc. Along with this, the authors have data about growth in these sectors by using AI systems, which make it cost-efficient with better performance than before.

In today's fast-growing world everyone is busy with their own stuff related to their life such as finding jobs which are suitable for them, learning many things on the internet, etc. Peoples want to do many tasks by their own but there are chances of errors and interventions by other peoples but how is it if they don't need to do it by their own and just have to command to their computers and it will do the task for them just in short period of time. That's always great to hear. For that, Artificial

DOI: 10.4018/978-1-6684-6418-2.ch009

Intelligence (AI) and Machine Learning (ML) came into existence, and they show great impact on the world by their efficiency to perform a task. Its earliest substantial work was done by the British logician and computer pioneer Alan Mathison Turing in the mid of 20th century, but John McCarthy is remembered as the father of AI. It is predicted that Artificial Intelligence can generate a revenue of $15.7 trillion to the global economy by 2035. It is estimated that China and United States will get the most benefit from it, which is estimated to be nearly 70% of the global impact.

In today's world, it is widely spread in many fields including IT-Sector, Agriculture, Healthcare, Transportation, Defense Services, Industries. All of these sectors have a great impact after the introduction of AI & ML. In this chapter, we will talk about applications of AI and ML in the Healthcare Sector and Transportation Sector.

Figure 1. Basic steps of implementation in AI & ML
Source: Lateef and Edureka (2021)

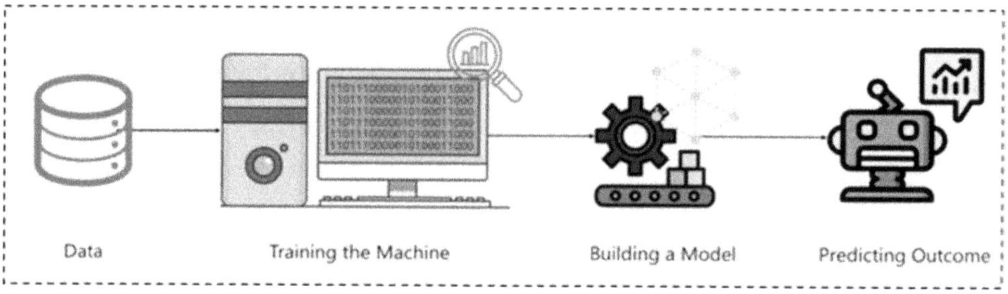

AI IN HEALTHCARE

Healthcare is a very important sector for every person's health and if a person is not physically well then it will affect their life as well as the lives of all the people associated with that person. In today's world, many techniques for treating a person are introduced, some are manual, and some are automatic. The machines and equipment used in this sector should be more effective and accurate in order to treat a person well such that the person will not be affected further in the future times. But sometimes due to manual interruptions and wrong implementation of the machines leads to life risks of the patients so the machines should be treated carefully and such a way that it won't affect the person who is being treated with that and sanitization after using that is also important so that if any serious patient is being treated by that then any traces of his/her body don't get absorbed there so the next person won't get affected by that.

To resolve these kind of issues, automatic machines are introduced in healthcare field, which are developed to give better accuracy and performance with low risk. Automatic Machines are based AI & ML and are known as smart machines, these are developed by providing data which are related to treat a person depending upon the use of that machine for specific disease detection, scanning or detection of some other internal issues. Some examples of these types of machines used in the healthcare field are MRI scanners, CT-scan, X-Ray machines, etc. Other than these AI & ML can be used in administration and operational purposes in healthcare sector like patient monitoring, clinical decision support, data entry for all patients in hospital's server so that data would be maintained. Every year, roughly 400,000 hospitalized patients suffer preventable harm, with 100,000 deaths.

Figure 2. Basic role of AI in healthcare sector
Source: DataFlair (2019)

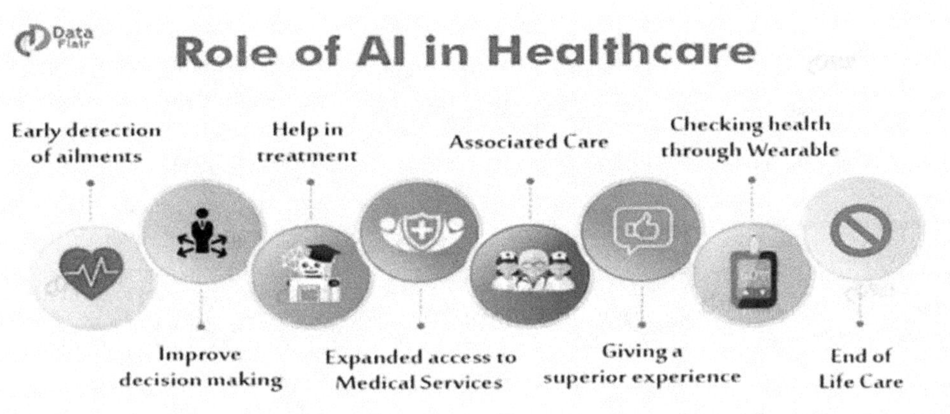

There are many ways in which AI can be used in the medical sector like robotic-assisted surgeries, nursing assistants, fraud detection, patient-monitoring devices, cybersecurity, dosage error reduction, connected machines, etc. All these operations are very important from their own perspective in the day-to-day life in the medical sector. Internet of Things (IoT) is one of the main advantageous factors which is being used in AI. It has been reported that it has been increasing in popularity continuously over the past years. Due to these improvements in communication technologies and data transfer speed, the ability to transmit large amounts of data has been increased drastically. In addition, more robust and advanced storage system and data processing capabilities provided by big data analytics and cloud technologies opened the doors for new applications and markets for IoT in real-time analytics and predictive modelling. As a result of these advancements in combining smart sensors, actuators, and data analytics in an IoT environment for real-time and continuous healthcare brings great promises to the healthcare industry. The fields that have come into existence are the Internet of Health Things (IoHT) or the Internet of Medical Things (IoMT) helps to create the balance in the healthcare management system (Nasr & Islam, 2021).

Estimation of funds or potential annual value of each of these parts by 2026 in billion dollars is given in Figure 3, which shows the estimated value of money that each part will get if everything goes right. People will also get good treatments with high quality machines and technologies which can be concluded with this estimated distribution which leads to the increase of machine dependency by the humans which have both pros. and cons. respectively.

Figure 3. Estimation of different usage of AI in healthcare sector
Source: Collier et al. (2018)

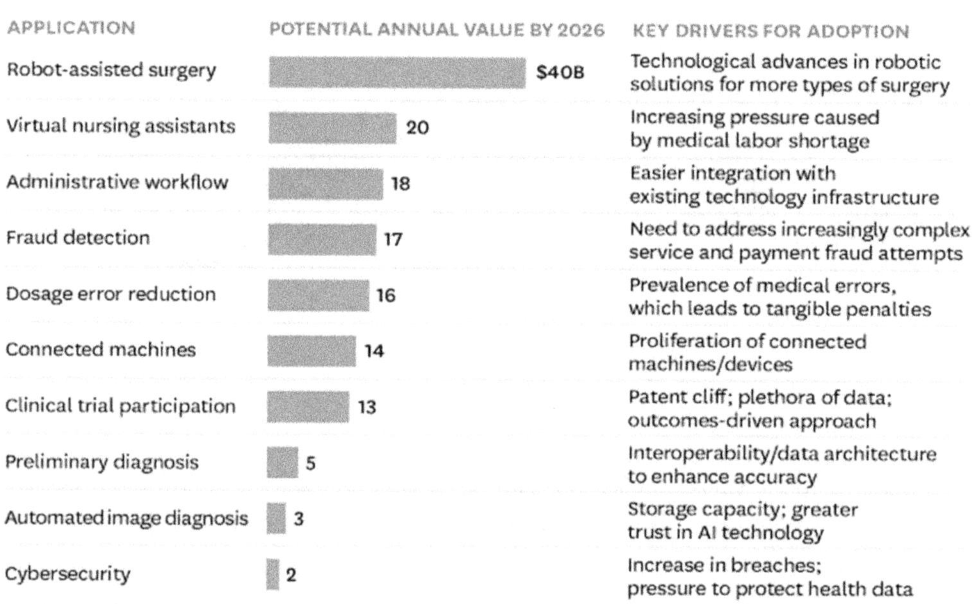

Benefits of AI for Healthcare

AI has a great impact on the healthcare sector, and it brings a new revolution in the field in medical which is very useful for the people as well as ease for the medical fields professionals and experts. The benefits of using AI in healthcare are as follows:

1. High quality of patient's lives.
2. Streamlined workflow.
3. High accuracy of treatment.
4. High employee efficiency.
5. Reduced costs.
6. Decreased labor demand.
7. Higher employee engagement with patients.
8. Reduced workload.
9. Enhanced protection against cybercrimes.

There are many other benefits than these which are important, but AI basically eases the way of working of the medical professionals and helps them to treat the patients easily without having issues.

Demerits of Using AI System

Everything in the world has benefits and demerits that are reduced by their alternatives and that way it becomes the new choice for people. Similarly, AI machines also have some demerits which are given below:

1. In the prediction system, there are chances of predicting wrong outputs as those are provided with data which is being introduced earlier or used before but requirement of a new type of data cannot be processed at that time so we can get output that is based on the previously happened cases.
2. Sometimes the machines behave unusual because they are using their own intelligence i.e. if a human do the task then that would go through many process like will it be the best way, effects after using this and some random questions but the machine will do the things that are either commanded or assigned that to do or by their own which may be different from a human's perspective.

These are two main demerits which may affect the whole system, but these can be overcome by monitoring and regular or in interval checkups of the systems and machines such that chances of these types of issues are less.

Tools and Technologies Used

In AI system, there are many things that are useful while some are not that useful for the specific purpose. Natural Language Processing (NLP) is the most useful technique in AI used in the medical sector. It includes forms of speech recognition and text analysis and then translation. Data processing, prediction, decision making, automatic operation algorithms are key domains in the AI usage in the medical sector. Deep Learning & AdaBoost are also techniques which are widely used in machines to perform tasks by their own without human intervention. Internet of Things (IoT) is one of the main technologies which can be used in the healthcare sector. Its benefits are defined as:

1. Simultaneous reporting and monitoring.
2. Data Analysis and Data Assortment.
3. End-to-End connectivity.
4. Remote Medical Assistance.
5. Tracking and Alerts (Mahalakshmi & Latha, 2019).

Challenges for AI in Health Industry

- The first challenge faced when using AI in healthcare is that it requires a large amount of data in order to give more effective and accurate results.
- The second challenge is that there is a risk of bias if the data used to train the algorithms is not representative of the population as a whole.
- Finally, it is difficult to compare results or combine data from multiple sources as there is lack of standardization across different AI systems.

Different Usage of AI in Healthcare

In this section, we will talk about the applications of AI in the Healthcare sector and the places where it can be used in future in more detail.

Chatbots

If a person needs an urgent information about their health or an instantaneous solution for any health issue and it will take more time to reach the nearest hospital so they can use online chat-based bots through which they can get their issues resolved in which data is added regarding the general health issues which can be treated instantaneously by any person. It uses AI & ML simultaneously to get better results with high accuracy. There is user-privacy designed for individual users who login to the system by completing the registration process and all other authentication processes specified in the system.

Basically, it replaces "live" customer service agent in past decades with a digital algorithm; a program that can be quite simple or quite sophisticated.

"Having one person call one person takes a lot of human capital" says Dr. Z. Brown of Northwell Health Solutions.

It is developed based on predictable conversation in advance, i.e. we have to think that what would be the best solution for the question asked by the user in advance so that when it will be asked then the chatbot give that response at that time.

What Does Every Patient Want?

1. The queries which may specify many kinds of patient's behavior would be dictated by custom.
2. Less on-site visit should be there.
3. To easily arrange meetings.

What Does the Front Desk Want?

1. Let the chatbot handle the issues according to the user's need.
2. It should confirm to the patient if there have been any changes in employment or insurance policy coverage.
3. The chatbot should be able to manage open time slots such that the patient doesn't face any issue.
4. If there is any time slot available, then the bot will fetch that just after the phone call or before the appointment; It would be easy if the employer and insurer has AI capabilities, otherwise, the medical assistant will have to arrange a phone call.

There are many hospitals across the country which have their own chatbot assistant for free services to provide a good experience to the people who can't get the medical facilities on time or don't get this at all like peoples in rural area. All the hospital administrative or control panel should check the bot within an interval to maintain the security and upgrade that by time to give users a better experience.

Virtual Nursing Assistants and Patient-Monitoring System

AI can also be used as an employee who can assist a patient regarding the health issues and other issues they will have at the hospital whether to showing the way to any specific place, room or any ward or predicting the medicines for any common disease. Also, the monitors showing the patient's health status are AI-based systems. These are available 24/7. It can save funds up to $20 billion annually for the healthcare industry.

Figure 4. Global healthcare chatbots market estimation
Source: Verified Market Research (2021)

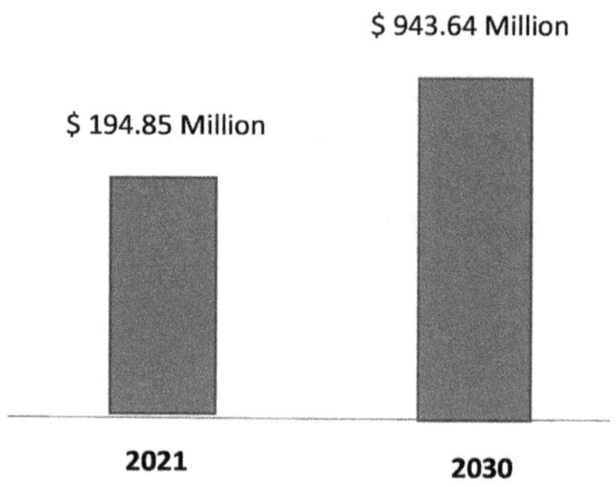

Figure 5. Sample of an AI chatbot system for healthcare
Source: The Medical Futurist (2021)

Maintaining a healthy lifestyle has become a challenge in the day-to-day life of people. So, smart health-monitoring system is taken as one of the best solutions for these challenges. The recent changes in industry 5.0 and 5G lead to the development of smart & economical sensors which can help in real-time health monitoring of peoples. The Smart Healthcare Monitoring (SHM) System has led to fast, economic and reliable health monitoring services from remote locations which was not possible earlier with traditional healthcare systems (Sujith et al., 2021).

Figure 6. Real-life virtual nursing assistants
Source: Smith (2020)

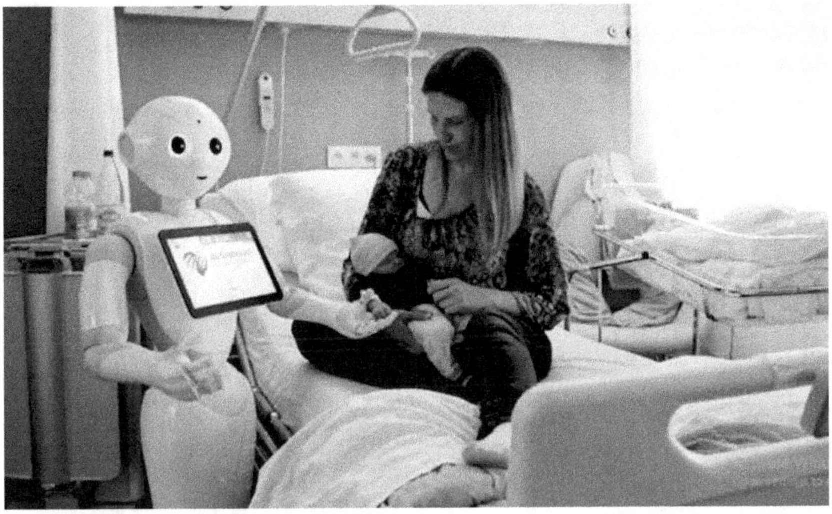

Surgeries by Robotic Systems

Using Collaborative robots based on AI technology has made a big revolutionary change in the way of study of medical surgeries. As a result, it can be seen that it can perform exceptionally well with high speed when making narrow and fine incisions. Generally, the results of a surgery, especially with a recent or complex one, its performance can differ with surgeon's ability to treat patients. Utilization of AI system can be reduced with variations in cases and improvement in the efficiency of doctors even the surgeons with high skills in the field at the same time. AI-based machines have precise performance, so, it can reduce the probability of shudders, or any uncertain incident caused by any bad movement during surgery.

AI-assisted robotic system can show a three-dimensional (3-D) magnification for articulation while performing the procedure with better accuracy and miniaturization which is known for execution of basic tasks like precise cutting and stitching.

Surgeons at the Maastricht University Medical Center, Netherlands, during surgery use AI-assisted robotic system to suture very fine and narrow blood vessels of size 0.03 to 0.08mm. Other than these operations, AI machines can be used to data from previous records of operations to present new surgical methods to treat different cases easily.

Figure 7. Example of real-time AI-assisted robots surgery in India
Source: Singh (2022)

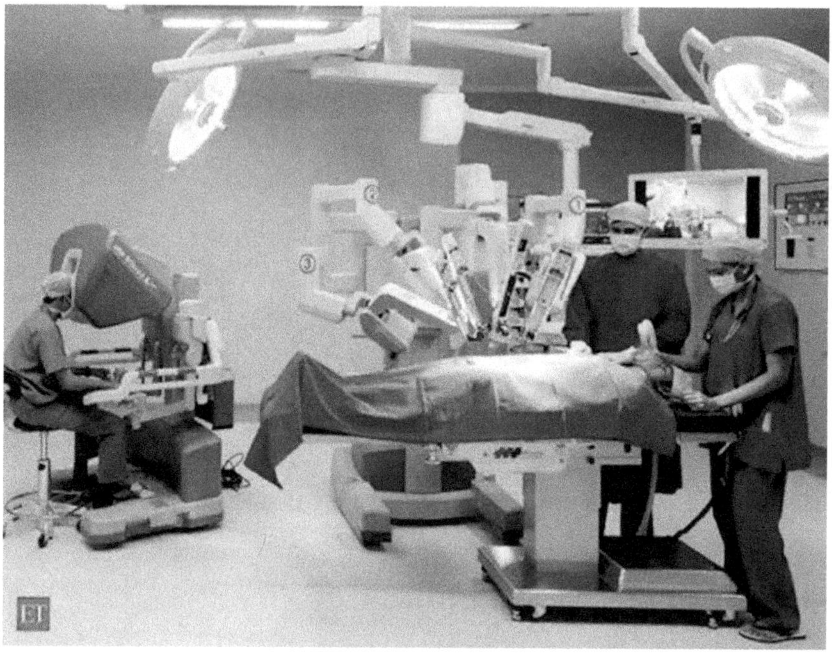

In a country like India where population is high, the medical sector should be strong in order to maintain the situation if any crisis happens such that there would be as low as possible number of life losses. AI supports the medical official to reduce the chances of error in treatment or surgery in order to get more efficiency and speed without causing side effects after the surgery.

Medicine Prescription System

It is known as one of the most valuable and popular applications of AI in healthcare. It is based upon gathering of large amounts of data from various unorder technological inventions such as advanced biotechnology, economical genome sequencing, and sensors for health monitoring used by the patients at home. It was done on advanced supercomputing algorithms with deep learning. A deep learning system has been developed by Intel, which is capable of identifying 23 patients at a time those have high cardiovascular disease risk chances which are not detected by traditional statistical methods, with an accuracy of 85%. Other than this, Google DeepMind and IBM Watson are also known for leading the list of data mining of medical record developers. Its main objective is to develop a good assistant provided with a high range of clinical (medical) knowledge, other than a logical, reasonable and analytical mindset. It will be helpful for the patients as they don't need to wait for their turn for a doctor's appointment in a hospital or stay in a queue to take an appointment just to get medicines.

Figure 8. Representation of data processing in precision medicines
Source: Wang et al. (2019)

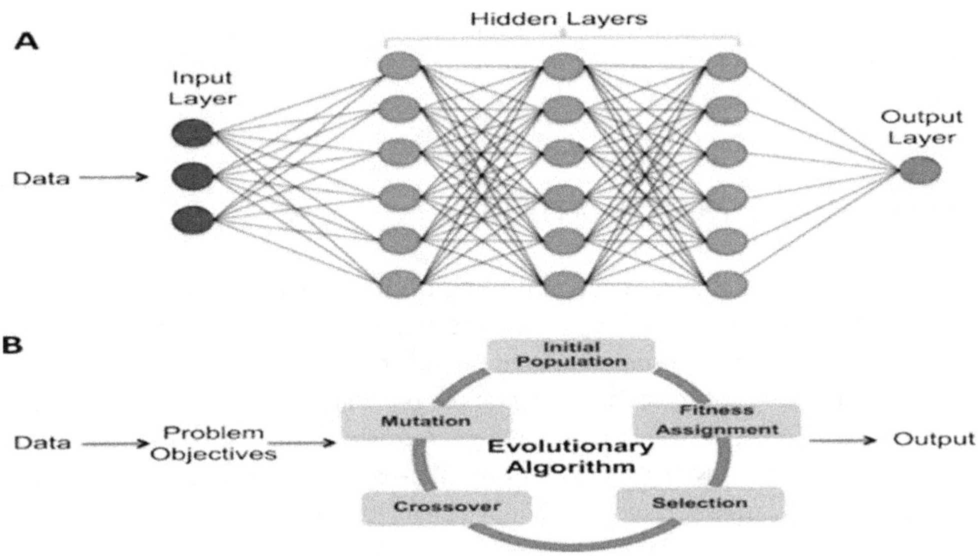

Administrative Workflow Assistant

AI can also be used for administrative work in healthcare. It includes assistance for the management in official records of patients, employees, monetary funds, and all other stuff. It ensures that care providers will give priority to urgent cases and work, medical officials like doctors, assistants and nurses, in order to save time for their daily jobs.

Out of the applications of AI used in various fields, AI can produce savings of about $18 billion in the health sector. Advanced technologies like voice-to-text conversion help on the administration side of health sector. It helps to automate non-patient care activities like prescribing medications, arranging tests, and writing chart notes.

Other Aspects about AI in Health Industry

The global market revenue of AI in the medical field is estimated to rise from $4.9 billion in 2020 to $45.2 billion in 2026. This breathtaking rise in this growth statistics indicates towards the needs of medical institutions to take up AI to offer new medical solutions to the peoples coming to the hospitals. Medical Institutions & Organizations should understand and explore the full powers of AI technology by straightening with medical technologies providing organizations who can help them to create strategies to make the right decision about investments in AI and guide them in return by field experts to enable their business to add good values for their customers in future.

AI FOR SMART TRANSPORTATION

In today's busy world, everyone is travelling from one place to another place for different purposes such as going to their offices, going out with their family and friends, etc. So, the transportation system should be good enough such that peoples won't get frustrated but in countries where the population density is more, there a huge amount of people travels everyday by their own means of travel or by using public transport vehicles. In the present time, fuel-based systems and non-automatic systems are used in transportation systems which may be hectic for people sometimes.

Smart systems can replace the present time using systems to overcome these issues. Basically, smart systems are based on AI systems. There are many areas where it can be used such as changing fuel-based vehicles to automatic electric-based vehicles, smart navigation systems and traffic control systems, etc. All these areas are important as these are areas where the peoples face issues in many ways whether issue with the vehicles, environmental issues aroused due to the vehicles and with the time to reach to a place due to different problems in their travelling routine whether it is with their vehicle or with the path they have choose to get to their destination.

Transport Management Systems (TMS) refers to the field of management of transportation systems, especially the operations used for maintaining transportation. The main purpose of these systems is to create efficient route planning, transparency, flexibility improvement and load optimization using large amounts of data. Transportation Strategies proposed for a city is attached to an informative system to maintain better administration which will be focusing on capturing data and process that data to transmitting it further and management of the data bus generated by the system. Due to the appearance of smart technologies over a couple of decades ago, various information systems for logistics and planning of routes have been introduced.

These systems provide improved data processing capabilities which are better to plan the transportation procedure effectively, which will lead to Intelligent Transportation Systems (ITS). The data which is generated by the users and vehicles can be used to develop an efficient ITS. Developing ITS ensures the improvement in performance exchange, acquisition and connection between vehicles, city infrastructure and other related activities (Iyer, 2021).

Figure 9. Basic roles of AI in smart transportation system
Source: Rogerson (2020)

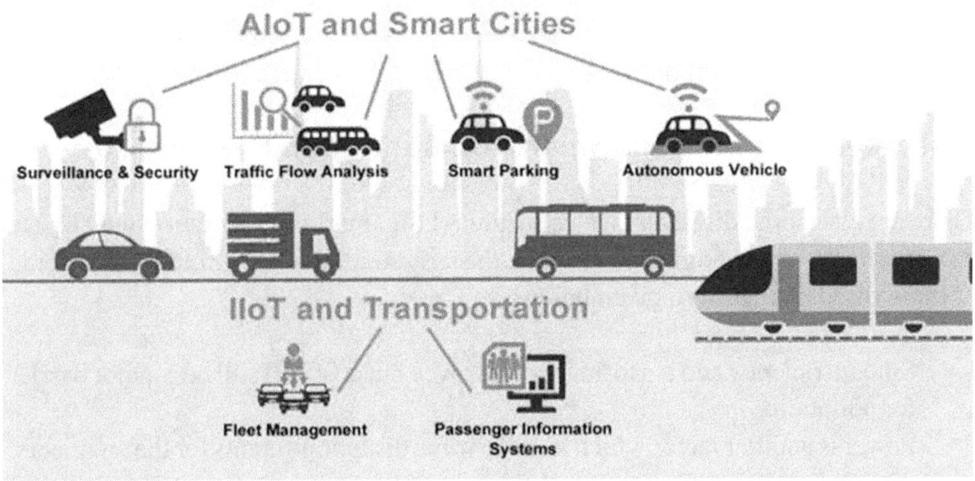

Advantages of AI for Smart Transportation System

Everything that comes into existence has advantages and disadvantages that would be overcome in their alternatives and that is with change of time not in a sudden. AI used in smart transportation system also have advantages and those are given below:

1. It will reduce traffic accident cases and ensure more safety for peoples. The number of accident cases which involve trucks at night is a big issue and it can be reduced with the help of automatic vehicles without drivers.
2. It improves the ability to process and predict data and results faster than human beings in a way such that travel or transport providers (operators) will schedule public and private transportation services in a considerably improved manner.
3. It may play a big role in developing new and innovative ideas which can be used to deal with environmental issue like air and noise pollution as well as

helping scientists and engineers to come up with more nature-loving methods to run vehicles and machinery for travel or transport.

4. Sometimes, it plays a big role in managing the route for emergency hospital cases. It can help in navigating the best way with low traffic and clear route so that the ambulance or peoples carrying any patient will reach the hospital in a short period of time without having any critical condition and it can be used to divide the path only for ambulance to pass through that by using automatic bars from underground so that they can be passed through that path to reach the hospital on time before having any type of critical condition.
5. It allows automation companies to decrease production costs and present new revenue system, including self-driving technology, route optimization and anticipating maintenance of the system. The long driving hours and waiting at traffic lights will not be an issue anymore with fully automated fleets.

Disadvantages of AI for Smart Transportation System

There are also some disadvantages of using AI for smart transportation which give an opportunity to the engineers to make the system more efficient in future. Some of these disadvantages are given below:

1. Since efficiency and performance improved but cost will still be a major barrier for adoption.
2. Power is another factor which leads to some disappointments for the engineers. AI requires more power for all the systems to work more intelligently and efficiently to get the maximum power of newer technologies in the system.

Other than these demerit points, AI is a very good revolutionary product that improves the transportation system by a very large factor in present times but there is a long way to achieve more good things in this.

Different Ways to use AI in Smart Transportation System

There are many ways to use AI in Smart Transportation System which will help people to easily travel from one place to another. Some of those are car navigation systems, driverless or automatic cars, delivery drones, smart traffic systems, smart automatic or auto-pilot flying vehicles like airplane, jets and many more.

Car Navigation System

Finding the best route to go to a place was difficult for a couple of years but nowadays it is very simple with the help of the internet and AI. AI gives the best result of it. Navigation systems are used in vehicles in order to get the best route and traffic status information through different routes with other information like total distance to be travel. Vehicle Navigation System uses AI to get the data of the route via geographical map study using machine learning and data processing so that it will give us the distance need to travel and the best way where the traffic is less so that we can get to our desired place in less time.

Figure 10. Example of car navigation system
Source: Bednarczyk (2020)

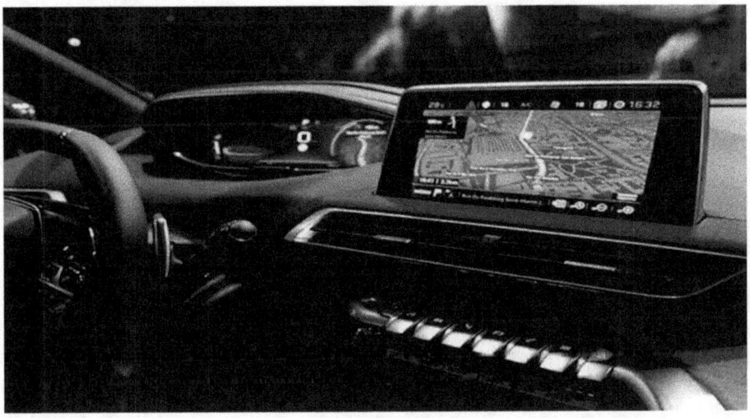

Nowadays, it is preinstalled in many cars, but some people buy this additionally to get their route data during their travelling. Its efficiency is very good depending upon the manufacturer company and system quality used in it. This can be used by using Wi-Fi connection or by Bluetooth via the devices through which we want to use it with whether it is our phone or any other electronic device.

Driverless or Automatic Vehicles

Sometimes, due to carelessness there are high chances of dangerous vehicle accidents, and it will affect to all the peoples travelling through the same route and to all the peoples who are associated with person who is being injured or loss their life due to this. To reduce the number of such accidents, driverless or automatic vehicles are introduced in order maintain the control even if the driver is not attentive due to any reason or anyone doesn't have a person the drive the vehicle now, but they have to reach some place on time.

This technique is introduced in many countries and many tech-giant companies have jumped into this to avail such technology to peoples after seeing the need of this type of technology whether it is due to extra use of fuel-based vehicle or due to other reason. We have seen many examples of this system like self-driving metros with sensors in many cities, self-driving electric vehicles made by many companies like Tesla, TATA motors and many more.

Figure 11. Comparison of self-driving vehicles, automated vehicles and connected vehicles
Source: Road Safety Facts (2022)

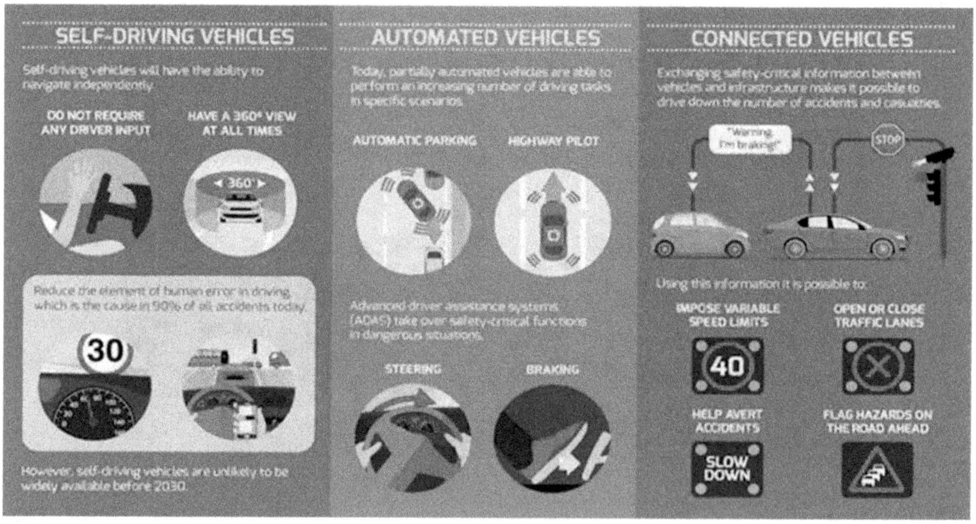

Self-driving and automated vehicles are used most often nowadays in private transportation vehicles while connected vehicles are used mostly in public transport vehicles like metro, buses.

Figure 12 shows the automatic car gear shifter which is a common example automated vehicle system operation in which gear is shifted according to the traffic on the route whether it is cleared or busy.

There are some other applications in this, like brain-controlled car for disabled peoples. Since these cars will rely on what the individual thinks, they will not require any physical movement on the part of the individual. This is done by using a bio-control system. The system is based on technology which was heard by many people in earlier times called brain-wave sensors.

Figure 12. Automatic car gear shifter
Source: Esquire (2020)

The bio-control system integrates signals from various other systems and compares them with originals in the database. It comprises of the following systems:

1. Brain-Computer Interface: EEG (Electroencephalograms) measured from the human scalp.
2. Automatic Security System
3. Automatic Navigation System (Shriaarthi et al., 2018).

Drones Used for Delivery

A machine performs various tasks very efficiently and with speed. There are various modes in which it can be used to provide service to the users and one of these types is drone system which can be used to delivery items to peoples where time to reach is high if we go through the roads. Drones are flying machines based on AI & ML and various electronic-based concepts. Nowadays, it is used very much in defense technologies to make various arm deployed weapons.

Drones are one of the greatest innovations in weapon system which leads to a great revolution for many countries like United States, Russia, China, India and many more, in terms of economic growth and security and arm strengths. But there are some disadvantages of drone systems which may lead to raising questions on its utility.

Figure 13. General drone with camera which can be used for delivery of items
Source: Majumder (2019)

Figure 14. Unmanned combat weapon drone system which can be used in wars
Source: Defense View (2020)

1. There are some chances that it may come in contact with any aircraft which may lead to a dangerous incident so, it should be of low height range, and it needs practice in order to avoid the dangerous consequences.
2. Some people may use it for bad purposes like spying which is not permitted and breach privacy of peoples.
3. It requires equipment of high cost in order to get good quality and efficient working.
4. It can be hijacked or manipulated.
5. It may trespass into authorized areas such as airports and military zones if not controlled properly.

There are four types of UAVs (Unmanned Armed Vehicles) which can be developed:

1. **Fixed-Wing:** Its range is up to 1850km and speed is up to 500km/h, flying time is up to 2000 minutes and carry payload up to 1500kg and its power is supplied using LiPo/fuel and its weight varies from 0.5 to 2500 kg but it is expensive and requires more space.
2. **Helicopters:** Its range is up to 400km, and speed is up to 120km/h, flying time 250 minutes and payload is 65 kg and power supply through LiPo/fuel and its weight varies from 25 to 200 kg but it is expensive and requires high maintenance.
3. **Multi-Coopers:** Its range is up to 100km and speed is up to 30km/h, flying time is up to 100 days and payload is up to 10kg and it has solar battery as power supply and its weight is up to 75kg but it is very expensive and very sensitive to wind.
4. **Loons:** Its range is up to 10km, and speed is up to 160km/h, flying time is up to 60 minutes and payload is up to 5kg and power supply is LiPo battery, and its weight is up to 25kg but it sensitive to wind and limited power supply (Rezwan & Choi, 2022).

Smart Traffic Control System

The number of vehicles being used in cities has increased badly in the past couple of years, which leads to many severe issues like traffic jams, accidents and many other serious issues as cities change to smart cities. In recent times, traffic jam is being counted as one of the biggest problems for designers and engineers to develop a smart traffic control and management system which can detect and reduce the total density of traffic in most of urban areas visited by drivers like offices, urban and institutions accurately and efficiently based on some modern technologies which includes Wireless Sensor Networks (WSNs), Surveillance Cameras and IoT (Internet of Things).

WSNs are something that can be used for collecting data about traffic on roads and about parking spaces available in an area of a city. The system allows driver to look through the traffic density figure and the count of available parking spaces in between the way to their final destinations by using an android mobile application in order to avoid unnecessary traffic jams and to take other possible routes so that they can avoid being stuck and it is also used to make it much easier for peoples to look for freely available parking spaces in order to avoid unrequired long travelling. The system joins three subsystems to each other in order to give better results and those are crossroad management, parking space management and mobile application, to connect people to a smart city.

Figure 15. Smart traffic control and management system
Source: Hilmani (2020)

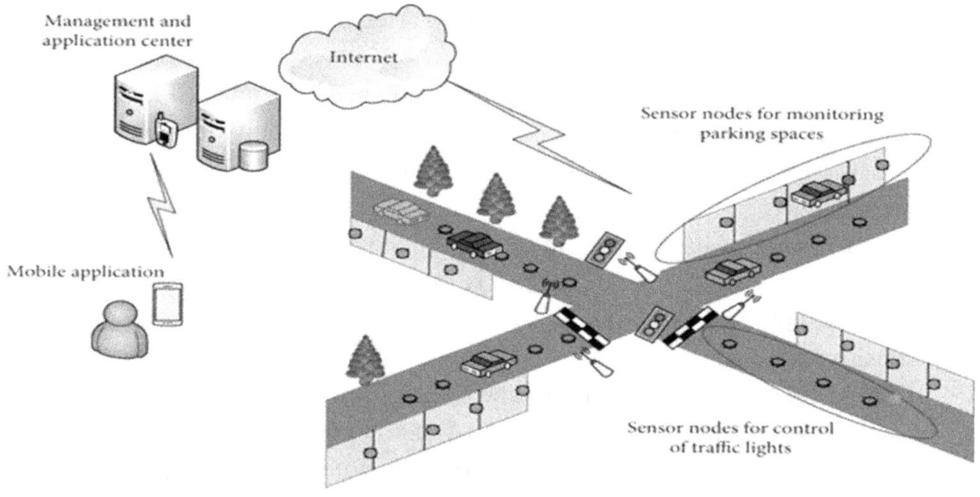

By using the above system, traffic control and management would be much easier in order to reduce traffic jams and reduce number of accident cases. Other than these, we can also use smart path divider whenever there is an ambulance near a traffic light so that it can pass through that path such that it doesn't have to wait for its turn to go ahead and reach the hospital on time.

Adaptive Traffic Signal Control (ATSC) as a key part of ITS, which plays a key role in reducing traffic congestion by real-time adaptation to dynamic traffic conditions. Moreover, these systems are integrated with IoT. MARL (Multi-Agent Reinforcement Learning) is used in ITS to improve the overall traffic in the city.

The ML methods such as neuro-fuzzy, immune network algorithms, neural networks (NN) and genetic algorithms have been used for ATSC in recent years (Damadam et al., 2022).

CONCLUSION

All the contents discussed are important points which should be known to people when they intend to go through any project related to AI in smart healthcare and smart transportation system. Most of the data given in this chapter are either theoretical concepts or facts based on study by different groups of peoples.

Artificial Intelligence System can be in many fields other than these two like agriculture, electronics, home appliances, wearable smart devices, etc. Its efficiency of doing tasks in various ways or more specifically in the best way, make this technology one of the best and popular innovation in the present time and its integration with different technologies make it more powerful and useful in many areas of development which will the peoples to live their life more easily and do their works efficiently in short period of time with less investment, which is more important in present time.

REFERENCES

Bednarczyk, A. T. (2022). *From good to great: the next step in embedded navigation.* https://www.tomtom.com/newsroom/explainers-and-insights/future-of-embedded-navigation/

Damadam, S., Zourbakhsh, M., Javidan, R., & Faroughi, A. (2022). An Intelligent IoT Based Traffic Light Management System: Deep Reinforcement Learning. *Smart Cities*, *5*(4), 1293–1311. doi:10.3390martcities5040066

DataFlair. (2019). *Importance of AI in Healthcare Sector*. https://data-flair.training/blogs/ai-in-healthcare-sector/amp

Facts, R. S. (2022). *How Can Automated and Connected Vehicles Improve Road Safety*. https://roadsafetyfacts.eu/how-can-automated-and-connected-vehicles-improve-road-safety/

Hilmani, A. (2020). *Automated Real-Time Intelligent Traffic Control System for Smart Cities Using Wireless Sensor Networks*. https://www.hindawi.com/jounarls/wcmc/2020/8841893/

Iyer. (2021). *AI enabled applications towards intelligent transportation*. doi:10.1016/j.treng.2021.100083

Kalis, B., Collier, M., & Fu, R. (2018). *10 Promising AI Applications in Health Care*. https://hbr.org/2018/05/10-promising-ai-applications-in-heath-care

Lateef, Z. E. (2021). *Introduction To Machine Learning: All You Need to Know About Machine Learning*. https://www.edureka.co/blog/introduction-to-machine-learning/amp/

Mahalakshmi & Latha. (2019). Artificial Intelligence with Internet of Things on Healthcare systems: A Survey. *International Journal of Advanced Trends in Computer Science and Engineering*, 2847-2854. doi:10.30534/ijatcse.2019.27862019

Mahmuod Nasr, M. (2021). Smart Healthcare in the Age of AI: Recent Advances, Challenges and Future Prospects. *IEEE Access : Practical Innovations, Open Solutions*, *9*, 145248–145270. doi:10.1109/ACCESS.2021.3118960

Majumder, A., & Standard, B. (2019). *Goods delivery by drones gets closer to reality, updated policy unveiled*. https://www.google.com/amp/s/wap.business-standard.com/article-amp/current-affairs/goods-deliver-by-drones-get-closer-to-reality-updated-policy-unveiled-119011600056.html

Rezwan, S., & Choi, W. (2022). Artificial Intelligence Approaches for UAV Navigation: Recent Advances and Future Challenges. *IEEE Access : Practical Innovations, Open Solutions*, *10*, 26320–26339. doi:10.1109/ACCESS.2022.3157626

Rogerson, S. (2020). *Sintrones fanless computer powers smart city applications*. https://www.iotm2mcouncil.org/iot-library/news/smart-cities-news/sintrones-fanless-computer-powers-smart-city-applications/

Shriaarthi, B., Suganya, B., & Sridevi, S. (2018). Brain Controlled Car for Disabled Using Artificial Intelligence. *International Journal of Engineering Research & Technology*, 69-71. doi:10.17577/IJERTCONV1IS03024

Singh, B. (2022). *India's first association of robotic surgeons launched in Guwahati*. https://www.google.com/amp/s/m.economictimes.com/news/india/indias-first-association-of-robotic-surgeons-formally-launched-in-guwahati/amp_articleshow/91219923.cms

Smith, J. (2020). *Can Artificial Intelligence Replace the Role of Doctors?* https://readwrite.com/can-artificial-intelligence-replace-the-role-of-doctors/

Sujith, Sajja, Mahalakshmi, Nuhmani, & Prasanalakshmi. (2021). Systematic review of smart health monitoring using deep learning and Artificial intelligence. *Neuroscience Informatics*, 1-6. doi:10.1016/j.neuri.2021.100028

The Medical Futurist. (2021). *The Top 12 Health Chatbots*. https://mediclafuturist.com/top-12-health-chatbots/

Tipan, E. (2020). *Manual vs. Automatic: Which Car Should You Get?* https://www.esquiremag.ph/wheels/drive/manual-vs-automatic-car-a2314-20200109-lfrm

Uddin, M., Wang, Y., & Woodbury-Smith, M. (2019). *Artificial intelligence for precision medicine in neurodevelopmental disorders*. https://www.nature.com/articles/s41746-019-0191-0

Verified Market Research. (2021). *Healthcare Chatbots Market Size and Forecast*. https://verifiedmarketresearch.com/product/healthcare-chatbots-market/

View, D. (2020). *Top 10 Best Unmanned Combat Aerial Vehicle*. https://www.defenceview.in/top-10-best-unmanned-combat-aerial-vehicle/

Chapter 10
Revolutionizing Agriculture With Cloud-Connected Irrigation Technology

Abhishek Kumar Singh
Govind Ballabh Pant Institute of Engineering and Technology, India

Pushkar Praveen
Govind Ballabh Pant Institute of Engineering and Technology, India

Devansh Tripathi
Govind Ballabh Pant Institute of Engineering and Technology, India

Ved Prakash Pandey
Govind Ballabh Pant Institute of Engineering and Technology, India

Pratibha Verma
Rajkiya Degree College, Puranpur, India

ABSTRACT

The current global water crisis, caused by factors such as population growth, urbanization, and climate change, has a significant impact on agriculture, which consumes 70% of the world's freshwater. One potential solution to this problem is the use of intelligent irrigation systems. Cloud-based smart irrigation systems offer a more efficient and cost-effective approach to water consumption management. This study examines the concept of cloud-based smart irrigation systems, their advantages, and their potential application in agriculture. To begin, the authors detail the various components of a cloud-based smart irrigation system, including sensors, controllers, and cloud computing. They then explore the benefits of cloud-based systems, which include scalability, cost-effectiveness, and real-time data

DOI: 10.4018/978-1-6684-6418-2.ch010

processing capabilities. They also delve into the different types of sensors used in smart irrigation systems, such as soil moisture, weather, and crop sensors, as well as their respective advantages and limitations. Furthermore, they discuss the various controllers utilized in smart irrigation systems, such as scheduling, sensor-based, and weather-based controllers. They elaborate on how these controllers function and the advantages they offer in terms of water conservation and crop yield optimization. Lastly, they examine the practical application of cloud-based smart irrigation systems in agriculture, including how to create, implement, and tailor the system to suit the specific requirements of a particular farm. They also consider potential barriers to adoption, such as cost and technological complexity, and how to overcome them. In short, cloud-based smart irrigation systems present a promising solution to the water crisis in agriculture. These systems can significantly reduce water waste, increase crop yield, and save costs by utilizing cloud computing technology to manage irrigation. With careful planning, installation, and configuration, cloud-based smart irrigation systems have the potential to be an effective tool for farmers in improving their productivity and sustainability.

INTRODUCTION

Water management and irrigation systems have been an integral part of human civilization for thousands of years. From the ancient civilizations of Mesopotamia to the modern-day societies, water management has played a crucial role in sustaining human life and fostering economic development. The history of water management can be traced back to the early agricultural societies, where farmers used simple techniques, such as diverting water from rivers and streams to irrigate their crops. In many parts of the world, water was considered a scarce resource, and the ability to manage it effectively was critical for survival.

As societies became more complex, so did their water management systems. Xiao Yun Zheng (2021) explains the water management in ancient Egypt, for example, the construction of canals and reservoirs allowed farmers to harness the power of the Nile River to irrigate their crops. The Maya civilization of Central America also developed sophisticated water management systems that allowed them to grow crops in areas with limited rainfall. The importance of water management and irrigation systems continued to grow as societies became more urbanized and industrialized. In the 19th century, the Industrial Revolution led to the development of large-scale irrigation systems that allowed for the cultivation of crops on a massive scale. The construction of dams and reservoirs also became increasingly common, providing a reliable source of water for both agriculture and industry.

Today, water management and irrigation systems are more important than ever. As the world's population continues to grow and climate change brings more unpredictable weather patterns, the need for effective water management has become critical. Governments and organizations around the world are investing in advanced water management technologies and infrastructure to ensure that future generations have access to clean water and food security. In short, the history of water management and irrigation systems is a testament to humanity's ability to adapt and innovate in the face of challenges. As we look towards the future, it is clear that effective water management will play a critical role in ensuring the health and prosperity of our planet and its people.

We all know about the scenario of water scarcity. It affects not only rural areas but also urban areas. This scarcity is further exacerbated by the uneven distribution of rainfall between locations and at different times of the year. As per information given by AQUASTAT, the global information system of the Food and Agricultural Organization of the United Nations (FOA), the agricultural industry uses 69% of the fresh water available on Earth. Water's significance and requirement may be seen in the agricultural production industry and for ensuring food security.

Almost 40% of the world's food is produced by irrigated agriculture. But farmers still irrigate their crops using conventional irrigation systems. These systems are time- and labor-intensive and ineffective, especially when several fields are dispersed over various geographical areas. For instance, in developing nations, farmers are required to constantly visit their fields and remain there when irrigation is taking place in order to manage and plan irrigation in the area in question. Moreover, conventional irrigation frequently results in the issue of overwatering, which uses a lot of water. Moreover, it may result in concerns with inadequate irrigation, which impacts crop quality. Farmers that only use rainwater for irrigation are concerned about this. As a result, it becomes extremely important to monitor and manage each crop's water requirements. Regardless of weather, soil type, or crop variety, this irrigation method must carefully match the water needs of the entire field. Due to evaporation and a lack of suitable resources, a large amount of irrigation water is squandered. Irrigation system management of water resources becomes crucial in agriculture to aid farmers in improving their produce. A smart irrigation system strategy has emerged to solve this issue. The system will be more practical and able to monitor a variety of characteristics, including temperature, humidity, and soil moisture, utilizing a variety of sensors thanks to the use of IoT.

The Internet of Things (IOT) is a concept in which every gadget is given an IP address that can be used to identify it by anyone online. Mechanical and digital devices that can communicate data without relying on networked human-to-human or human-to-computer communication contain Unique Identifiers (UIDs). It essentially began with the "internet of computers." According to research, the number of "things" or gadgets connected to the internet will skyrocket. The "Internet of Things" network was born out of this (IoT). Different devices can interact with one other thanks to recent advancements in technology that enable wireless control of environments using things like Bluetooth and Wi-Fi. H.

H. Meena et al. (2020) proposes the implementation of a smart monitoring and controlling automated irrigation system based on Arduino for effective management of water resources in India, which is an agro-based country with a large population. The system uses a microcontroller to turn ON/OFF the motor and measures various parameters like temperature, humidity, and soil moisture. It also sends alerts/messages through GSM to keep the user updated about the field status and has an LCD screen for monitoring the system. The system enables farmers to monitor their crops from anywhere at any time through SMS, which helps in cost-cutting and produces good results. The paper also briefly discusses the application of IoT technology in the field of agriculture.

The water flow through the sprinkler system is currently controlled by a timer system in many garden centres. As there is no way to check the soil's moisture level, this practice is prone to overwater plants. For instance, even though a garden center's plants received enough water from the rainfall, if there was a strong downpour just before the sprinkler system was scheduled to turn on, the system would still turn on at the scheduled time and water the center's plants. Customers end up buying overwatered plants, which is bad for the garden center's reputation.

The goal of this project was to develop a prototype that could illustrate the idea of watering plants based on measurements of soil moisture. The project's solution needed to be competitively priced to draw a variety of garden centres, not only big corporate centres. The item must be as dependable as possible though. Many plants in the garden centre won't receive the water they require if the product doesn't work, and their health will start to deteriorate.

LITERATURE REVIEW

A system that makes use of Internet of Things (IoT) technology to automate and optimize the irrigation process for crop cultivation is required for overcome the issues related to water management systems. For this purpose different scientists and researchers have given their ideas and technology for eliminates these problems. In which IoT based system is also designed to short out the issues related to irrigations. T. Thaher (2020) explained the crucial role of water in economic development and the agricultural sector, which is the largest consumer of freshwater. Traditional irrigation systems are time-consuming and ineffective, especially when managing several fields distributed in different geographical regions. To address these issues, the paper proposes a smart irrigation system using a Wireless Sensor Network (WSN) and Internet of Things (IoT) cloud services. The system comprises a WSN, a control unit, and cloud services. The sensors data is wirelessly gathered using Arduino Uno and XBee ZigBee modules, and YL-69 sensor monitors soil moisture. Raspberry Pi is used to process data, provide decisions, and transfer them to ThingSpeak IoT cloud. The system makes use of a variety of sensors to gather information on environmental factors including temperature, humidity, and soil moisture before sending it to a cloud server for processing and analysis. Based on crop-specific metrics and environmental factors, the scientists have built a machine learning system that evaluates the data to establish the best irrigation schedule. The device also comes with a mobile app that gives customers access to real-time data and enables remote monitoring and management of the irrigation process. The results of the authors' implementation of the suggested system and testing it in a greenhouse environment demonstrate that the system can greatly increase crop output and water usage efficiency while consuming less water. The benefits of employing a cloud-based strategy, including scalability, flexibility, and cost-effectiveness, are also covered in the study. This system enables farmers to monitor crop status and make necessary decisions.

L. Gracia et al. (2020) discusses the importance of water management in countries facing water scarcity, especially in agriculture, where a significant amount of water is consumed. With global warming leading to water scarcity, studies focused on reducing water usage in irrigation have gained momentum. Traditional commercial sensors are expensive, making them difficult for smaller farmers to afford. However, the emergence of low-cost sensors and IoT and WSN technologies has made it possible to develop affordable irrigation management systems. The paper summarizes the state-of-the-art smart irrigation systems, discussing the parameters monitored in irrigation systems, the nodes and wireless technologies utilized, and the challenges and best practices for implementing sensor-based irrigation systems.

S. Ismail et al. (2022) shows the use of IoT for managing and monitoring natural resources such as water is an upcoming and sustainable solution. Various researches have focused on developing smart IoT-based water management systems for applications in agriculture, industry, residential, and crude oil exploration sectors. Additionally, they proposes a new optical water management system to fill a gap in the literature, specifically for the oilfield industry, where significant amounts of water are required to stimulate oil production. This system uses an advanced optical technique for sensing and transmitting data. Overall, this paper serves as a guide for future research efforts that aim to integrate IoT in water management and monitoring.

The system is made to continuously monitor and modify irrigation based on environmental factors like temperature, humidity, soil moisture, and light intensity. The ideal irrigation schedule is determined using the authors' innovative fuzzy logic-based irrigation scheduling algorithm, which takes into account a number of environmental variables and crop-specific parameters. The device also comes with a smartphone app that lets customers monitor and control the irrigation process remotely while giving them access to real-time data and alarms. The results of the authors' implementation of the suggested system and testing it in a greenhouse environment demonstrate that the system can greatly increase crop output and water usage efficiency while consuming less water.

An algorithm-based low-cost automated irrigation system with a soil moisture sensor is described in Agarwal and Agarwal (2017). This system uses a soil moisture sensor to measure the soil's moisture content and choose the best irrigation schedule. The system consists of a water pump, a soil moisture sensor, and a microcontroller-based circuit. When the soil moisture level drops below a predetermined threshold, the authors' algorithm analyses the soil moisture data and instructs the water pump to start irrigating the crops. The method can drastically lower water consumption while preserving crop output, according to the authors' testing and implementation of the suggested system in a small-scale farming setting. The benefits of employing a low-cost automated system, including accessibility, usability, and energy efficiency, are also covered in the article.

K. Taneja and S. Bhatiain (2017), in their paper, "Automatic Irrigation System Using Arduino UNO," presented an automated irrigation system that employs an Arduino UNO microcontroller board to control the irrigation process based on soil moisture data. A soil moisture sensor, a water pump, and an Arduino UNO board with a relay module make up the system. The authors have created a programme that analyses data on soil moisture and, when the moisture level drops below a predetermined level, activates the water pump to irrigate the crops. A LCD monitor that presents real-time statistics and status information is also part of the system. The system can greatly increase water use efficiency, reduce water consumption, and preserve crop output, according to the authors' testing and implementation of the suggested system in a small-scale farming setting. The benefits of adopting an Arduino UNO-based system, including their accessibility, adaptability, and affordability, are also covered in the article.

Al-Bahadly et al. (2015) proposed a garden watering system that uses a moisture sensor to calculate the ideal watering schedule for. A soil moisture sensor, a water pump, and a microcontroller-based circuit make up the system. When the soil moisture level drops below a predetermined threshold, the authors' algorithm reads the soil moisture data and activates the water pump to irrigate the plants. The technique can greatly increase water use efficiency while sustaining healthy plants, according to the authors' implementation and testing in a small garden setting. The benefits of adopting a moisture-based watering system, including water conservation, energy efficiency, and less labour, are also covered in the study.

D. Divani et al. (2016) describes an automated plant watering system that uses a soil moisture sensor to ascertain the watering requirements of plants. A microcontroller-based circuit, a soil moisture sensor, a water pump, and a water tank make up the system. When the soil moisture level drops below a predetermined threshold, the authors' algorithm reads the soil moisture data and activates the water pump to irrigate the plants. A LCD monitor that presents real-time statistics and status information is also part of the system. The technique can greatly increase water use efficiency while maintaining healthy plants, according to the authors' testing and implementation of the suggested system in a small-scale farming setting. The benefits of adopting an automated plant watering system, including enhanced plant health, less labour, and water conservation, are also covered in the study.

B. Narayanavaram et al. (2020) proposed an automated agriculture system that employs an Arduino microcontroller board to control several agricultural factors. In this paper several sensors and actuators are part of the system, including a soil moisture sensor, a temperature sensor, a humidity sensor, a water pump, and a fertiliser dispenser used. According to predetermined criteria and thresholds, the authors have created a programme that reads sensor data and regulates actuators. A smartphone application that offers real-time data and control functionalities is also part of the system. The proposed approach was evaluated and put into practise by the authors in a small-scale farming setting, and the results indicate that it has a considerable impact on labour efficiency and production. The benefits of employing an Arduino-based system, such as flexibility, affordability, and usability, are also covered in the article.

In an article titled "Intelligent and Smart Irrigation System Utilizing Edge Computing and IoT" by M. Safdar Munir et al. (2021), they offer an irrigation system that makes use of edge computing and the Internet of Things (IoT) to increase water use efficiency and decrease labour. The system is made up of a number of sensors and actuators, including valves, a water pump, a soil moisture sensor, a temperature sensor, and a humidity sensor. The authors have created a programme that uses machine learning methods and pre-established rules to read sensor data and control actuators. A mobile application and a cloud-based platform for data analysis and storage are also included in the system. The technique can greatly increase water use efficiency while maintaining healthy plants, according to the authors' testing and implementation of the suggested system in a small-scale farming setting. The benefits of IoT and edge computing, such as real-time monitoring, data analysis, and remote control, are also covered in the study.

The "Smart Irrigation System" by Asmae EL Mezouari et al. (2022) describes an irrigation system that maximises water utilisation in agriculture by combining Internet of Things (IoT) and machine learning algorithms. The system is made up of a number of sensors and actuators, including valves, a water pump, a soil moisture sensor, a temperature sensor, and a humidity sensor. The authors have created a programme that reads sensor data and manages actuators using machine learning techniques that consider a variety of elements, including soil type, crop kind, and weather conditions. A mobile application and a cloud-based platform for data analysis and storage are also included in the system. The technique can greatly increase water use efficiency while maintaining healthy plants, according to the authors' testing and implementation of the suggested system in a small-scale farming setting. The benefits of IoT and machine learning in the agricultural industry, such as real-time monitoring, data analysis, and decision assistance, are also covered in the study.

The "IoT-Based Smart Irrigation System" by P. Roopa Ranjani et al. (2018) describes an irrigation system that makes use of the Internet of Things (IoT) to automate and optimise water consumption in agriculture. The system is made up of a number of sensors and actuators, including valves, a water pump, a soil moisture sensor, a temperature sensor, and a humidity sensor. According to predetermined criteria and thresholds, the authors have created a programme that reads sensor data and regulates actuators. A mobile application and a cloud-based platform for data analysis and storage are also included in the system. The technique can greatly increase water use efficiency while maintaining healthy plants, according to the authors' testing and implementation of the suggested system in a small-scale farming setting. The benefits of implementing IoT in agriculture, including real-time monitoring, remote control, and decision assistance, are also covered in the article.

The article titled "IoT-Based Smart Irrigation System" by Srishti Rawal (2017) presents a smart irrigation system that utilizes IoT technology to monitor and control the water supply to crops. The system includes sensors that measure soil moisture, temperature, and humidity, which are then transmitted to a central unit using wireless communication. The central unit processes the data and uses it to control the water supply to the crops. The system is designed to be energy-efficient and cost-effective, with the ability to save up to 80% of water compared to conventional irrigation systems. The paper concludes that the smart irrigation system has the potential to revolutionize the way we farm and produce food, leading to increased sustainability and efficiency in agriculture.

The article titled "Smart Irrigation System Based on the Internet of Things and the Cloud" by M. Mustafa et al. (2021) presents a smart irrigation system that uses IoT technology and cloud computing to optimize water usage in agriculture. The system includes sensors that measure soil moisture, temperature, and humidity, which are then transmitted to a central cloud-based server. The server processes the data and uses it to control the water supply to the crops, ensuring that they receive the right amount of water at the right time. The system also includes a mobile application that allows farmers to monitor and control the irrigation system remotely. The paper concludes that the smart irrigation system has the potential to improve crop yields, reduce water waste, and increase efficiency in agriculture, making it a promising solution for sustainable farming.

The article titled "Low Power Cost Effective Automatic Irrigation System" by S. Nallani and V. Berlin Hency (2015) presents an automatic irrigation system that is designed to be low power and cost-effective. The system utilizes a microcontroller and sensors to measure soil moisture levels and control the water supply to the crops. The system is also designed to be wireless, allowing farmers to control and monitor the irrigation system remotely using a mobile application. The paper concludes that the automatic irrigation system has the potential to improve the efficiency and sustainability of agriculture by reducing water waste and improving crop yields. Additionally, the low power and cost-effective design make it accessible to small-scale farmers who may not have access to expensive irrigation systems.

The article titled "Sensor-Based Automated Irrigation System with IoT: A Technical Review" by K. Kansara et al. (2015) presents a review of the use of IoT technology in sensor-based automated irrigation systems. The review highlights the importance of monitoring soil moisture levels and the use of sensors to detect these levels. The paper also discusses the use of wireless communication and cloud computing to enable remote monitoring and control of the irrigation system. The authors conclude that the use of IoT technology in automated irrigation systems has the potential to increase efficiency and reduce water waste in agriculture, leading to improved crop yields and more sustainable farming practices.

METHODOLOGY

In order to make better informed judgements about when and how much to water plants, a cloud-based smart irrigation system employs cloud computing to collect data about weather patterns, soil moisture levels, and other environmental elements. The following are some essential actions in putting in place this kind of system:

Data Gathering: Many different sources, including weather stations, soil moisture sensors, and satellite imaging, are used to gather environmental data.

Data Processing: To find patterns and trends, the acquired data is processed and examined in the cloud.

Decision-Making: The system determines when and how much to water plants based on the data analysis, and then issues instructions to the irrigation system.

Feedback Loop: To maximize plant growth while consuming the least amount of water, the system continuously evaluates the outcomes of its decisions and modifies its algorithms as necessary.

In short, a cloud-based smart irrigation system uses real-time data and cutting-edge analytics to make accurate irrigation decisions, leading to more effective water use and healthier plants.

Feedback loop: To maximize plant growth while consuming the least amount of water, the system continuously evaluates the outcomes of its decisions and modifies its algorithms as necessary.

In conclusion, a cloud-based smart irrigation system uses real-time data and cutting-edge analytics to make accurate irrigation decisions, leading to more effective water use and healthier plants.

Figure 1. Flowchart of a cloud-based smart irrigation system

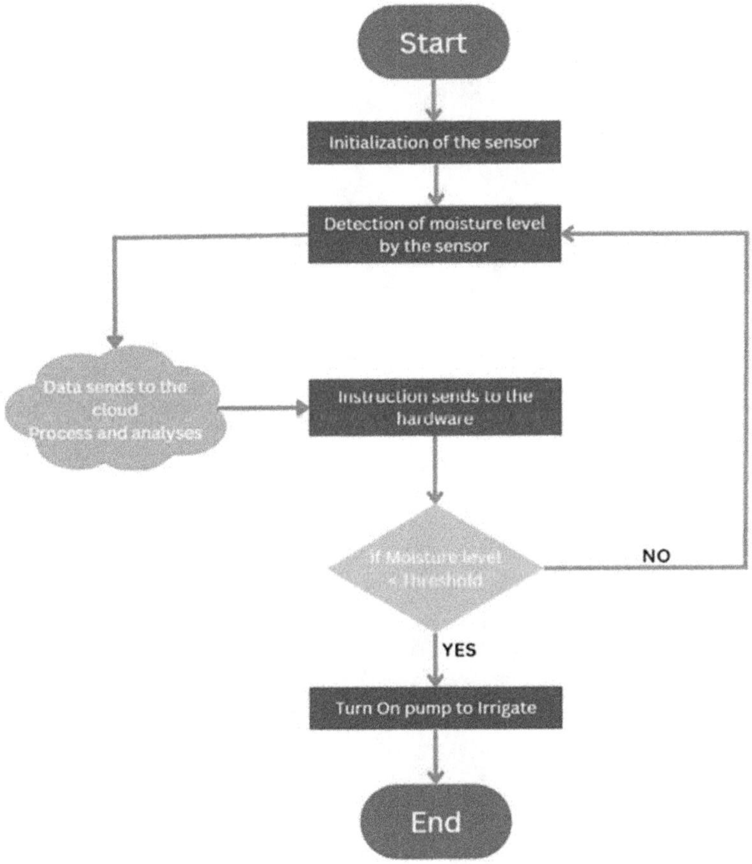

IMPLEMENTATION

Revolutionizing agriculture with cloud-connected irrigation technology has the potential to transform the way we farm and produce food. This innovative technology leverages the power of the cloud to monitor and manage irrigation systems remotely, giving farmers more control over their crops and reducing waste. By using sensors and data analytics, cloud-connected irrigation systems can optimize water usage, reduce energy costs, and improve crop yields. This technology also enables farmers to access real-time information about soil moisture, weather patterns, and plant health, allowing for more informed decision-making. Ultimately, the implementation of cloud-connected irrigation technology has the potential to increase sustainability and efficiency in agriculture, leading to a more secure food supply for the growing population.

Flowchart of the Model

An illustration of the flowchart model is shown in Figure 1.

Figure 2. Block diagram of a cloud-based smart irrigation system

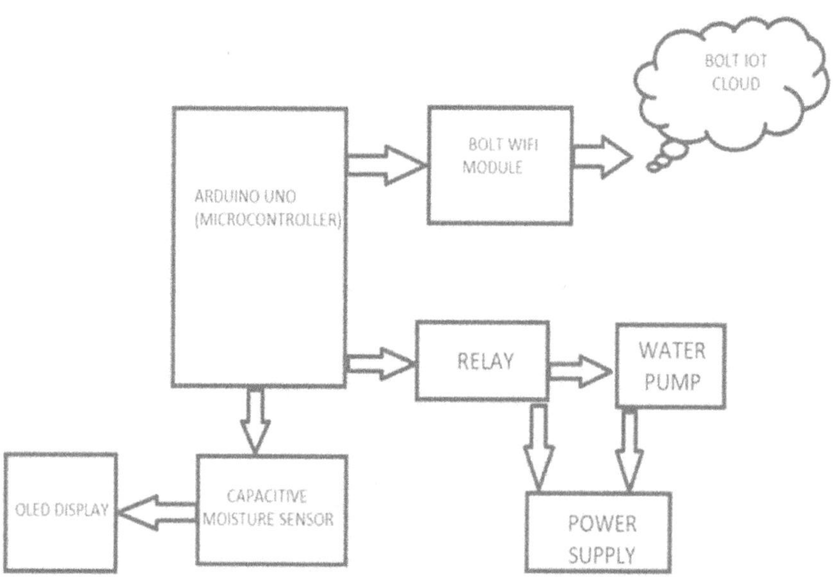

Schematic Diagram

The block diagram up top shows how the entire project functions. The microcontroller, or Arduino UNO, is the main system controller. The capacitive moisture sensor measures the amount of moisture in the soil and transmits a signal to the Arduino UNO. The data is subsequently sent to the bolt wi-fi module and shown on the OLED display concurrently. The microcontroller and an application can communicate wirelessly thanks to the Wi-Fi module. The Arduino uses a relay to activate or deactivate the appliance when it receives the signal. Physical connections exist between the water pump, relay, and Arduino, which are all powered by the same power supply unit. The Arduino UNO unit transmits the data to the bolt wi-fi module and shows it on the OLED display once the capacitive moisture sensor evaluates the soil moisture level and provides a signal to it. The Wi-Fi module aids the microcontroller in connecting to a Wi-Fi network and sending data to a wireless application. After receiving the signal, the Arduino uses a relay to switch the appliance on or off. There is a physical connection between the Arduino, relay, and water pump. The pump, the relay, and the microcontroller are all powered by a single power supply unit.

Circuit Diagram

A cutting-edge approach to effective and efficient irrigation management can be seen in the circuit diagram of a cloud-based smart irrigation system. The irrigation process is automated by this cutting-edge system, which also enables real-time monitoring and control from any location with an internet connection. The circuit diagram offers a visual depiction of the main hardware elements and connections, explaining how the system operates and how the various elements cooperate to produce the best irrigation outcomes. With this cloud-based smart irrigation system, users can precisely manage the amount of water supplied to crops based on real-time meteorological data and soil moisture levels, which helps users save water, cut expenses, and increase crop yields. Commercial farms, greenhouses, and urban gardens looking for a cutting-edge, effective irrigation management solution should consider this system.

This is the simulation of the hardware circuit in Proteuse. The circuit showing in Figure 3 is in Off condition and the circuit showing in Figure 4 is in ON condition.

An experiment was carried out by reading the water requirements of irrigated plants in a small area. The result of the experiment is shown in the graph obtained from the bolt-IoT cloud platform. It displays the timestamps when the moisture readings were less, i.e., when the pump is automatically switched on to supply water and vice versa.

Figure 3. Circuit is in OFF condition

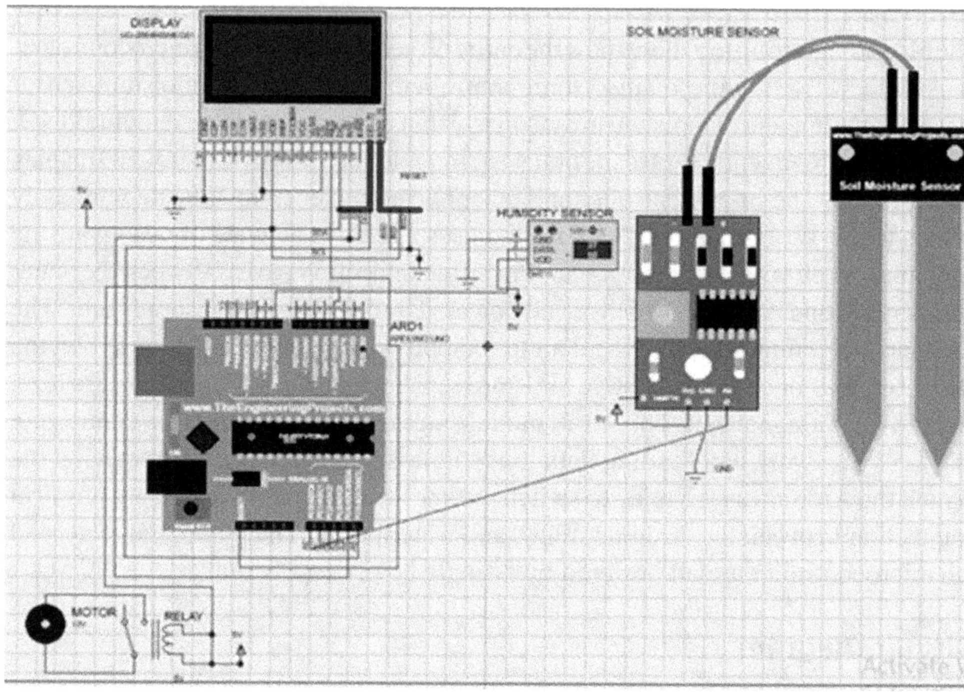

Figure 4. Circuit is in ON condition

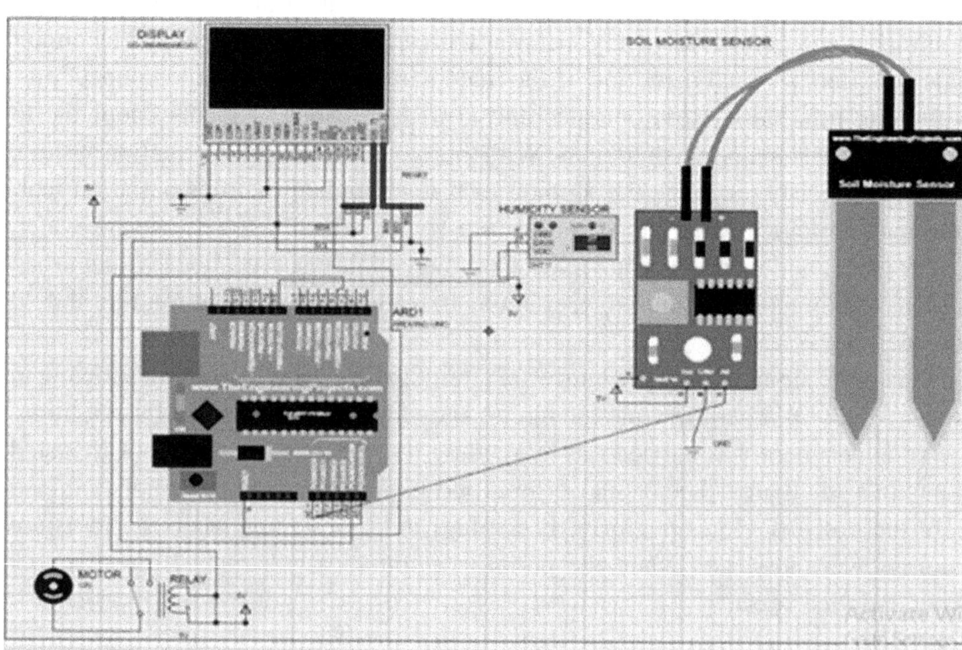

DESCRIPTION OF DIFFERENT TECHNICAL COMPONENTS USED

Software Used

The detailed description of the software that has been used in this project is as follows:

IDE for Arduino

The Arduino Integrated Development Environment (IDE), sometimes known as the Arduino Software (IDE), provides a graphical user interface for basic C programming of microcontrollers along with a few more keywords known as the Arduino language. A text editor for writing code, a message box, a text console, a toolbar with buttons for frequently used features, and a number of menus are all included. To download applications and communicate with Arduino hardware, it establishes a connection with it. Two primary functions are setup() and loop (). Although loop() defines the primary working function, setup() defines the initialization of variables.

BOLT Cloud

IoT is the main goal of Bolt Cloud's design. The Bolt platform allows for the online control of Arduino, Raspberry Pi, and other devices through IOS and Android applications. It can do a lot of things, such save and visualize data, display sensor data, and remotely control devices.

Hardware Used

Table 1 shows the standard market price of the hardware or equipment that has been used in this project.

Table 1. Hardware components details

No.	Component	Quantity	Cost (INR)
1.	ARDUINO UNO	1	600
2.	BOLT Wi-Fi MODULE	1	1500
3.	CAPACITIVE MOISTURE SENSOR	1	180
4.	OLED DISPLAY	1	175
5.	WATER PUMP	1	120
6.	5V- 1 – CHANNEL RELAY	1	70

The detailed description of the following components are as follows:

Arduino Uno

Popular microcontroller board called Arduino Uno is based on the ATmega328P microcontroller. It is frequently used to build a range of electronic projects by professionals, students, and amateurs. The Arduino Uno's salient characteristics are as follows:

Processor: Powerful 8-bit microcontroller with 32KB of flash memory and 2KB of RAM, the ATmega328P.

Inputs and Outputs: The Arduino Uno includes six analogue inputs, 14 digital input/output pins, and a USB connection for power and programming.

Power: The board features an integrated voltage regulator to deliver a steady 5V output, and it can be powered by either a USB cable or an external power source.

Programming: The Arduino Integrated Development Environment (IDE), a free software tool that makes it simple to write, compile, and upload code to the board, can be used to programme the Arduino Uno.

Expandability: The board can be expanded using shields, which are add-on boards that provide additional functionality such as Wi-Fi, Bluetooth, motor control, and more.

Overall, the Arduino Uno is a flexible and user-friendly microcontroller board that makes it simple to build a variety of electronic projects, from straightforward robots and home automation systems to sophisticated robots and blinking LED lights.

Figure 5. Arduino module

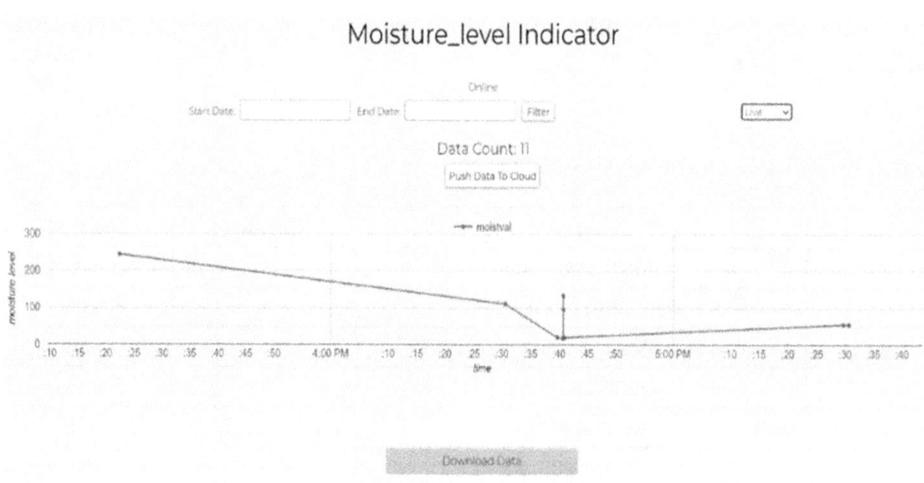

Wi-Fi Module for Bolt

The Internet of Things (IoT) applications can be easily integrated with the internet using the portable and reasonably priced BOLT Wi-Fi module. The Inventrom-created BOLT Wi-Fi module offers a user-friendly framework for creating Internet of Things (IoT) applications that need wireless connectivity. It is created using the Arduino Integrated Development Environment and is based on the ESP8266 Wi-Fi module (IDE). The BOLT Wi-Fi module is an excellent option for novices and amateurs because it is simple to use and includes pre-built libraries for a variety of sensors and components. Additionally, it offers cloud communication with IFTTT, Twilio, and Temboo platforms, enabling users to create sophisticated Internet of Things (IoT) applications with little effort. The BOLT Wi-Fi module is a well-liked option among makers and hobbyists for creating cutting-edge IoT projects because of its inexpensive cost, simplicity of use, and solid functionality.

With BOLT v1, a total of 12 Digital I/O pins are provided.

- Specifically, pins 0,1,2,3,4,5,6 and 7, as well as A0, A1, A2 and A3.
- All components function at 5V logic levels, where HIGH = 5V = Binary 1 and LOW = 0V = Binary 0.
- Each pin in output mode has a 40mA maximum sinking capacity, although users are advised to use no more than 20mA at a time.
- When used as an input (reading the voltage on a pin), anything less than 2.5 volts is considered low voltage and anything greater than 2 volts is considered low voltage.
- 5 volts is considered high voltage.

Figure 6. Bolt wi-fi module

Capacitive Moisture Sensor

An electronic device that gauges soil moisture is the capacitive soil moisture sensor. It is based on the concept of capacitance, which describes how two conductive surfaces can store a charge when they are separated by a non-conductive substance. Two electrodes that are put into the soil make up the sensor, which also has an integrated circuit that measures the capacitance between the electrodes. The capacitance between the electrodes changes along with the soil moisture level, and the integrated circuit can track this change. The very precise and dependable capacitive soil moisture sensor can be used in a variety of fields, such as horticulture, agricultural, and environmental monitoring. Also, many sensors include built-in libraries that can be used with well-known microcontrollers like Arduino or Raspberry Pi, making it very simple to use. The capacitive soil moisture sensor has grown to be a popular option for anyone wishing to monitor soil moisture levels in real-time due to its simplicity, accuracy, and versatility.

Figure 7. Capacitive moisture sensor

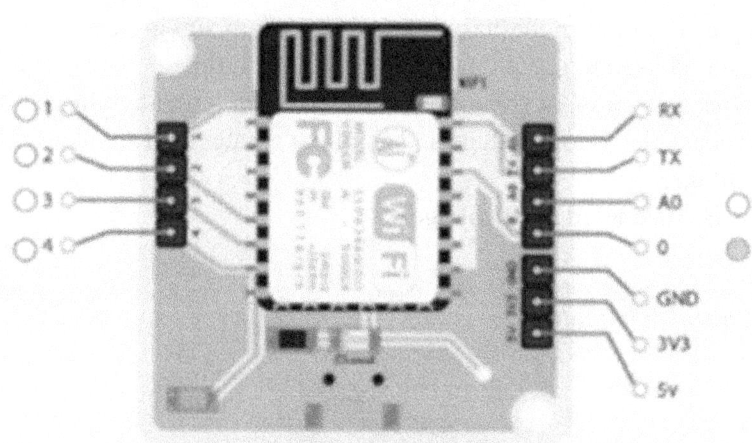

OLED Display

The OLED (Organic Light Emitting Diode) display is a type of display that emits light when an electric current is applied using organic molecules. Wide viewing angles, excellent image clarity, high contrast ratios, and low power consumption are all characteristics of OLED displays. Also, they offer higher refresh rates than other display technologies, which makes them perfect for use in virtual reality and gaming.

OLED displays are commonly used in mobile phones, tablets, televisions, and other electronic devices. They are also popular in the maker community, where they are used to display real-time data in projects like smart home automation and environmental monitoring.

One of the key advantages of OLED displays is that they are very thin and flexible, allowing them to be used in a wide range of applications. They also have a high pixel density, which means that images and text are sharp and clear, even at small sizes.

Overall, OLED displays are a highly versatile and popular display technology that offers a range of benefits over traditional display technologies. They are widely used in many different applications and are expected to continue to grow in popularity in the coming years.

Figure 8. OLED display

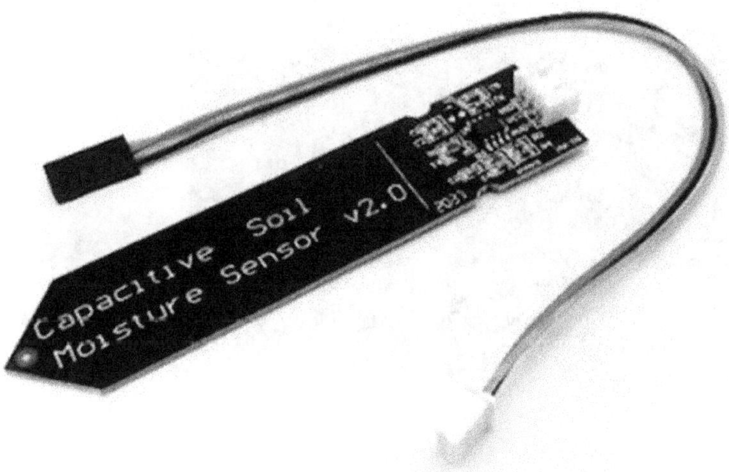

Water Pump

A water pump is an essential device used for moving water from one place to another. It is commonly used for various purposes such as irrigation, water supply, drainage, and circulation. The water pump works by using mechanical force to create suction that draws water into the pump and then pushes it out through a pipe or hose.

Water pumps are available in various types, including centrifugal, positive displacement, and submersible pumps, each with its specific application. Centrifugal pumps are the most common type and are used for applications that require a high flow rate of water, such as irrigation or swimming pool circulation. Positive displacement pumps are used for applications that require a precise flow rate, such as dosing chemicals or mixing fluids. Submersible pumps are designed to operate while submerged in water and are commonly used for draining flooded areas or pumping water from wells.

The size of the pump, the motor's output, and the kind of impeller being utilised are some of the variables that affect the effectiveness and power of a water pump. To ensure effective and dependable functioning, the appropriate pump must be chosen for the application.

Water pumps are crucial tools utilised in a variety of situations, from household to industrial, and their significance cannot be understated.

Figure 9. Water pump

1 Channel, 5 Volt Relay Module

A module for a 5V-1-Channel Relay is an electronic device used to control high-voltage devices using a low voltage signal. It consists of a relay switch, a transistor, and a diode. The relay switch acts as a switch that can turn on or off a high-voltage device, while the transistor is used to amplify the low voltage signal required to activate the relay switch. The diode is used to protect the transistor from back EMF (Electromotive Force) generated by the relay.

The 5V-1-Channel Relay module is a common component in electrical applications, particularly in robotics and home automation where it is used to manage lights, motors, and other high-voltage equipment. It is simple to use and is compatible with development boards and microcontrollers like Arduino, Raspberry Pi, and others.

Overall, the 5V-1-Channel Relay module is an essential component in many electronic projects that require the control of high-voltage devices. It is a versatile and reliable device that is widely used by hobbyists, makers, and professionals alike.

Figure 10. 5V relay module

WORKING MODEL

A sophisticated, automated system called a cloud-based smart irrigation system makes use of cloud computing to monitor and manage irrigation operations from a distance. In order to calculate the ideal amount of water required for plants, this system makes use of a variety of sensors and devices to assess soil moisture levels, temperature, humidity, and meteorological conditions. Additionally, it connects to the cloud via a Wi-Fi module and sends data to a central server for processing and analysis.

This working model of a cloud-based smart irrigation system provides an efficient and convenient way of managing irrigation operations, saving water and reducing energy costs. The system can be controlled from a mobile device or computer, allowing farmers and gardeners to easily monitor and adjust the watering schedule based on real-time data. With this model, users can set up customized irrigation schedules, receive alerts when there are deviations from optimal conditions, and optimize water usage to achieve maximum crop yield. Overall, this working model of a cloud-based smart irrigation system is a valuable tool for sustainable agriculture and water conservation.

Figure 11. Hardware in OFF condition

Figure 12. Hardware in ON condition

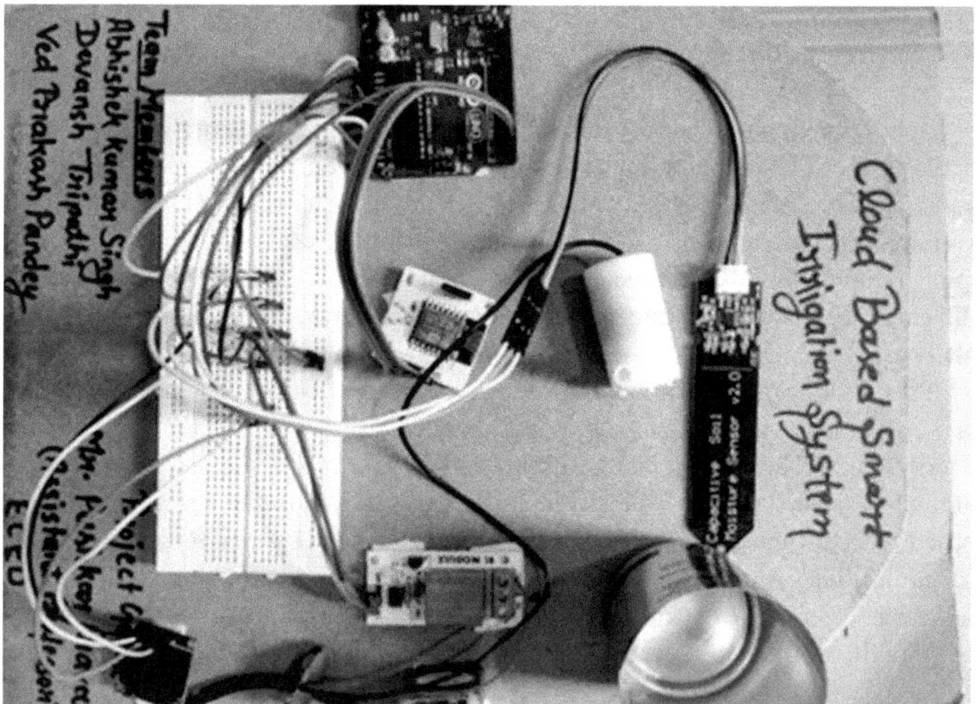

CONCLUSION

In conclusion, the adoption of cloud-based smart irrigation systems can significantly help agriculture by offering a financially advantageous, effective, and environmentally responsible approach to manage water use. These systems collect and interpret sensor data using cloud computing, enabling real-time irrigation monitoring and control. By using weather and crop sensors, as well as controllers that optimize irrigation based on these factors, cloud-based systems can reduce water waste and improve crop yield.

Additionally, cloud-based systems offer scalability, allowing farmers to expand or adjust their irrigation systems as needed. Although there may be initial barriers to adoption, such as cost and technical complexity, the long-term benefits of reduced water usage and improved crop yields can outweigh these challenges.

All things considered, cloud-based smart irrigation systems have the potential to change how water is managed in agriculture, resulting in a more effective and sustainable use of this vital resource.

In the future, the project can be adjusted and improved taking into account the atmospheric temperature and humidity via the DHT22 or DHT11 sensor. This data will allow the creation of different soil moisture thresholds, so that it can be used for seasonal irrigation. Machine learning algorithms (such as predictive algorithms) and artificial intelligence models can also be implemented in projects.

REFERENCES

Agarwal, S., & Agarwal, N. (2017). *An algorithm based low-cost automated system for irrigation with soil moisture sensor*. International Conference on Computer Communication and Informatics (ICCCI -2017), Coimbatore, India. 10.1109/ICCCI.2017.8117741

Al-Bahadly, I., & Thompson, J. (2015). Garden Watering System Based on Moisture Sensing. *International Conference on Sensing Technology*. 10.1109/ICSensT.2015.7438404

Divani, D., & Patil, P. (2016). Automated Plant Watering System. *International Conference on Computation of Power, Energy Information and Communication (ICCPEIC)*. 10.1109/ICCPEIC.2016.7557245

El Mezouari, El Fazziki, & Sadgal. (2022). Smart Irrigation System. *IFCA Papers Online, 55*(10), 3298-3303.

Gracia, L., Parra, L., Jimenez, J. M., Lloret, J., & Lorenz, P. (2020). IoT-Based Smart Irrigation Systems: An Overview on the Recent Trends on Sensors and IoT Systems for Irrigation in Precision Agriculture. *Surveys of Sensor Networks and Sensor Systems Deployments*, *20*(4), 1042. doi:10.339020041042 PMID:32075172

Ismail, S., Dawoud, D. W., Ismail, N., Marsh, R., & Alshami, A. S. (2022). Iot-Based Water Management Systems: Survey and Future Research Direction. *IEEE Access : Practical Innovations, Open Solutions*, *10*, 35942–35952. Advance online publication. doi:10.1109/ACCESS.2022.3163742

Kansara, K. (2015). Sensor Based Automated Irrigation System with IOT: A Technical Review. *International Journal of Computer Science and Information Technologies*, *6*(6), 5331–5333.

Meena, H., Nandanwar, H., Pahl, D., & Chauhan, A. (2020). IoT based perceptive monitoring and controlling an automated irrigation system. *11th International conference on computing, communication and networking technologies (ICCCNT)*. 10.1109/ICCCNT49239.2020.9225455

Mustafa, M., Abbas, A., & Bsoul, Q. (2021). Smart Irrigation System Based on the Internet of Things and the Cloud. *International Journal for Modern Trends in Science and Technology*, *7*(9), 19–24.

Nallani & Berlin Hency. (2015). Low Power Cost Effective Automatic Irrigation System. *Indian Journal of Science and Technology, 8*(23).

Narayanavaram, B., Manoj Kumar Reddy, E., & Rashmi, M. R. (2020). Arduino based Automation of Agriculture A Step towards Modernization of Agriculture. *International Conference on Electronics, Communication and Aerospace Technology (ICECA-2020).* 10.1109/ICECA49313.2020.9297546

Rawal, S. (2017, February). IoT Based Smart Irrigation System. *International Journal of Computers and Applications*, *159*(8).

Roopa Ranjani & Sravya. (2018). IoT Based Smart Irrigation System. *International Journal of Research, 5*(12).

Safdar Munir, Bajwa, Ashraf, Anwar, & Rashid. (2021). Intelligent and Smart Irrigation System Using Edge Computing and IoT. Wiley.

Taneja, K., & Bhatia, S. (2017). Automatic Irrigation System using Arduino UNO. *International Conference on Intelligent Computing and Control Systems ICICCS.*

Thaher, T., & Ishaq, I. (2020). Cloud-based Internet of Things Approach for Smart Irrigation System: Design and Implementation. *International Conference on Promising Electronic Technologies (ICPET).* 10.1109/ICPET51420.2020.00015

Xiao, Y. Z. (2021). *The Cultural Dynamics in Water Management from Ancient History to the Present Age*. IWA Publishing.

Chapter 11
COVID-19 Diagnosis and Treatment Challenges in Rural India

Shanti Chandra
Indian Institute of Information technology, India

Shivaji Kanojiya
Indian Institute of Technology (BHU), Varanasi, India

Agya Ram Verma
Govind Ballabh Pant Institute of Engineering and Technology, India

Yatendra Kumar
M.J.P. Rohilkhand University, India

ABSTRACT

COVID-19 was first identified in Wuhan, China in December 2019. India is the second most affected country of the world. About 516 million people have been affected worldwide by it so far, and approximately 6 million deaths have been caused. In India, more than 40 million people have been infected according to official data released by the government. Early detection of COVID-19 infection is the first part to deal with this pandemic. However, detection of COVID-19 infection is the biggest problem in rural India because of bad connectivity and poor medical facilities. After a certain period of time there are waves of increasing numbers of cases because of its variants. Rural India faced the problem of shortage of medical resources because of a spike in the number of cases in a region. The goal of this study is to develop a deep learning model that can help in detecting COVID-19 cases using chest x-rays, thus leading to inexpensive and fast early detection of COVID-19 to prevent its spread.

DOI: 10.4018/978-1-6684-6418-2.ch011

INTRODUCTION

COVID-19 (coronavirus disease 2019) is a highly transmissible disease caused by a virus named SARS-CoV-2. The World Health Organization (WHO) declared it as a pandemic on 11th of March 2020. It has affected more than 200 countries with more than 61.7 Cr confirmed cases and more than 6.5 million confirmed deaths. According to studies there could be annual surges in COVID-19 infections through to 2025 and beyond. As it is a highly spreadable disease its early detection plays a significant role. Some of the already used techniques are Rapid Antigen Test, RT-PCR Tests etc. Reverse-transcription polymerase chain reaction in short RT-PCR is the standard method which is widely used for detection of COVID-19 disease. But it still has many cons as it is very expensive, requires experts to conduct it and time consuming. Whenever there is spike in no of cases there is always a high demand for rapid detecting method that is accurate, fast, and cheap.

CT-Scan can be considered as an alternative approach as they are easily accessible and fast to obtain but again, they are too expensive to afford. This is not an issue for the low-income countries only, but many developed countries are also struggling to overcome this rapid testing. Today many doctors, researchers, and scientists from all over the world are working day and night to fight this pandemic. Many of them have published series of papers indicating the solution for COVID-19 detection from chest X-Ray images (Aggarwal, 2022; Chowdhury, 2020). However, there is need to easily classify the Chest X-rays as normal or Covid-19. Currently, Artificial Intelligence (AI) based solutions are being used in many biomedical health problems. Today cutting edge deep learning based techniques have allowed image recognition and classification with great speed and accuracy. To be specific, Convolutional Neural Network (CNN) has been proved to play an extremely beneficial role in feature extraction and learning from images. Therefore, an approach to classify Chest X-rays as Covid-19 or normal has been proposed in this study which can play an important role in early detection of Covid-19 in a very fast, cheap, and easily accessible manner.

LITERATURE REVIEW

Artificial Intelligence based techniques have given outstanding results and accuracy in studies related to medical image analysis. Since the arrival of covid-19 in December 2019. Researchers are analyzing deep learning-based approaches to classify Chest X-Rays as Covid positive or normal. Not only covid Because of its proven results CNN is to detect the Pneumonia, lungs diseases, tumor detection etc. which includes image recognition and classification. Several studies have achieved great results, but the findings still lack high accuracy and sensitivity which has prohibited to use this approach for production use to detect Covid-19 (Ahsan, 2021). In the paper the author proposed Covid-Net architecture. In the study the given model was trained using an open resource called COVIDx, with the help of more than 13,000 X-ray images, The accuracy of about 93.3% was achieved (Ahsan et al., 2020). Another relevant study which uses the same kind of method Transfer- and Compose-type deep CNN for COVID-19 detection using chest X-ray image classification. Which provides the results with very good sensitivity of 97.91%, and comparatively low specificity of 91.87% and have an accuracy of 95.12% (Chowdhury, 2020). Many other studies have worked on the CNN based approach to detect Covid-19 emphasizing that this can be reliable tool for doctors. Research gap related to image padding; image augmentation can be seen in most of the studies. Also, neural networks are known to require a large amount of dataset to get more accurate results. In the proposed methodology a customized CNN have been used which has layers of image padding, and model is trained including image augmentation on a large number of datasets (Aggarwal, 2022). Later in this paper accuracy has been calculated, which seems higher than the previous techniques used.

METHODOLOGY

Data Preprocessing

In this study, two databases were collected and merged for experimental purposes. Both were publicly available. One contained 1073 images which was collected from Kaggle (Rahman, 2020). It had 423 Lung opacity CXR images which were collected from RSNA (Radiological Society of North America), and 345 Viral Pneumonia images including rest other normal images.

Second source of data was open-source GitHub project in (Sarki, 2022). It contained 473 images, 143 were infected and the rest were normal. Two examples of X-Rays of are described below.

For this study, data augmentation and image padding methods were applied. Using data augmentation variety of data fed to the classifier for classification can be enhanced. In the case of Image augmentations widely used types are horizontal flip, rotation, width shift and height shift on all the extracted images from the original dataset. Vertical flip cannot be applied in case of chest X-rays because they are not vertically flipped. Along with image augmentation and enhancement, images were scaled to specific resolution of 256×256 pixels. Gather data was then split into ration 80:20 for training and testing purposes. As described above this study included rich datasets collected from trusted sources for research purposes

Figure 1. (a) Normal, (b) COVID-19 positive

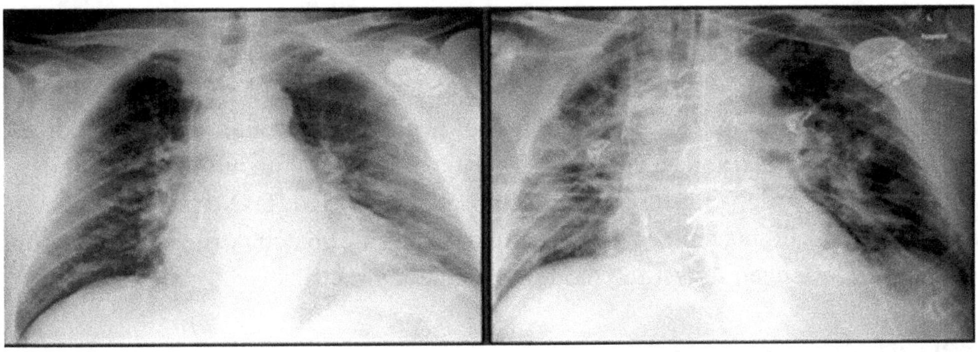

and they were preprocessed in a way to train the model in better way in terms of quality and quantity.

CNN Model

This proposed CNN architecture which is modified model of standard MobileNet has four convolution layers that take a chest X-Ray image of 256×256 as its input (Sarki, 2022). These convolution layers consist of a filter set or also known as kernel. The filters are convoluted against the input image and then it extracts features by creating a new layer thus reducing the size of the matrix. In the first most convolution layer kernel filters of 5×5×2 having a batch size of 32 are used. Then the second layer is max-pooling layer with stride of 1×1. In this layer the input is reduced to half of its size i.e., 128 × 128. Additionally, image padding layer was added in between in all the layers of the model, the output from the pooling layer is passed through a Relu activation function. The standard Relu activation function returns 0 if the input is negative or the value itself if it is positive. Mathematically,

$$R(x)= \max⟦(0,x)⟧ \tag{1}$$

The nonlinear output obtained from the previous layer is now passed into the next layer which is again convolution layer with 5 × 5 × 64 having exact 128 filters, and its stride value is the same i.e., 1 × 1. As done previously, the output gathered is again passed to a max-pooling layer with strides of 2 × 2. It again reduces the size of received input to half. That is now it becomes 64 × 64. Again, passing the output through ReLU activation, the third convolution layer with 256 filters and the kernel size 5 × 5 ×128 with 1 × 1 stride comes in action. The obtained matrix is passed to max-pooling layer, from which a tensor of shape 32×32 is obtained (Bayar, 2018).

As mentioned earlier after applying ReLU activation function it is fed into the fourth and the last convolution layer which has about 512 filters and having kernel size 5×5×256 and stride 1×1. Now it is max pooled to a size of 12×12. The obtained tensor is flattened with 26,100 neurons. At the end, to avoid network overfitting a dropout layer is applied. After that model needs to be trained with two indices signifying Covid as 0 and Normal as 1. The above CNN architecture was inspired by the standard MobileNet. MobileNet are subset of modified ResNets that are known to achieve very high accuracy whenever there is lack of batch normalization. After the training, we need to test the model for accuracy metrics.

Performance Analysis

Performance of a model can be analyzed using confusion matrix. Model can be evaluated based on accuracy, specificity, precision, and recall. In this case of binary classification, confusion matrix is composed of 2*2 matrix which has True positive are the cases of COVID-19 that are accurately predicted. False positive is used for cases of COVID-19 that were wrongly classified as normal. False negative (FN) points to instances of COVID-19 that were inaccurately categorized as normal. True negative (TN) are instances that are correctly categorized as normal (Chicco, 2021).

Using the following equations, we can calculate the accuracy, precision, specificity, and recall respectively with the help of TP, FP, FN, TN.

Accuracy is defined as the ratio of correctly identified Covid – 19 and Healthy cases from all samples. The accuracy of a model determines how correct the values can be anticipated (Alakus, 2020).

$$Accuracy = \frac{TP + TN}{TP + FP + TN + FN} \tag{2}$$

Precision or Positive Predictive value (PPV) is defined as the ratio of correctly identified Covid – 19 cases from all samples. Precision is used to determine how many instances predictions made are accurate (Alakus, 2020).

$$Precision\,/\,Positive\ Predictive\ value\left(PPV\right) = \frac{TP}{TP + FP} \tag{3}$$

Recall or Sensitivity is defined as the ratio of correctly identified Covid – 19 cases from sum correctly identified Covid – 19 and incorrectly identified Covid – 19 cases as healthy. Recall shows the model's ability to detect positive samples (Alakus, 2020).

$$Recall\,/\,Sensitivity = \frac{TP}{TP + FN} \tag{4}$$

F1 score is the overall measure of the accuracy of the model using precision and recall (Alakus, 2020).

$$F_1 = 2 \times \frac{Precision \times Recall}{Precision + Recall} \tag{5}$$

Specificity is defined as the ratio of correctly identified healthy cases from sum correctly identified healthy cases and incorrectly classified healthy cases as COVID-19, i.e., all the original healthy samples (Alakus, 2020).

$$Specificity\left(SP\right) = \frac{TN}{TN + FP} \tag{6}$$

Negative predicted value (NPV) is defined as the ratio of correctly identified healthy cases from all samples classified as healthy (Alakus, 2020).

$$NPV = \frac{TN}{TN + FN} \tag{7}$$

Results

Figure 2 displays the confusion matrix obtained from the designed model. The confusion matrix shows that the CNN model test employs the dataset's chest X-ray images, of which 642 images belong to the COVID-19 positive class and 1003 are of normal lung images.

From the above matrix calculated accuracy, precision, recall, and F1 score are as follows.

Table 1. Percentage of accuracy, precision, recall, and F1 score

Accuracy	96.2%
Precision /Positive Predictive Value (PPV)	94.4%
Recall	95.8%
F_1 Score	95.094%

Figure 2. Confusion matrix of the model

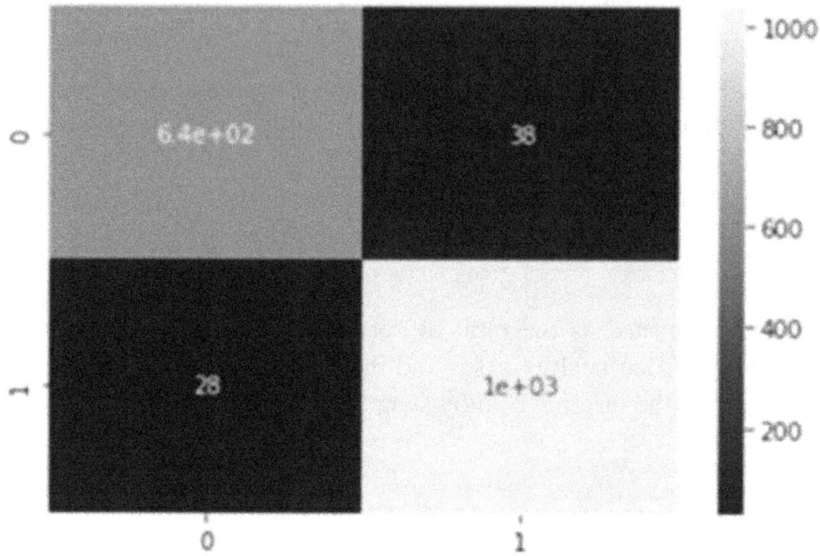

From Table 1, it can be seen that the accuracy, precision, recall and F1 score of this model is about 96.2%, 94.4%, 95.8%, and 95.094% respectively. Other parameters such as specificity, NPV can also be calculated for future purposes if required. A brief analysis on the compilation time of the above model was also made. It took approximately 319 seconds for about 1500 images.

CONCLUSION AND FUTURE WORK

In this study, CNN model was trained and presented how it shows outstanding results in the detection of COVID-19 using Chest X-Rays images. Although to improve the model a large number of images are needed, it already has achieved an accuracy of 96%. By improving the networks in the CNN model, we can achieve more accuracy. The future work for this study can be to develop a real-world API which can be further integrated with Webpage, mobile app for public can use. This Model also proves the importance of Convolutional Neural Networks in the medical field specially in image recognition. The findings can assist scientists and doctors to choose the suitable models which is required for various image recognition and classification methods, which will be important when in future there will be very limited resources and time in a pandemic situation like the present COVID-19.

Furthermore, the proposed methodology can be used for extending this binary classification problem to multi classes which may include lung-based diseases such as detecting lung cancer, Pneumonia etc. which can lead to detect any of these diseases in cheap and easy way using the existing laboratory setup especially in poor areas.

REFERENCES

Aggarwal. (2022). COVID-19 image classification using deep learning: Advances, challenges and opportunities. *Computers in Biology and Medicine*, 105350.

Ahsan. (2021). COVID-19 detection from chest X-ray images using feature fusion and deep learning. *Sensors, 21*(4), 1480.

Ahsan, M. M., Gupta, K. D., Islam, M. M., Sen, S., Rahman, M. L., & Shakhawat Hossain, M. (2020). Covid-19 symptoms detection based on nasnetmobile with explainable ai using various imaging modalities. *Machine Learning and Knowledge Extraction*, 2(4), 490–504. doi:10.3390/make2040027

Alakus. (2020). Comparison of deep learning approaches to predict COVID-19 infection. *Chaos, Solitons & Fractals, 140*, 110120.

Bayar. (2018). Constrained convolutional neural networks: A new approach towards general purpose image manipulation detection. *IEEE Transactions on Information Forensics and Security, 13*(11), 2691-2706.

Chicco. (2021). The Matthews correlation coefficient (MCC) is more reliable than balanced accuracy, bookmaker informedness, and markedness in two-class confusion matrix evaluation. *BioData Mining, 14*(1), 1-22.

Chowdhury. (2020). Can AI help in screening viral and COVID-19 pneumonia? *IEEE Access, 8*, 132665-132676.

Rahman. (2020). *COVID-19 radiography database*. https://www. kaggle. com/ tawsifurrahman/covid19-radiography-database

Sarki. (2022). Automated detection of COVID-19 through convolutional neural network using chest x-ray images. *PloS One, 17*(1).

Chapter 12
Air Pollution Effects on the Habitat of Insects and Strategic Control to Improve Productivity

Loveneet Mishra
Department of DEECE, Galgotias University, India

Usha Chauhan
Department of DEECE, Galgotias University, India

S. P. S. Chauhan
SCSE, Galgotias University, India

Harshit Singh
SCSE, Galgotias University, India

ABSTRACT

Urbanization has significantly worsened the ecosystem by releasing harmful substances and gases into the air through air pollution, which has an impact on the population demographics of insect herbivores. A pest outbreak and the formation of new pest biotypes are caused by the careless application of wide spectrum pesticides, which decimate the population of important pollinators and foragers. The potential efficiency of phloem-feeding homopteran insects such aphids is increased by pollutants like sulphur dioxides, oxides of nitrogen, fluorides, and heavy metals in the environment. Industrial melanism in moths is caused by a decline in the light-coloured lichens as a result of pollution. However, trophic amplification caused by high toxicant concentrations reduces the susceptibility of the parasitise & predators.

DOI: 10.4018/978-1-6684-6418-2.ch012

The population growth of invading alien pest species & tropical disease vectors is amplified by air pollution and the acceleration of global warming. However, steps have been taken to reduce air pollution and boost agricultural output. The fabric filtration method uses filter cakes to remove air contaminants in this manner. Successful electrostatic precipitation strategies remove charged droplets of solid or liquid from syngas where the suspended particles exist. However, methods like thermal incineration & catalytic oxidation, which essentially use a catalyst activator to transform organic chemicals in waste gas streams into CO_2, liquid water vapor, as well as inorganic-gases at extreme temps, are in use. Additional to these methods Due to their low operating and capital expenses, biofilters could be employed in conjunction with the other treatment technologies. Thus, successful reduction of air pollution can restore the health of agricultural plants and lessen food shortage.

INTRODUCTION

In essence, pollution is defined as unwelcome elements present at the incorrect location, concentration, and time. In a nutshell, pollution is the introduction of various energy forms and substances into the environment that are mostly the result of human activities and are likely to cause harm to the ecosystem or its ecological constituents, structure, and functions. Sewage, fertilizers (both organic and inorganic), siltation, pesticides, heavy metals (such as cadmium, mercury, and lead), polycyclic aromatic hydrocarbons, industrial atmospheric emissions (such as sulphur dioxide & NO_x, it may contribute to acidic precipitation; CO2 and methane are examples of greenhouse gases.), radiation (from Chernobyl disaster), heat (like the release of toxic gases due to power plant heat pollution), activities such as mining, and pesticides are just a However, because the by-products of such catastrophes are not intentionally incorporated into the ecosystem, natural tragedies cannot ever constitute the proper source of pollution. Arthropods, which include insects and other hexapods, are impacted by pollution in the same way that other living organisms are. This is because the phylum Arthropoda is the largest in the animal kingdom. Insects, however, are used in biomonitoring to reflect the broader collections of creatures in the ecosystem and serve provide surrogates for the impacts of pollution (Houlden et al.,). This allows insects to play a dual role in the monitoring process.

Effects of Air Pollution on the Insects

Insects and humans have a close relationship, whether it be in the context of agricultural production or in the context of sharing a home with insects. Humans and insects have a significant link. Insects that are able to find their way inside our houses have a major bearing on air quality that we breathe (Farnoud, 2008). It is well known that certain insects, such as German cockroaches (Blattella germanica) and house dust mites (different species of Dermatophagoides), exhibiting ubiquitous behavior, create antigens that trigger allergic reactions in people. On the other hand, the antigens Der p1 and Der f1, which are generated by dust mites, are around year-round and are not as susceptible to seasonal changes as other allergens. However, the decline of insects such as gigantic water bugs, dung rollers, copper and blue butterflies is due to the high levels of pollution caused by vehicles and the high concentrations of contaminants found in water bodies. Because giant water bugs (family Belostomatidae) and other species of water bugs feed on the larvae of frogs and mosquitoes, this has a significant impact on the ecological system's ability to remain in a state of equilibrium. In this way, these bugs manage the diseases that are transmitted by mosquitoes. The indiscriminate use of broad-spectrum herbicides and insecticides has led to a decrease in the population of blue and copper butterflies, both of which feed on the weeds that are found in agricultural fields. A recent meta-analysis found that whereas the population of herbivorous insects increases in polluted areas, the species of predatory invertebrates decreases. This is in contrast to the increase in the number of herbivorous insects. Because predatory insects are more susceptible to pollution than herbivores, pollution leads to the selective elimination of the predators, which in turn enables herbivores to grow in the enemy-free region that is devoid of biotic environmental resistance (Imperial College at Silwood Park, 1990). Experiments have also been done out to measure the level of intelligence possessed by insect parasitoids in their quest for prey.

The findings indicate that the hunting behavior & efficiency of Braconid parasitoids are greatly reduced due to O3 fumigation, but not due to NO2 or SO2 fumigations. Rather, these data point to O3 as the culprit. The presence of these pollutants in the air can interfere with olfactory receptor responses of the parasitoids, as suggested by the fact that O3 fumigation may result in a large reduction in the percentage of parasitism caused by insect parasitoids. Pollutants such as ozones (smog) and nitrate radicals formed as a result of exhaustion bind with volatile scent molecule emitted by the flowers, which chemically alters as well as degrades the molecules so that they no longer possess a sweet fragrance and are unable to attract pollinating insects to the plants. This causes the flowers to fail in their ability to produce fruit and vegetables. Therefore, air pollution has a direct influence on these pollinators and foragers since it makes it more difficult for them to locate their food, which ultimately leads to a greater mortality rate for them. The process of pollination is hampered, which in turn leads to a decrease in agricultural productivity (Farnoud, 2008).

The presence of sewage in aquatic bodies causes the expansion of more tolerant insect organisms such as blood worm larvae, dronefly maggots, and aquatic earthworms, which actually supersede the less tolerant species. Aphids, with the exception of Acyrthosiphon pisum, that feed on Vicia faba show an increased mean relative rate of growth of individuals when they feed on the food fumigated with air pollutants such as acid mists rather than clean food. This is because of an increase in both the nymphal & population growth rates in aphids. On the other hand, the environmental stress brought on by pollution, such as sulphur dioxide, oxides of nitrogen, or fluorides, encourages the growth of aphids on herbaceous & woody plants (Imperial College at Silwood Park, 1990). The higher C/N ratio in the plants is a direct consequence of the increased CO2 concentration in the atmosphere. The population of herbivorous insects will drop as a result of a lower amount of nitrogen in the plant leaf, while the aphid population will increase. The evolution of novel phenotypes in insects has been attributed to the presence of air pollution. Reasoned that of the industrial revolution and dumping of industrial waste into the environment, insects began to take on a darker, blacker appearance. As a result of the increased pollution in the air caused by coal-fired industries, a smaller population of pale phenotypes led to the evolution of black phenotypes. This phenomenon is known as "Industrial melanism," and it occurs when an increase in the melanin pigments occurs on the insect species that are living in the habitat of industrial pollution. Additionally, the changes in the percentages of dark and pale phenotypes that are determined by genes reflect natural selection. This might be a fantastic defense strategy for insects against other animals that eat them, such as predators (AWMA, 1992; Baun et al., 2004).

Control of Air Pollution.

Fabric filters, electrostatic precipitators, thermal incinerators, and biofiltration technologies are some of the tools that can be utilized in an effort to properly reduce air pollution. The following is a condensed explanation of each of the many technologies.

Fabric Filters

A baghouse, also known as a bag filter as well as fabric filter, is a type of air pollution prevention system that remove particles from the surrounding air as well as gases that are emitted as a byproduct of industrial activities. Even with extremely minute particles, the fabric filters have a collection efficiency of 99% for any and all particulate matter that passes through them (Beachler et al., 1995). Fabric filters are frequently utilized by factories producing electricity, steel, and medicine, corporations that produce both food and chemicals in order to regulate the emission of air pollutants. Fabric filters function in a manner analogous to that of a conventional vacuum cleaner. The fabrics used to make these filters can be made from a wide variety of materials, from paper and cotton to Nomex polyamide, glass fibre, Tefl, and sometimes even alloy steels. After being adsorbed on the fabric surfaces, dust-laden gases are allowed to go through the fabric bags, which may be compressed into a filter chamber with anywhere from one's to several hundred bags in each compartment. These dusts are disengaged from the fabrics by the pressure of the air & fall into the collecting hoppers, where they are removed from the environment.

Inertial collection of dust particles occurs when particulates strike aligned fibers in the direction opposite the gas flow. Interception occurs when particles, instead of crossing the liquid streamline, experience fibers because of fiber proximity (Deshusses & Cox, 2003). Maximization of where the filtering takes place is utilised by developing a system of long, thin fabric tubes (bags) to serve as the fabric filter media. Brownnian movement occurs when the particles move through the fabric filter media. The mixture of these elements leads in the Stacking of Dust Particles, which eventually lead to an increase in the resistance to the flow of gas; as a result, the filters need to be cleaned on a regular basis.

Baghouses are able to be categorized as (i)mechanical shakers because there is no positive pressure inside the shaker while it is operating, and even low pressure has the potential to impede the cleaning process. (ii)Reverse Air, in which the dusty hoarder is flushed by circulating clean air through it in the opposite direction, it creates pressure causing cracking of the dust cakes; (iii)Pulse jet, in which Over a line of bags, pressurized air is injected in brief bursts through a shared pipe.; both of these methods are used to clean the dust collector. In semi-dry desulfurization systems, baghouses may be employed as a component of a multistage gas cleaning system. In this role, they may perform the functions of a reactor in addition to those of particulate removing devices. Baghouses may be fitted with calalytic reactors, which perform the function of a chemical reactor, or they may be fitted with sonic horns, both of which are designed to complement the cleaning energy available. However, even at lower temperatures, acid gases can sometimes condense, causing the fabric filter casing to deteriorate, as well as leading to bag blindness, which prevents air from passing through the bags. The usage of refractory minerals or metallic fabrics, which can be quite pricey, can be an option for temperature regulation in environments with high temperatures. Additionally, the lifespan of the fabric may be reduced due to an increase in temperature as well as the presence of acidic and alkaline components of the gas (Cohn et al., 2006; Courtenay & Bryant, 2008).

Electrostatic Precipitator (ESP)

Electrostatic precipitators have a low-pressure loss and excellent collecting efficiency, making them ideal for usage in both indoor and industrial settings. In the manufacturing of paper, cement, power, and metal, it is put to use to filter out solid particulate matter or tiny mists from the contaminated air stream. As a result of the fact that these instruments have been in use for a considerable amount of time, they are in need of being upgraded in order to increase their performance and dependability. In general, an ESP is made up of electrodes that release electrons, They are commonly referred to as a "Corona chamber"," which is where these contaminated air flow is ionized, a collection plate, which is where the ionized suspended particles migrate, rappers or mallets type hammers, which are where accumulated particulate matter from the collecting plate is removed at periodic intervals, and hoppers, which collect the particles that have been removed at the base of the ESP and dispose of them (DIN EN 1948: Teil 1-3, 1996).

The three different types of ESP units are the dry negatively corona units, the wet negatively corona units, and the wet positive corona units. Dry negative corona units have a voltage that is intrinsically superior, and as a result, they are utilized quite frequently. The fundamentals of electrostatic precipitation are the charging of particles, the transit velocity of electric charges, and the ratio of particulates. It is possible for the performance of the ESP to suffer as a result of back corona when dealing with high-resistivity dusts, re-entrainment when dealing with low-resistivity dusts, and corona quenching when dealing with fine dusts. Pulsed energisation is beneficial for the control of high resistivity dusts, which in turn leads in reduced power usage. Pulsed energization can be used to generate non-thermal plasma, which can then be used to initiate chemical reactions that remediate gaseous pollutants such as nitrogen oxide and volatile organic chemicals (EPA, 1992; Manfrida, n.d.).

The polluted flue gas is forced to flow between two electrodes that are in the form taken by metal bars, wires, and plates within a conduit in order for electrostatic smoke precipitators to work well. A strong negative voltage charge can be found at the first electrode. Due to the fact that the particles of dirt acquire a negative charge as they go through it. The positively charged metal plates that make up the second electrode are located in the upper portion of the pipe. This part of the pipe is known as the cathode. Because opposite charges are attracted to one another, As the soot particles, which are negatively charged, are drawn to the positively charged plates, that get stuck to the plates because of the positive charge. Shaking the collecting plates periodically to remove the soot is something that can be done either by hand or by a machine, depending on your preference. Some power plants burn coal with a lower sulphur content, which results in more ash being produced, in an effort to reduce the amount of sulphur dioxide emissions that leads to acid rain. Because coal has less energy than other fuels, it must be burned at a higher rate to produce the same amount of energy, which results in an increased amount of ash. The ashes that are created from coal with a low sulphur content have a higher electrical resistivity than those produced from coal with a higher sulphur content, which makes the

ESP less effective. Since low-sulphur coal often burns at reduced temperatures and produces flue gas with a high-water content, the capacity of ESP is also impacted by these factors. This is because low-sulphur Coal requires a cooler flame than other fuels. When the particles in the gas stream are electrostatically charged, the ESP is able to work (Gammage & Berven, 1996). The charged particles are subsequently drawn to the plates or collecting devices where they are deposited. The purified air is allowed to leave the precipitator and enter the atmosphere via a stack after it has been released. When there are sufficient particles stored in the collecting devices, such particles are removed by mechanical rappers after they have been stored for a certain amount of time. Wet or dry particle bodies are collected in the hopper, and a conveyor system then transports them to an area where they can be disposed of or recycled (Pardon et al.,).

The Wet Electrostatic Precipitator, commonly known as a WESP, has the ability to simultaneously remove dusts and gaseous contaminants from the environment. However, as contaminants such as dioxins, micron particles, heavy metals, acid mists, and liquid droplets are removed, their effectiveness increases. Some examples of these contaminants include sulfuric acid mists and oil mists. WESP, despite the fact that they are extremely unreliable, can be used with air streams that are water vapour saturated (with 100% relative humidity), and It works best in conditions when the gases are wet, flammable, or sticky, and free of moisture.

Thermal Incinerations

Thermal incinerators, also known as oxidizers, are combustion devices that oxidize gaseous or vaporous organic substances, carbon monoxide, & carbon dioxide but also water from other volatile toxic gases. This enables the devices to reduce emissions of these potentially harmful air pollutants. In many ways, catalytic oxidizers and thermal oxidizers, which employ heat rather than a catalyst to speed up oxidation reactions, are extremely similar. The design is made in such a way that it contributes to the ignition of the organic constituents in the water stream through high temperature and sufficient residence time, both of which enable the complete combustion process to take place (). Additionally, the design contributes to the thorough combustion air and exhaust gas combining.

However, the Ferrox process allows for safer disposal of air pollution control leftovers produced by municipal solid waste incineration units. Because an iron oxide matrix is used in the process, soluble salts can be removed and heavy metals can be held in place. The thermal treatment that takes place within the combustion chamber results in an additional level of stability. Incineration involves the ignition of organic substances in the waste products, resulting in ash, exhaust gas, and heat. The ashes that are released are the inorganic ingredients that have the potential to be transported by the exhaust gases as particulates. The exhausted gases are cleansed of any contaminants before being released into the atmosphere, and the heat that is produced is used to generate electricity. It is possible that thermal incineration with energy recovery is comparable to a number of other processes, including gasification, pyrolysis, and anaerobic digestion. The energy recovered by incineration is in the shape of high-temperature heat, whereas the main source of energy that is created through gasification is often combustible gas, but the processes are otherwise quite similar. The history and development of municipal incinerator air pollution control shows that Milwaukee, Wisconsin has been a pioneer throughout the process of creating a reliable pollution prevention system for municipal incinerators where they use Green Bay Avenue Incinerator and Lincoln Avenue Incinerator. This is evidenced by the fact that Milwaukee uses Green Bay Avenue Incinerator and Lincoln Avenue Incinerator. However, in order to remove the particulate matter and inorganic acids from the flue gas, air pollution control equipment such as scrubbers are also utilized.

Bag houses, venturi scrubbers, and inonizing wet scrubbers are some of the components or devices that may be introduced into the waste treatment system in order to manage the particles. These components and devices are selected in accordance with the elements of the flue gas. Flue gas can have its temperature lowered by quench systems, conditioning towers, and heat exchangers (Heredity, 2013). Absorbers and scrubbers are responsible for cleaning the flue gas once it has been contaminated by acid mists.

There are two different reaction mechanisms that can be used in incinerators to minimize the amount of nitrogen oxides in exhausted gas. These are known as selective non catalytic reduction (SNCR) & selective catalytic reduction (SCR). In SNCR method, NH3 or, for systems of a smaller scale, urea could be employed as reductants. In this process, urea transforms into NH3 at temperatures between 850 and 1050 degrees Celsius. It is possible to represent the equations as:

NH_2-CO-NH_2 + H_2O ® $2NH_3$+ CO_2; $4NO$+$4NH_3$+O_2® $4N_2$+$6H_2O$; $6NO_2$+$8NH_3$® $7N_2$+$12H_2O$

A solid catalyst is utilized at temperatures spectrum from 200°C - 400°C in the presence of NH3 throughout the SCR process. Because acidic media catalysts become inactive at higher temperatures, the SCR module cannot be fitted until after the particulate materials and acidic components that were previously present in the flue gas have been eliminated. The total reaction can be broken down like this:

$4NO$+$4NH_3$+O_2® $4N_2$+$6H_2O$; $6NO$+ $4NH_3$® $5N_2$+ $6H_2O$; $4NO_2$+$8NH_3$+O_2® $3N_2$+$6H_2O$

Biofilters

If the pollutant is biodegradable in the allotted amount of time, at the appropriate concentrations, and If the ambient air temperature is adequate for sustaining life, then a biofilter, also known as a bio-oxidation system, is a relatively straightforward piece of equipment that functions and provides a cost-effective solution (Industrial Melanism, 2012; Kidd, 1991). When it comes to the control of pollution across a vast volume of air, a biofilter might be the only approach that is both cost-effective and efficient when there is no secondary contamination. The removal of air pollutants by microorganisms, which are used in biofilters, contributes to the filtration and purification of air. In this case, the pollutant is transmitted to a slim biofilm on the outside of the packing material. as the air that is loaded with the particle components travels through a packed bed. In the process of degrading the pollutant into more manageable forms such as water, heat, and biomass, the biofilm traps and encases microorganisms such as bacteria and fungi, which are then immobilized.

hazardous organic compound + O_2®CO_2 + H_2O + Heat + Biomass

Toxic gases and trace amounts of volatile organic compounds (VOCs) can be eliminated with the use of a biofilter (VOC). Biodegradation is a method for reducing air pollution by breaking down toxins through the activity of microorganisms trapped in a biofilm of a thin layer of water. Biodegradation of pollutants is a method of reducing air pollution that is sometimes associated with the term "biofilter" (Quina et al., 2011). The treated substances are often a mixture of volatile organic molecules and a wide variety of sulphur-containing compounds. Maintaining the correct moisture concentration and temperature throughout the entire system is the primary obstacle that must be overcome in order to achieve optimal performance from the biofilter.

Before the air is allowed to enter the bed, it is often humidified using a water spraying mechanism, a humidification chamber, and bio scrubbers. Toxic pollutants such as hydrogen sulphide (H2S), diminished sulphur compounds (RSCs), and biodegradable chlorinated compounds (BCCs), additional byproducts include innocuous sulphate or chloride. In comparison to biofilters, biotrickling filters are more complicated, but they are typically more successful in acidic byproducts of particle chemical processing, such as hydrogen sulphide.

Industries that employ biofilters to reduce air pollution include those dealing with food and animal products, pharmaceuticals, wood goods, paint & coating application, and manufacture, as well as resin manufacturing & implementation. Biofilters are also used in the pharmaceutical industry, the wood product industry, the paint and coating industry, and the resin industry (Proceedings of the 11th International Symposium on Insect-Plant Relationships, n.d.). These biofiters can be built in a wide variety of forms, in a largely unregulated open area, with the necessary piping and delivery systems hidden beneath the ground; these are the primary advantages of biofilters. Other advantages of biofilters include reduced capital costs, lower operating costs, lesser use of chemicals, and no combustion source. The treatment efficiency of biofilters can be more than 90% even for lower concentrations of the contaminants they are designed to remove, making them useful for the removal of odorous gases, hazardous chemicals, and volatile organic compounds. The process of biofiltration can be carried out using a variety of different kinds of media, microbial species, and operational systems. Nevertheless, biofilters have a number of significant limitations. Because of the filtration process and the growth of the biomass, a substantial amount of material accumulates in the filtration, leading to clogging and flow channelling (Hsu, 1999). Therefore, to a large extent, it depends on the nature of the application and the medium that is used for the development of microorganisms, clogging can be reduced using either physical or chemical means. It is also possible to use chemicals such as those that are oxidizing (peroxides, ozone), or biocides.

CONCLUSION

Therefore, a successful control of air pollution by the technologies discussed above would lead to a better environment for living organisms, particularly arthropods, which will help in the general preservation of the environment and the normal functioning of food chains from higher trophic levels to lower. The regulation of air pollution would, in the long run, be helpful in putting a stop to global warming, which has been showing worrying signs in recent times in the form of secondary pest and disease outbreaks, the emergence of invasive or alien pests, pest resurgence and resistance, and, most importantly, biotype development in the pests and diseases that are the major factors in the decline of agricultural production.

REFERENCES

AWMA. (1992). *Air & Waste Management Association, Air Pollution Engineering Manual*. Van Nostrand Reinhold.

Baun, D. L., Bergfeldt, B., Mogensen, E. P. B., Vehlow, J., & Christensen, T. H. (2004). Thermal treatment of stabilised APC-residue in a waste incinerator pilot plant. Part 2: Leaching characteristics of bottom ashes. *Waste Management & Research*, *22*, 58–68. doi:10.1177/0734242X04042284 PMID:15113115

Beachler, D. S., Joseph, J., & Pompelia, M. (1995). *Lesson 2: Fabric Filter Bag Cleaning*. North Carolina State University.

Cohn, R. D., Arbes, S. J. Jr, Jaramillo, R., Reid, L. H., & Zeldin, D. C. (2006). National prevalence and exposure risk for cockroach allergen in U.S. households. *Environmental Health Perspectives*, *114*(4), 522–526. doi:10.1289/ehp.8561 PMID:16581539

Courtenay, J., & Bryant, M. (2008). *Pleated cartridges provide increased baghouse capacity and improved filter performance.* Academic Press.

Deshusses, M., & Cox, H. H. J. (2003). *Biotrickling Filters for Air Pollution Control Chapter*. doi:10.1002/0471263397.env105

DIN EN 1948: Teil 1-3. (1996). *Bestimmung der Massenkonzentration von PCDD/F*. Determination of Massconcentration of PCDD/F. Deutsche Norm. VDI/DIN Handbuch Reinhaltung der Luft, Band 5, Beuth Verlag.

EPA. (1992). *Control Techniques for Volatile Organic Emissions from Stationary Sources*. EPA-453/R-92-018.

Farnoud, A. (2008). *Electrostatic Removal of Diesel Particulate Matter*. ProQuest.

Gammage, R. B., & Berven, B. A. (1996). *Indoor Air and Human Health* (2nd ed.). Lewis Publishers.

Heredity. (2013). *The peppered moth and industrial melanism: Evolution of a natural selection case study.* Academic Press.

Houlden, G., McNeill, S., Aminu-Kano, M., & Bell, J. N. B. (n.d.). *Air pollution and agricultural aphid pests. I: Fumigation experiments with SO_2 and NO_2*. Academic Press.

Hsu, S-L. (Ed.). (1999). *Brownfields and Property Values*. United States Environmental Protection Agency.

Imperial College at Silwood Park. (1990). Environmental Pollution. Academic Press.

Industrial Melanism. (2012). College of William and Mary. doi:10.1002/9780470015902.a0001788.pub3

Kidd, N. A. C. (1991). The implications of air pollution for conifer aphid population dynamics: A simulation analysis. *Journal of Applied Entomology*, *111*(1-5), 166–171. doi:10.1111/j.1439-0418.1991.tb00307.x

Manfrida, J. P. (n.d.). Insects, Allergy and Indoor Air Quality. *The Environmental Reporter, 9*(2).

Pardon, G., Ladhani, L., Sandstrom, N., & van der Wijngaart, W. (n.d.). Aerosol sampling using an electrostatic precipitator integrated with a microfluidic interface, Sensors and actuators. *B, Chemical, 212*, 344-352. doi:10.1016/j.snb.2015.02.008

Proceedings of the 11th International Symposium on Insect-Plant Relationships. (n.d.). *Series Entomologica*.

Quina, M. J., Bordado, J. C. M., & Quinta-Ferreira, R. M. (2011). Air Pollution Control in Municipal Solid Waste Incinerators, The Impact of Air Pollution on Health, Economy, Environment and Agricultural Sources. Academic Press.

Chapter 13
Automation of Street Light of Adarsh Gram With Manual Intervention and Automatic Luminosity Control With Traffic Flow

Rahul Kumar Singh
Indian Institute of Information Technology, Design, and Manufacturing, Jabalpur, India

Anju Bala
Dr. Bhim Rao Ambedkar Engineering College of Information Technology, India

Pratibha Verma
Rajkiya Degree College, Puranpur, India

ABSTRACT

Sixty-five percent of India's population still lives in the villages. Even after years of independence, the condition of villages are not good for living. The basic facilities and amenities like drinking water, hospital, and electricity are not available in most villages in India. The streets of villages are mostly dark either due to no light or malfunctioning lights. This leads women and children facing problems at night while going outside. The authors try to overcome this problem with the concept of electrification of street lights with minimum power consumption by automatic control along with intensity control with simple, cheap, durable, and easily reparable components with and without manual intervention and further expanding into solar power.

DOI: 10.4018/978-1-6684-6418-2.ch013

INTRODUCTION

An intuitive program obsessed with the idea of our Father of Nation "Mahatma Gandhi" on comprehensive rural development, Smart Village (Adarsh Gram) is a concept widely accepted and recognized over the whole national level, including state union territories, driven with the help of local governments of India (India. gov, n.d.; Pmagy.gov, n.d.-a; Pmagy.gov, n.d.-b). It is adapted from Mahatma Gandhi's 150th birth anniversary as an ideal of Adarsh Gram and Swaraj in Figure 1. During this initiative, the prime objective of the Model Village Scheme is that every Member of Parliament and Minister to endorse a rural village by 2019 by working on social cultural development and bringing awareness among the people of villages and transforming it into a model under the Sansad Adarsh Gram Yojana guidelines (Adafruit, n.d.; Light Dependent Resistors, 1997). The target setup under the Sansad Adarsh Gram Yojana is to forge the rural village from the root level as an integrated village development deploying the strategy that takes into account the personality, sanitation, plantation, water supply, electrification, internal roads, health, financial, human, social, and economic spheres.

Figure 1. Model for smart village under Sansad Adarsh Gram Yojana
Source: Pmagy.gov (n.d.-a)

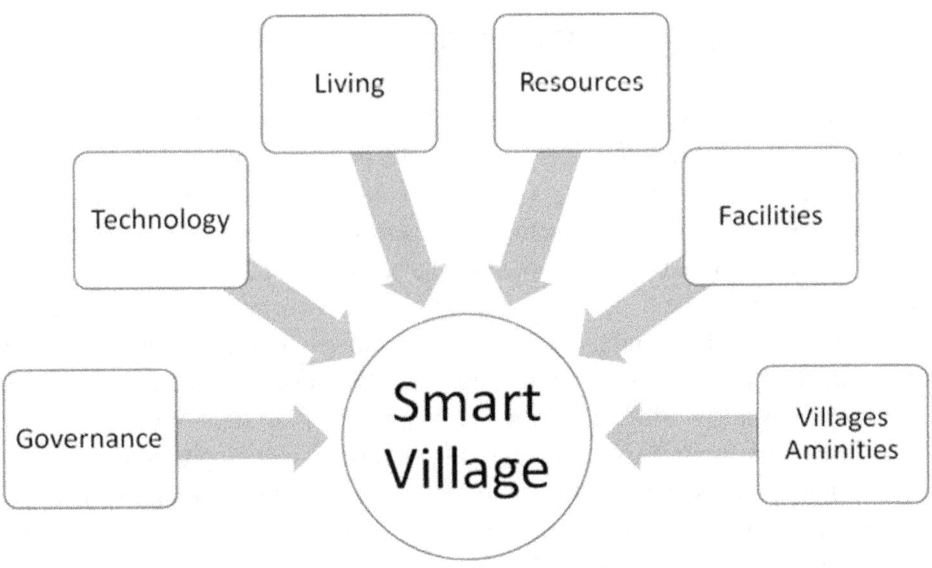

Mahatma Gandhi's put his idea and long-term view on Adarsh Gram (a model village) and Gram Swaraj (village self-rule/independence) which served as the inspiration for Smart Village India and his idea was brought to realistic on his 150th anniversary. Gandhi advocates for integrated rural development to benefit the bulk of the population as the main initiative following India's independence in 1947 in his two works, Hind Swaraj and Gram (Village) Swaraj (Smart Village, n.d.).

As part of the Smart Villages Initiative under Sansad Adarsh Gram Yojana guidelines, various self-running local and global organizations, countries, global organization, business tycoons and businesspeople, celebrities have come forward for improving the status of rural areas by adopting villages and making efforts to promote sustainable development by constructing internal roads, planting trees, conserving water, and providing basic amenities like sanitation and street lighting (Passive Infrared Sensor, n.d.; Photorisistor, n.d.). During the day times villages have access to the roads and streets but during night non-availability of street lights malfunctioning street lights or poor light facilities or no electricity in the village, the street remains dark in night. These problems are faced mostly by females. To overcome we have come up with the concept of electrification of street light with minimum power consumption by automatic and simple cheap durable easy reparable components with and without manual intervention. In case, if villages are not electrified this can be extended to solar powered panel.

LIGHT DETECTOR RESISTOR MODULE

A light detector resistor which is known by various names like Photocell, photoconductor, photoresistor, short name LDR, or photo-conductive cell (Stefan, 1879; Wien, 1897). This device fall under the category of passive component that has a non-linear property of resistance variation with the respect to falling light over its conduction junction or window or the sensitive surface Table 1. All these photo conductive cell devices and materials have this variation in resistance with increase in incident light intensity; which is commonly coined as the property of photoconductivity. Using this property and resistance variation with luminosity, the concept is widely and worldwide utilized in the production of light-sensitive detector circuits, light darkening activated, and light activation switching modules working in the category of a resistance semiconductor Looking at the working methodology if we look it's during no light or absence of light (dark), the resistance of an LDR is in the range of several mega ohms (MΩ) Figure 2 on the other hand in the presence of light, the resistance of an LDR falls as per the property of resistor semiconductor in a range of as a few hundred ohms. The sunlight falling on a conductive window or sensitive surface passes through a defined frequency, and the energy is absorbed by the photon present within, afterward; this photon loses its energy to the bound electron and these energy-absorbed electrons jump into the conduction band as per band theory. This conduction band-free electron and their whole partners will start the conduction of electricity, hence creating a lower resistance path and resistance of the device will come down. Since different materials have different conduction properties and sensitivity properties to frequency. Unique photoconductive cells respond significantly differently to photons in certain wavelength bands Figure 3.

Table 1. Table for various luminous sources

Available Light Source	luminous Flux per Unit Area (Lux)
Full Moon light	0.1
60W normal Bulb at 1 mtrs distance	50
Fluorescent tube/ Tube light /CFL	500
Full day sunlight	30,000

Figure 2. Variation of resistance with luminosity

Figure 3. Spectral response

Figure 4. A commercial most commonly used CdS photocell from a streetlight

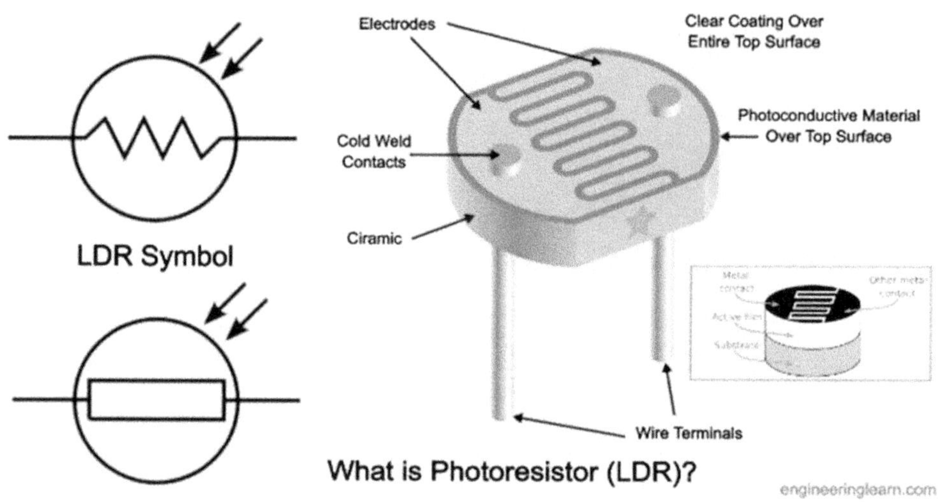

As explained above, under the category of Resistor Semiconductor a LDR is comparatively less light-sensitive than a photodiode or a phototransistor, but its application to detect sunlight is acceptable. If we compare the response time of photoresistors/LDR devices it has a certain extent of latency between sensitive surfaces exposed to daylight and the subsequent decrease in the resistance due to the behavior of the valence band and conduction band. This concept of energy absorption by photon and subsequent losing energy to electrons which was earlier explained lead to this latency and this value came around 10 milliseconds. The lag time when going from lit to dark environments is even greater than, often as long as one second in (Table 2).

Table 2. Source datasheet of LDR

Electrical characteristics

Parameter	Conditions	Min.	Typ.	Max.	Units
Cell resistance	10 lux 100 lux	20 -	- 5	100 -	kΩ kΩ
Dark resistance	10 lux after 10 sec	20	-	-	MΩ
Spectral response	-	-	550	-	nm
Rise time	10ftc	-	45	-	ms
Fall time	10ftc	-	55	-	ms

Implementation

The intramural elements of a photoconductive cell of LDR for a typically automated streetlight Figure 4 are depicted. The photoresistor is kept facing directly toward the sunlight or ambiance light where the maximum intensity of light is falling without any obstruction and this falling light controls the current flow through the transistors figure, which further operates the relay that opens the main power contacts to switch on/off the lights. At night, the LDR detectors have no sunlight, forcing the resistance of the photodetector to vary as per its property of a resistor semiconductor. Thus, closing the power contacts by the transistor leads to electrifying the streetlight through the same Relay Figure 5. This is one of the parts of the working module of our work in this project.

Figure 5. What is photoresistor?

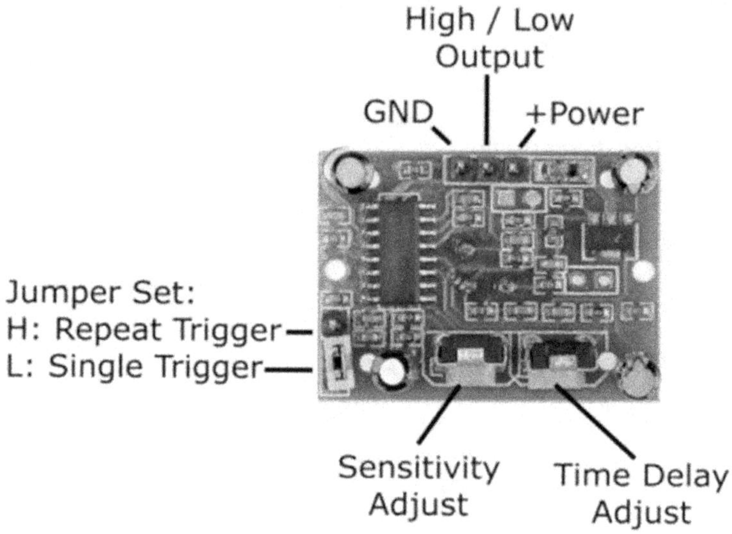

LDR Usages

1. Light Detector Resistors are cheap, low powered, and easily available, abundantly available found its application in mostly automation of streetlight to control its ON/OFF of lights. The daylight/Sunlight incident on the conductive surface of LDR module turns off the street light during daytime. By this, no manual intervention is required and the streetlight is turned off and electric energy is saved by ensuring the light is only turned ON as per the requirement.
2. Arrays of photo resistors are utilized in Security systems which used laser light.
3. LDR found their usages in Audio compressor circuits. Audio compressors are used to reduce the gain when it is found that the amplitude of the incoming signal is more than the required set value. This generally amplifies the low and soft sounds without clipping the loud sounds.
4. These are found in infrared astronomy using infrared cameras
5. Used in medical equipment and places where light failing circuit are installed to detect the failures.
6. These sensors are used in smoke detection using the luminosity concept.
7. Modern television and mobile screen used these LDR sensors for automatic contrast and brightness control.

8. optical coding or optical communication is a field where LDR found its vast applications.

PASSIVE INFRARED SENSOR MODULE

The Passive Infrared Sensor module coined commonly as PIR sensor via HCSR-501 generic module in Figure 6 is an electronically designed passive module containing a sensor that measures the infrared (IR) light radiating from the subject its field of sight or vicinity. The most common application and widely used is for the detection of motion and utilization as a PIR-based motion detector. These devices find their way in low-cost devices with accuracy, mostly in security alarms, burglar alarms, unattended area intruder detectors, security breaches, and automatic lighting applications in Figure 7.

Figure 6. Passive infrared sensor

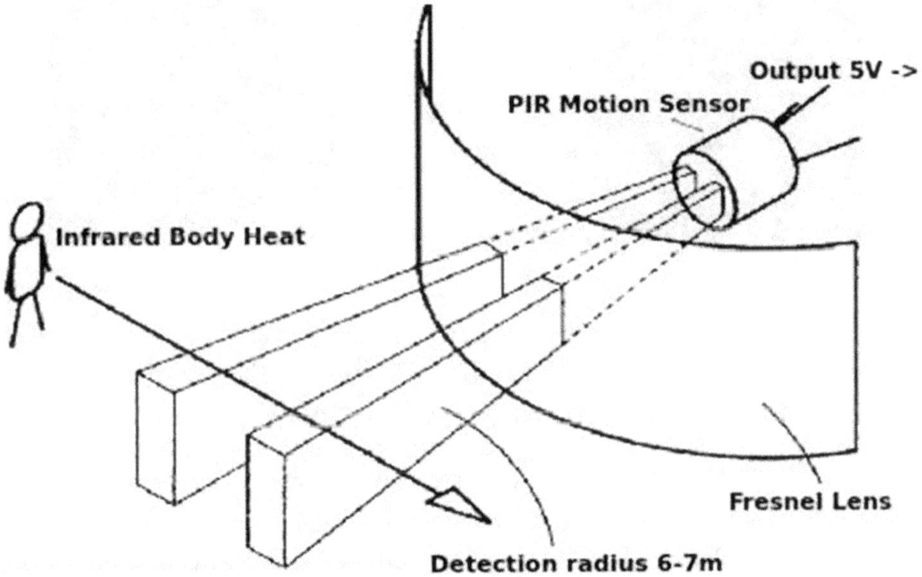

A Passive infrared detector helps in detecting the movement in its area of surveillance or vicinity, while it is unable to track the type of object or subject that has passed in that zone, or it is unable to identify the exact what has been intruded in that area. To get the exact information of the object/ subject that has passed through the area of vicinity an imaging-based IR in Figure 8 sensor like a thermographic camera, thermal camera, or thermal images can be utilized. Further, these thermal imagers or thermal-based camera is like ordinary camera except that ordinary camera used visible light for their operations. While the thermal camera utilized the wavelength from the range of 1,000nm to 14,000nm.

Figure 7. Human detection using PIR module

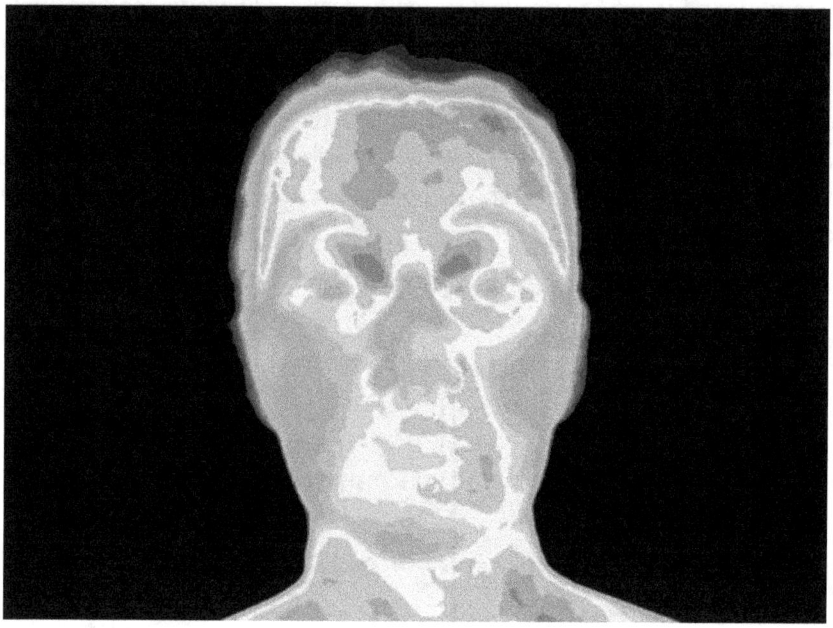

A PIR sensor is coined as "PIR", or somewhere "PID", for "passive infrared detector". The term "passive", and "infrared", refers to the information as these devices do not have the property of radiation as the Active component does rather than fully working on the detection method using the Fresnel lens attached to the sensitive surface. The infrared discovered in 1800, derived its name from the Latin prefix for "below". They work on the concept that each and every object radiates a certain amount of black body radiation as a function of temperature (Kirchoff, 1859; Mehra & Rechenberg, 1982). The four main laws that govern these properties and working of InfraRed Radiation and thermal imaging: are Stefan–Boltzmann law, Kirchhoff's law of thermal radiation (Fresnel, 1821), Wien's displacement law, and Planck's radiation law (Bernhard, 2019).

Figure 8. Thermal image of a human face

Operation

A Passive Infrared or pyroelectric-based sensor (used here HCSR501) is able to detect the changes in the amount of infrared radiation falling through the pyroelectric sensor module which is collected and converged at a point by the Fresnel lens in Figure 9. The amount of radiation as explained about varies accordingly and it depends on the temperature and surface characteristics of the subjects which are present in the area of vicinity of the pyroelectric sensor. If an object, such as a person or bullock cart of females, in our project for street light activation, passes over the street or road in the Fresnel angle area covered by the PIR module, the temperature which was constant due to no intrusion detected till now, suddenly at that point in the pyroelectric sensor's field of sight will start to increases from ambient temperature to body temperature of above said subjects/objects. The sensor collecting this radiation through the Fresnel lens converts this resulting change in the total collected infrared radiation as a change in the output voltage via module, and this triggers the detection of objects. This detection will further trigger our Relay module for the automation of street lights. Since this radiation pattern or black body radiation is a characteristic property of an object may have a different infrared emission pattern and the same emission pattern to be detected by PIR module with respect to the surrounding around that object, thus it will trigger the pyroelectric detector.

Figure 9. The inner structure of PIR module Fresnel lens

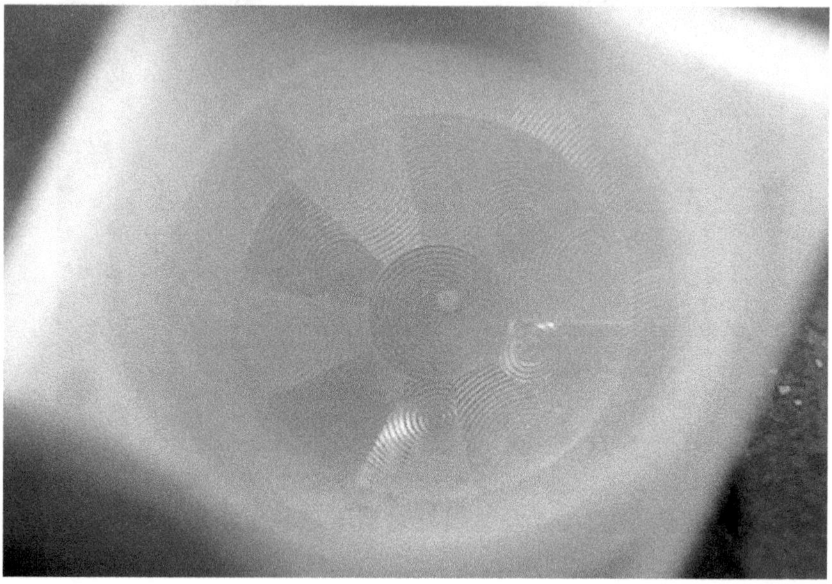

Figure 10. Fresnel lens

Fresnel Lens Figure 10 is a type of lens that falls under the category of the composite lens. It was designed during year 1788-1827 by French physicist Augustin-Jean Fresnel. Henceforward its name is coined as Fresnel lens. It was designed basically for lighthouse application. After the invention of this lens, it has been popularly named the "invention that saved a million ships" in Figure 11.

Figure 11. Ships

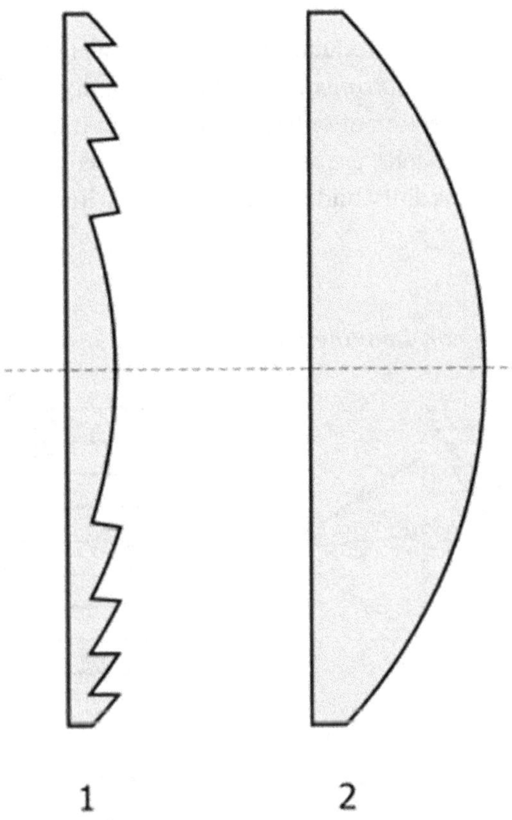

A composite lens as per its structure design has a peculiar property to the conventional lens which has a larger aperture and small focal lengths which used to have a large mass and volume for its construction and design. The overall weight and thickness of a Fresnel lens can be reduced drastically as it was used in earlier conventional lens. This was great usage for the mobility and installation in light house.

In the design of a Fresnel lens, the full length of lens is divided into a set of concentric annular sections with these annular section reducing the weight and mass. As per theory of Fresnel lens, the ideal number of concentric Fresnel lens is infinite. This design will reduced the continuous surface as in the overall standard conventional lens to stepwise a discontinuous surface within them maintaining the same curvature. If there is constrain in regard of available surface the curved surface is reduced to flat surface with different angles along various sections. These lens acts like prism from inner surfaces with the advanced in the technology now more compact and complex monolithic structure of Fresel lens can be 3-D printed and CNC milled.

Since Fresnel lens allows the reduction in mass and weight by overall change the design and structure thus compromising the image quality but herein our projects we are restricted to the collection of radiation rather than imaging thermal detection camera. Still conventional lens are used for better imaging. To distinguish the application of Fresnel lens falls under two categories imaging and non-Imaging Lens in Figure 12.

Figure 12. Imaging and non-Imaging lens

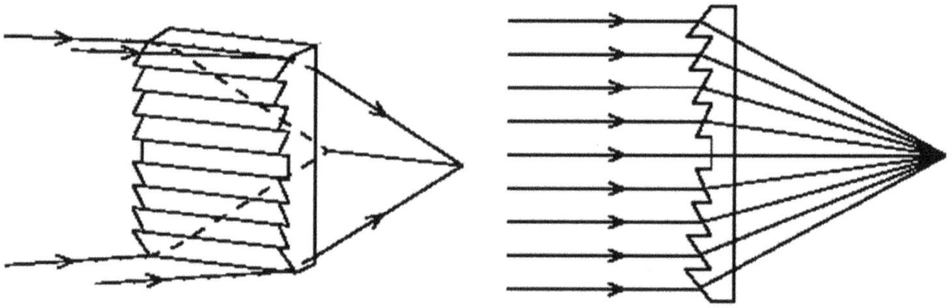

In our project we have utilized the Fresnel lens (HCSR501) module which is pre-equipped with the Spherical dome shaped Lens. The lens has effective range of about 10 meters (30 feet), and a field of sight which fall under the 180°. Modules ranging from the field of sight range 360° also available for application. Here our operation region is restricted to the road which is enough under the region of 180°. As explained above about the Fresnel lens it is found that effective range totally depended on the type of Fresnel lens, with single segment mirrors design PIR module we can effectively increase our range for detection likely up to 100 feet or 30 meters. Can be utilized further to detect fire and carnivorous in village premises. Changing the current orientation or utilizing the reversible orientation mirror that will allow using it's as a broad coverage or restrict to very narrow coverage. The overall change in the mirror orientation leads us to shape the coverage area or change the angle of view.

PIR (HC-SR501) Module

HC-SR501 in Table 3 is a pyro electric-based module that is based on Infrared Radiation technology. It has an automatic control circuitry that comes with, high sensitivity, high reliability, and ultra-low-voltage operating mode, highly durable, and easily replaceable. This module mostly found its application in a remote location, unattended location, and area for auto-detection for equipment rooms, in the region where devices can be run over battery or low power PIRs modules have a pyroelectric sensor (that can be identified as the round metal under the Fresnel lens housed inside the rectangular frame *that* is used to detect the infrared radiation. As per black body radiation every object emits some level of thermal radiation, the hotter the object emits more and the colder the object emits the lesser the radiation. But here are more concerned to detect the motion of objects rather than detecting the IR levels hence the module is split into two halves. The two halves are such wired to the circuitry that so that they cancel each other out each other in normal mode. While if one half detected the radiation as compared to the other half whether it is less or more the output will trigger from swing low to swing high and vice versa.

Table 3. Specification

Sr no	Parameter	Values
	Voltage	: 5V – 20V
	Power Consumption:	65mA
	TTL output:	3.3V, 0V
	Lock time:	0.2 sec
	Trigger methods:	L – disable repeat trigger, H enable repeat trigge
	Sensing range:	less than 120 degree, within 7 meters
	Temperature:	– 15 ~ +70
	Dimension:	32*24 mm
	Lens dimension in diameter:	23mm

Whether we are looking the hobby projects or commercial products that are required to detect when an intruder person, PIR modules are of great use.

Table 4. Absolute max. rating

Absolute max. ratings

Description	Condition	Range	Unit
Supply voltage	--	3 ~ 5	V
Input voltage	--	Vss-0.3~ Vdd+0.3	V
Output current	Vdd=5V	10	mA
Operating temperature	--	-20 ~ +70	°C
Storage temperature	--	-40 ~ +125	°C

PIR module Application

1. It is used for the automatic sense of light for the mezzanine floor.
2. These are installed in lavatory rooms for lighting purposes, auto flush of the toilet.
3. PIR sensors found their application in metro undercroft area.
4. PIR module is installed in the basement area, warehouse, garage, etc. in house where lighting are automatic ON/OFF as per person entering and leaving.

PIR (HC-SR501) Module Features

1. **Automatic Initiation:** If you enter the area of the vicinity the output goes to high and as persons leave the sensing ranges the output goes to low.
2. **Photosensitive Control:** Can be set as per the intensity of light.
3. **Temperature Control:** Can be set as per ambience temperature.
4. **Triggered methodology:** Jumper triggerable.
   a. **Non-Repeatable Trigger:** When the Pyrosensor output goes to high, and the delay time is over, the output is driven automatically from higher level to lower level.
   b. **Repeatable Trigger:** When the Pyrosensor output goes to high, along with the delay period, also there is person/subject over the road in its area of vicinity, the output will holds its value of high until the people/

subject left after the delay on the higher level goes low (the sensor module in this repeatable trigger mode will detects a time delay period, with automatically stretch out for every intruder activity).

1. PIR sensor comes with operating voltage range: default voltage DC4.5V-20V.
2. **Power Consumption:** During ideal mode or static mode the current consumption <50 microamps.
3. **Output High Signal:** The module comes with plugin module and PINS add the features to extend its operation utility.

RESEARCH ELABORATIONS

In this system of automatic light control of village we have keep following points while implementing this idea:

1. Villages still have to be economical as most pf villages still have an issue of load shedding. so, light will glow only when its only getting dark using the light detector sensor. The light detector sensor will be activated only when it is getting dark. This concept sounds promising, but it comes with a traditional drawback of power wastage on the streets where full power of light is not required.
2. In our paper we have overcome this by reducing the power consumption by lighting only 30 percentages of LEDS from panel and keeping rest LED OFF. A smart feedback system is utilized for this for this dimming LED in street. This dimmed street light provides the visibility in street so that no street will be under dark zone.
3. If there is an intruder detected in the area of the vicinity of the Street light. The PIR sensor installed on every alternate street light will increase the intensity of the street light to the 100% for that duration of time to light the street to 100%. since the alternate PIR sensor are used so that light intensity will available up to 3 lights with no lag and discontinuous in light on street Figure 13.
4. The villages are remote places where early maintenance and Since we have utilized dual LED strip in a single panel for each street light, so we can have a maintenance free type LED strip. This will extra security over which if any led panel failed or get fused out the other strip from panel can be used.

BLOCK DIAGRAM

Figure 13. Working block diagram of circuitry

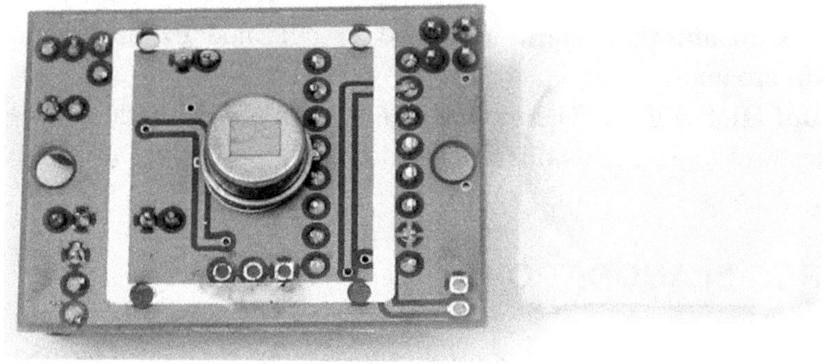

Figure 14. Kicad model of the circuit

RESULTS

The results were taken References and following finding were incurred LDR were quiet effective while detecting the sunlight and working well in dark.

PIR sensor or thermal sensor was tested to detect the intrusion for the person in its are for light intensity to rise and found its response a little delay this delay is incorporated as it will be sure that light wont light up unnecessary.

Figure 15. 3D prototype module of the circuit

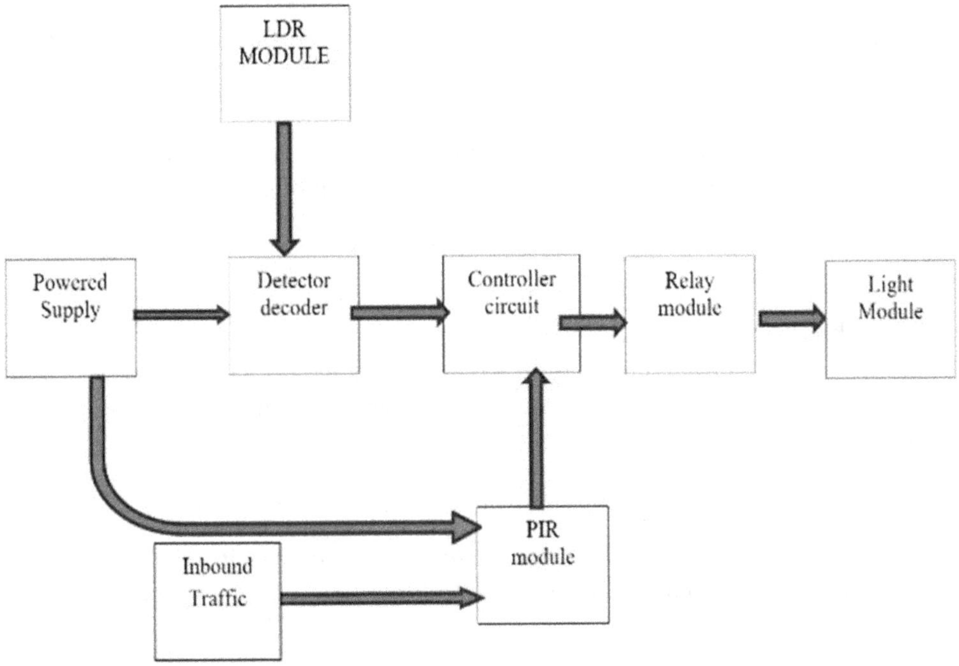

FUTURE

This circuit Figure 14 is designed using basic and readily available component that does not require any specific knowledge of the electronics and component so it can be easily repairable and changeable.

All components are available in module so any module that goes defective can be replaced easily Figure 15.

Since LED are driven from DC supply hence this circuit can be further extended over to solar power supply in respect of traditional Bulb that require AC power supply.

APPLICATION

Street lighting in villages.
Farm lighting using solar powered.
Terrace lightup when intruder detected.

CONCLUSION

1. In this system of automatic light control of village we have used two control units to drive led source so that power consumption is less and tradition bulb method is eliminated.
2. All components are low powered component so no issue for power wastage and can even work in villages with low voltage supply.

REFERENCES

Adafruit. (n.d.). https://cdn-learn.adafruit.com/downloads/pdf/pir-passive-infrared-proximity-motion-sensor.pdf

Bernhard, A. (2019). *The invention that saved a million ships*. BBC.

Fresnel. (1821). Note sur le calcul des teintes que la polarisation développe dans les lames cristallisées. *Annales de Chimie et de Physique, 17*, 102–11. doi:10.5281/zenodo.4058004

India.gov. (n.d.). https://www.india.gov.in/spotlight/saansad-adarsh-gram-yojana

Kirchoff, G. (1859). *Kirchoff laws of thermal radiation*. Monatsberichte der Akademie der Wis-senschaften zu Berlin.

Light Dependent Resistors. (1997). *NORP12 RS stock number 651-507 NSL19-M51 RS stock number 596-14*. https://components101.com/sites/default/files/component_datasheet/LDR%20Datasheet.pdf

Mehra, J., & Rechenberg, H. (1982). *The Historical Development of Quantum Theory* (Vol. 1). Springer-Verlag. doi:10.1007/978-1-4612-5783-7

Passive Infrared Sensor. (n.d.). In *Wikipedia*. https://en.wikipedia.org/wiki/Passive_infrared_sensor

Photorisistor. (n.d.). In *Wikipedia*. https://en.wikipedia.org/wiki/Photoresistor

Pmagy.gov. (n.d.-a). https://pmagy.gov.in/new-dashboard/Districts-Villages-covered

Pmagy.gov. (n.d.-b). https://pmagy.gov.in/new-dashboard/Docs/PMAGY-CSMC-Meeting-12022020.pdf

Robitaille, P-M. (2009). *Kirchhoff's Law of Thermal Emission: 150 Years*. Department of Radiology, The Ohio State University.

Smart Village. (n.d.). https://www.smartvillage.biz/

Stefan, J. (1879). Über die Beziehung zwischen der Wärmestrahlung und der Temperatur [On the relationship between heat radiation and temperature]. *Sitzungsberichte der Mathematisch-naturwissenschaftlichen Classe der Kaiserlichen Akademie der Wissenschaften*, *79*, 391–428.

Wien, W. (1897). On the division of energy in the emission-spectrum of a black body. *Philosophical Magazine*, *43*(262), 214–220. doi:10.1080/14786449708620983

Chapter 14
Advanced Detection:
Fungi-Bacterial Diseases in Plants Using Modified Deep Neural Network and DSURF

Pooja Singh
SCSE, Galgotias University, India

Usha Chauhan
DEECE, Galgotias University, India

S. P. S. Chauhan
SCSE, Galgotias University, India

Harshit Singh
SCSE, Galgotias University, India

ABSTRACT

In the present scenario, due to climate change, farmers crops get fungi due to bacteria since soil temperatures change very rapidly according to sudden climate changes for which the crop is getting spoiled. At this advanced era, disease can be detected early so that crops are safe. Different types of fungi-bacterial disease will be detected and prevented by machine learning-based predicted deterministic probabilistic and artificial technology-based CNN for colour changes in plants. This chapter described machine learning techniques and proposed modified algorithms to identify and classify plant diseases. Deep neural network (DNN) models and algorithms are used to improve object accuracy and entropy to reduce the complexity of computational processes and improve the features during deep learning processes (e.g., modified deep neural network [MDNN]). Additionally, they support dynamic feature extraction DSURF and classifier combinations for creating image clusters with the help of clustering and deterministic probability.

DOI: 10.4018/978-1-6684-6418-2.ch014

INTRODUCTION

In this world, plants are the core part of the survival of living creatures like human beings and animals because they provide us food. In short, food is a necessity and basic need of each and every human being so that they can get good nutrition's to keep them healthy and energetic. There are several categories and qualities of foods that can get from plants that can be utilized by living creatures like fruits, vegetables, meat, pulses, dairy products and so on. Quality of food is the foremost important factor in today's competitive food industry. The demand of food is directly proportional to the quality of food. As excessive the food's high-quality is, the higher probabilities of demand for that particular food in marketplace nowadays (Choudhary et al., 2022). So, as soon as the people focus on quality of food, the greater the monitoring of food quality was enhanced also. FAO (Food and agriculture report of the United Nations) evaluations shows that every year around 20 to 40 percent of crop productions are damaged globally due to fungi and other diseases found in plants (Brooks et al., 2022). Annually, illness and other diseases of plants cost's the world's economy around $220 billion (Rs 22 crore approximately) and fungi-bacterial insects around US$70 billion (Rs 7 crore approximately). It was found that annually around such crores rupees spends into the illness treatment of plants globally. It was shown that 10 out of 100 human beings sick due to food they eat which is getting from plant suffered from fungi and bacterial diseases results death of person also. According to the WHO, around 4,20,000 people die every year due to the consumption of bacterial fungus foods getting from diseased plants. This study gives us the reason to take necessary steps on the improvement of curing towards global spread of plants bacteria and diseases in order to decrease mortality rate in the world (Yan et al., 2022). So, its obligatory to cognizance on the appropriate standard of edibles which begins from the very simple and primary step, wherein the food is kept and produced (food warehouses). Food quality can be assessed on several parameters like visual appearance, size, color, shape and texture of that food (Chakrabarti et al., 2022). There are some specialized food inspectors also in food industries who are taking care of the quality of food manually, but this process is very time consuming and costly. Many times, food inspector found the diseased food (foul smell and bad taste) due to fungi bacterial diseases of plants. So, it's our responsibility to find out some new and advanced methods that decline the mortality rate and health issues also which ensures the safety of food comes from the fungi diseased plants. Artificial intelligence is a technique which imitates the functioning of as same as human brain. We can use the techniques of man-made intelligence like computer learning and (DL) deep learning in the market of food industry for numerous purposes like sorting and preparation of food, service delivery operation of edibles, improvement in the safecty and the quality of food because all ill food comes from fungi and diseased plants (Xing et al., 2022). Machine learning and deep learning models are using nowadays by several government agencies in order to make predictions to get high efficiency in complex processes like safety of food and quality of food. Many derived conclusions such as convolution networks neural(CNN), also artificial networks neural (ANN) are frequently applined in industry of edibles to solve several problems arises by fungi and diseased plants. ANN is found to be an excellent tool for assessing the food which is produced from diseased plants. We can use advanced detection techniques like deterministic probability-based data fusion in finding fungi and bacterial diseases in plants (Ali & Abdin, 2021). The aim of this paper is to find classified and optimistic solutions using CNN, deep learning, ANN, ML, clustering and deterministic probability-based data fusion in order to find fungus and bacteria in plants so that humans can get fresh food not diseased food. Figure 1 shows the basic steps required to classify fungi-bacterial diseases in plants.

Figure 1. Basic steps for plant disease detection and classification

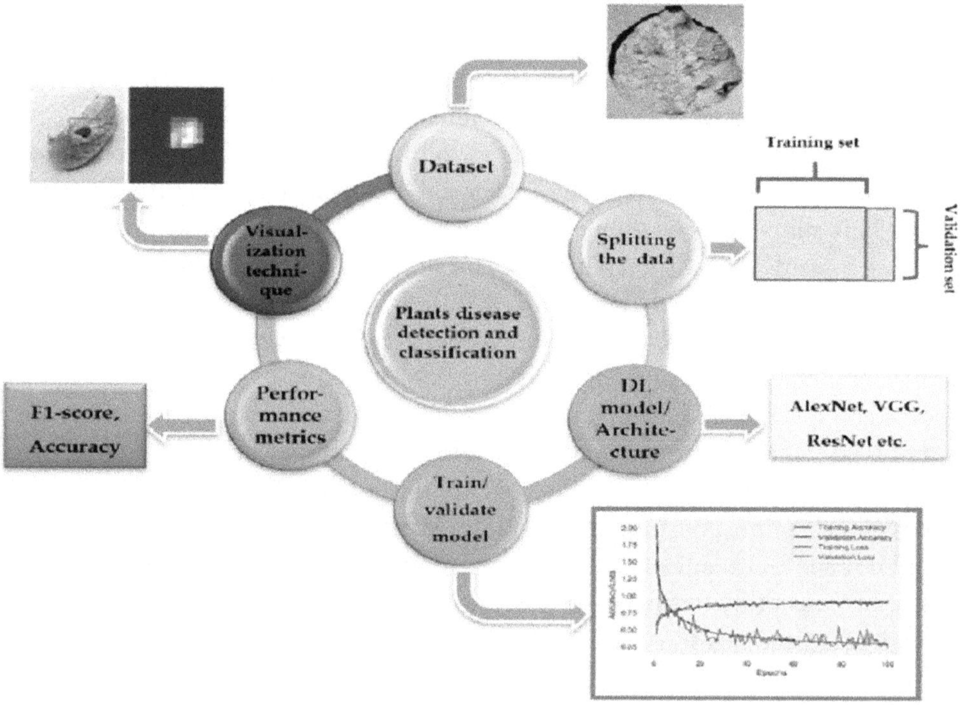

Literature Review

Authors of paper (Khan, Zhang, Albattah et al, 2021) show the challenges and problems in order to detect automated plant diseases in the field of agricultural science. From the last few decades, lot of research has been done in order to detect numerous diseases which can be transferred between humans and animals. These types of diseases can spread very fast and leads to mortality rate also. Using artificial intelligence methods in automatic recognition of plant diseases has been shown in this work in order to become it pandemic. The work proposed in paper (Chowdhury et al., 2021) is based on a popular approach of CNN by using the concept of deep learning known as Efficient Net in order to assess the plant diseases. Comparative analysis has been presented of two binary classification models (U-Net and Modified U-Net) on the basis of healthy and unhealthy leaves. Classification (on the basis of segmented images) has been done on 6 class and 10 class methods and at last Efficient Net B4 ten class classification has achieved highest accuracy of 99.89%. Algorithms has been proposed in paper (Medus et al., 2021) for the detection of condition of sealing food tray whether it is normal or foul smell so that consumers could be safe. On the basis of hyperspectral images which is used in deep learning approach can be used for the inspection of food tray. Their experiments shown by using the food trays dataset obtained from food industry obtained an overall accuracy of 90.1% by using the concept of Deep Belief Network (DBN), Extreme Learning Machine (ELM). Researchers (Liu et al., 2021) used the applications of CNN for recognizing and evaluating problematic food matrices and seeing CNN as a powerful tool for real time recognition of plant diseases in future. A survey has been done by scientists (Zhu et al., 2021) to represent a scenario on the old-style ML and DL techniques employed for the processing of food and shown latest methods, gaps and challenges in the field of food industry. The efficacy of food industry has greatly improved due to intensive research that have been carried out by various researchers in this field using the AL and ML concepts. Paper (Zhang et al., 2021) discussed the latest gaps and conventional chemometric techniques's challenges and also the evolving deep learning methods for the spectral evaluation of agro-products. This review work helped in enhancing the generalization and accurateness for quality assessment of food products. For improving efficacy and security and set standard parameters of quality control a lot of work has been presented in research (Banus et al., 2021) based on CNN and DL methods. Quality is a chief and prime factor in any field which decides the standard of particular product. Experimental results illustrated that there are many models like Dense-Net 61 are available to secure the occurrence of problematic products to users and remove them from the edibles production belt. Authors of research (Salim et al., 2021) has discussed several methods of food assessment on the basis of numerous approaches and effective different variables. Their study depicted that deep learning is much better than other assessment strategies like fixed ML methods and manual detection. An AI based framework has been proposed for farmers in paper (Almadhor et al., 2021) to avoid all upcoming usual production losses, so that they can take necessary steps in advance. Authors show the use of several image recognition techniques to ensure the food's qualty (Khan, Kumar, Dhingra et al, 2021). In paper (Katiyar et al., 2021), artificial bee colony algorithm has been proposed in order to distribute food without quality loss. A model vibhishan has introduced by the researchers of paper (Khan, Tyagi, & Chauhan, 2021) for safety of food warehouse. Some modelling techniques have proposed to improve the quality of food using AI techniques (Sahni et al., 2021). This study presents the automatic detection of flora's leaves diseases on the basis of color and text features of leaves using machine learning classifier (Ozdemir & Polat, 2020). In the study (Bhole & Kumar, 2020) a model named squeeze-Net has been represented to evaluate quality of mangoes using deep learning centric method. Results shown that accuracy achieved after classification is approx. 93% using RGB and thermal images. A systematic study was performed by researchers (Nayak et al., 2020) using ANN applications for food processing and then further assess the range value of its popularity and demand in market. ANN showed the best approach in terms of accuracy in food industry after experimenting the raw foods. Machine learning methods are used in paper (Lam et al., 2020) for image classification and segmentations type issues in order to assess flora diseases on publicly available database. The data sets are healthy and unhealthy leaves, spotted leaves, rotten leaves and infectious leaves. Figure 2 shows the basic steps required to classify plants Diseases (Hasan, Yusuf, & Alzubaidi, 2020).

Problem Formulations

The difficulty with this traditional method is that you have to select the main features of a given image. The more scheduled activities, the more difficult it will be to remove features. A computer vision engineer's decision and a lengthy process of trial and error must determine which task best describes the categories of different objects (Debauche et al., 2020; Gobalakrishnan et al., 2020; Hasan, Yusuf, & Alzubaidi, 2020; Loey et al., 2020; Misra, 2020). Additionally, each job description must take into account several parameters, each of which must be carefully adjusted by a computer vision engineer. Deep Reading (DL) introduces the concept of reading from start to finish. Here, the machine is provided with a set of image data with annotations with classes of objects on each image. Thus the DL model is 'trained' in a given data, in which neural networks detect subpatterns in the image categories and automatically apply descriptive and critical features in relation to each phase of each object (Simon, 2020). It has been proven that DNNs perform much better than traditional algorithms, in terms of computer requirements and training time. The integration layer is usually followed by a conversion layer to speed up the learning process and remove negligence input elements, Figure 3 (Xie et al., 2021) is used to reduce the amount of memory used by the network. For example, a large union effectively reduces important pixels in an image by moving the window above the input and automatically removing the size of that window (Alava et al., 2019; Alfian et al., 2019; Guine & Raquel, 2019; Stangierski et al., 2019).

Figure 2. Flow chart for plant disease detection and classification

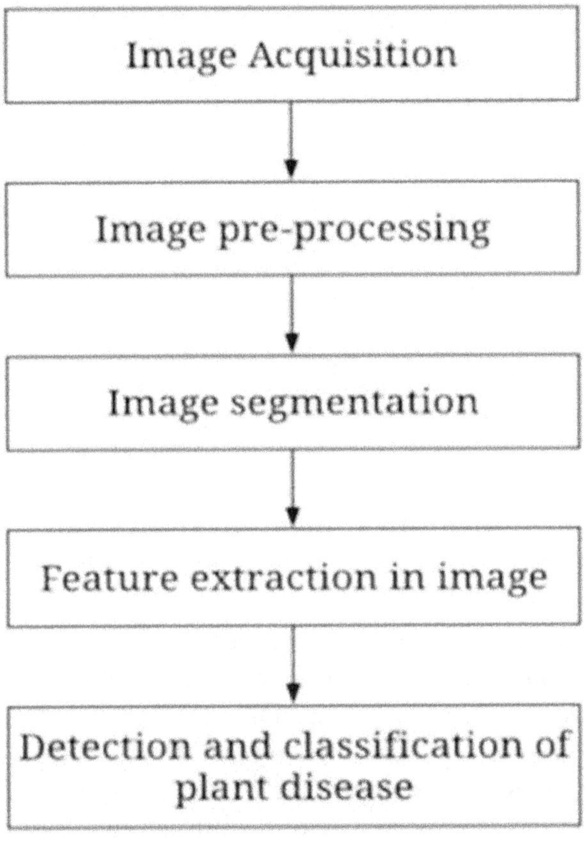

With all the latest and up-coming methods in Computer Vision using this approach, the workflow of a Computer Vision engineer has changed dramatically as the knowledge and expertise of extracting handcrafted features has been replaced by repetitive knowledge and expertise with in-depth learning structures as demonstrated in various AI techniques (Annabel et al., 2019; Ben et al., 2019; Nachtigall et al., 2016; Selvaraj et al., 2019; Zhou et al., 2019). such as CNN, ANN and DL have been used successfully in diagnosing Rice, Wheat, Maize, Cotton, Tomatoes, Peas, Potatoes, Cucumbers, Cassava, Berries, Peaches, Grapes, Olives, Mangoes, Bananas, Apples, Sugar paper, Tea, etc (Gungor et al., 2010; Meng et al., 2013; Nachtigall et al., 2016). This paper discusses AI strategies and other modifications such as MDNN (modified DNN) and SURF algorithm added with flexible features as DSURF is developed and used in agriculture to diagnose plant diseases and stages and their future to achieve agricultural accuracy.

Figure 3. Deep neural network workflow
Source: Xie et al. (2021)

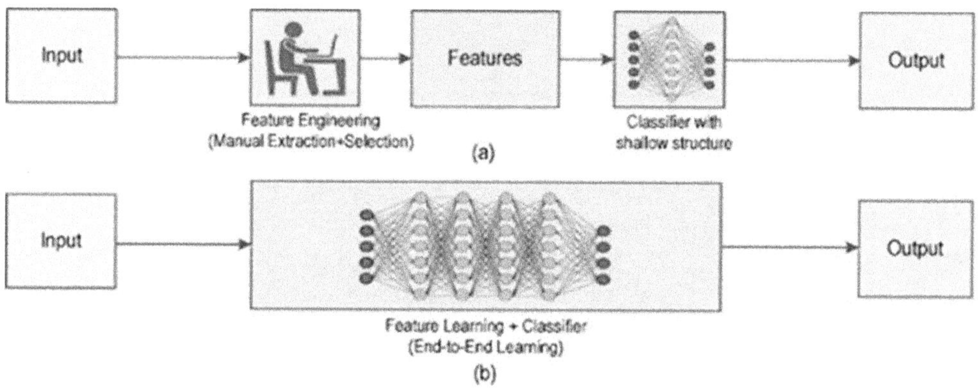

Methodology

A novel interpretation of the solution proposed in the given research work is to improve the dieses classification and detection process with overall better accuracy in network performance through various models.

Here in this paper, try to improve MDNN (Modified-DNN) using DSURF(Dynamic-SURF) and compare the results of MDNN and DSURF.

The SURF algorithm is a type of state-of-the-art algorithm with multiple working layer to simplify computational constraints compared to existing image detection techniques. SURF transforms image data into local spatial feature vectors which is known as descriptors. This vector is later used with key points collected and by running the algorithm at all different scales and Gaussian blurs.

These features have infinite geometric scaling and rotation, so they can produce results with high strength and precision regardless of shape. The algorithm is divided into four main steps.

1. **Image Space Detection:** Image spatial detection is useful for processing images in a shear and scale invariant manner. a formal principle that deals with photo structures at distinctive scales through representing photos as a one parameter group of smoothened photos as shown in equations 1-3.

$$L(x,y,\sigma)=G(x,y,\sigma)*I(x,y)G(x,y,\sigma)=\frac{1}{2\pi\sigma}e-(x2+y2)/2\sigma 2 \quad (1)$$

$$G(x,y,\sigma)=\frac{1}{2\pi\sigma}e-(x2+y2)/2\sigma 2 \quad (2)$$

Where, L(x, y, σ): the scaled image (result of the convolution process).

This will performed between the Gaussian function G (x, y, σ) and the image I (x, y).

$$D(x,y,\sigma)=L(x,y,K\sigma)-L(x.y,I(x,y,\sigma) \quad (3)$$

Where,

L(x,y,kσ) is the previous image convolution.
kσ: Image scale
G(x,y,kσ): Gaussian blur.

2. **Image Localization on Image Key Point:** Image localization refers to finding one or more object in a photo and drawing a large rectangle surrounding the boundary. Object detection combines these two operations to find and classify one or more objects in an image as shown on equation 4.

$$D(x)=D+\frac{\delta Dr}{\delta x}x+1/2xr\frac{\delta 2D}{\delta x2}x \quad (4)$$

Here, we used difference Gaussian (DoG) to remove unwanted extrema found in the image. This is done using the Taylor extension of the scaled space function.

3. **Assignment Orientation:** The direction angle of the feature point is calculated according to the gradation of the image. Then the distance between the key

point orientation angles is converted, so the next step is to orient each key point. This orientation ensures rotational invariance. The more immutability, the better as shown in equation 5 and 6.

$$M(x,y)=\left(L(x+1,y)-L(x-1,y)\right)2+\left(\left(L(x,y+1)-L(x,y-1)\right)2\right)1/2 \tag{5}$$

$$\Theta(x,y)=\tan-1\left(\left(L(x,y+1)-L(x,y-1)\right)\right)/\left(\left(L(x+1,y)-L(x-1,y)\right)\right) \tag{6}$$

Where,

m(x,y) is the magnitude of the image gradient, and

θ(x,y) is the direction precomputed using the pixel difference.

4. **Descriptor Generator for Key Point:** It uses the data gathered in the previous step of the algorithm, and like this last step, you can change the level of local shape distortion and lighting changes.

Algorithm for MDNN and Dynamic SURF are as follows.

Algorithm 1: Modified -MDNN algorithm

1. Input: Feature points prepared from Training Data (Tri D), Target as Disease Trays (Tar R) and Neurons Weights (NW)
2. Result: Separate and classify plant disease
3. Launch MDNN and set parameters
4. → No. of Iterations/ Epochs (E)
5. → Neurons as a network component (NC)
6. → Performance test training: parameters: Cross Entropy, Functions Gradient, Error and Validation → Apprentice training: Scaled Conjugate Gradient (Training)
7. → TriD Category: Ran.dom
8. For every collection of Tri
9. If data for training (Tri D) ε The first phase in the feature collection
10. Group (1) = CategoriesofTriD
11. Otherwise if data for training (TriD) ε Phase 2 of the feature collection
12. Group (2) = CategoriesofTriD
13. If not if data for training (TriD) ε Section 3 of the feature set
14. Group (3) = CategoriesofTriD

15. If not if Training Data (TriD) ε Section 4 of the feature set
16. Group (4) = CategoriesofTriD
17. Conclusion
18. Launch MDNN in PDDS using TriD and also with the help of their TarR
19. Net = patter.nnet (N) // where MDNN is initiated and saved on the network as a structure
20. Collection training training requirements as per the needs and train the program
21. Net = Train (Net, Tri D, clusters)
22. Split Effects = Imitate (Net, tested leaf's features)
23. If Div Results = True
24. Show the results separated by leaf disease
25. Calculate performance parameters
26. Conclusion
27. Recovery: Separated Outcomes
28. Conclusion

On the other hand, DNN algorithms can learn complex structures from large amounts of data and demonstrate the performance of various machine learning tasks in recent years. The most popular features of many computer-readable methods are the use of high-resolution data and signal magnitude, frequency, phase dispersion, and wavelet coefficients.

Therefore, in this paper, we use cumulate as a training tool and input it into a modified DNN (MDNN) algorithm. Various research assignments focus on machine learning methods rather than feature analysis used as input. Therefore, in this study, to reduce the computational complexity and speed up signal recognition when using the basic algorithm of the DNN structure, the proposed algorithm uses only the features that most directly affect the performance of the segment.

The proposed method is more efficient than the existing method using the same information and communication coefficients when selecting features. Here, we have compared the behavoir of the proposed method of importing various leaf samples using SURF and SURF together with a conventional DNN. MDNN with DSURF.

Algorithm 2: DynamicSURF (DSURF and fast feature)

1. Input: The split ROI of a flora leaf photo
2. Output: DSURF key points as feature vector Calculate ROI size [Matrix mXn]
3. For → 1 in every Line
4. By n → 1 for all Col.
5. Image ROI Scale = Measurement (plant leaf (m,n), scale size)

6. Loc_Keypoints = Local Performance (Image ROI Scale (m,n))
7. k_Keypoints = Shape (Loc_ Keypoint (m, n), angle)
8. DSURF_Keypoints = Filter (k_Keypoint (m.n), Square Filter)
9. Finish
10. Finish
11. Return: Feature vector = DSURF_Keypoints as feature points
12. Finish

Result Analysis

This segment describes the outcomes and evaluation acquired after the assessment of the proposed assignment. For testing, QoS parameters consisting of recall, Precision, Accuracy, F measure, and errors are taken into consideration. Figured 4-7 represents recall, F measure, error, precision, and accuracy.

Results of Recall F measure, precision, accuracy and Error are also shown in Tables 1-5. Results of Tables 1-5 and Figures 4-7 are compared as DNN with standard SURF and MDNN with DSURF.

In Figuress 4-7, x axis represents the number of samples and y axis represents the accuracy values obtained. Green bars are illuminated on DNN with SURF, and orange bars indicate that MDNNs have DSURF.

Figure 8 and Table 5 show 97.53 median value of MDNN and DSURF. There is a 0.763% improvement in the model that is presented for accuracy.

Figure 4. Precision computation graph

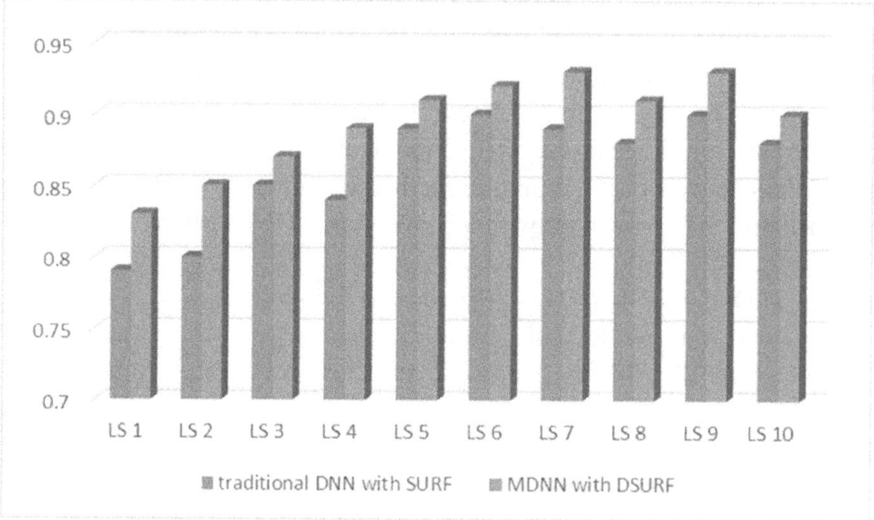

Table 1. Computation of precision

Number of Leaf Samples (LS)	Traditional DNN With SURF	MDNN With DSURF
LS 1	0.79	0.83
LS 2	0.8	0.85
LS 3	0.85	0.87
LS 4	0.84	0.89
LS 5	0.89	0.91
LS 6	0.9	0.92
LS 7	0.89	0.93
LS 8	0.88	0.91
LS 9	0.9	0.93
LS 10	0.88	0.9

Figure 5. Recall computation graph

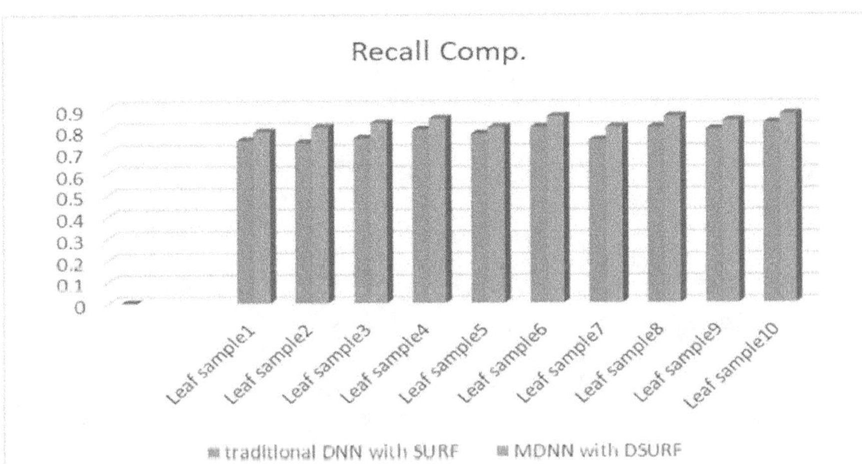

Figures 8 and 9 and Table 5 show the comparative accuracy of the proposed activities. A evaluation was made to reveal the effectiveness of the proposed assignment. 1.1% is an growth in accuracy by means of Bin Lu et al. 4.31% accuracy improvement with Zhang Chuanlei, 1.74% accuracy development with G. Nachtigall Lucas. It's been observed that the proposed method is extremely effective with accuracy in comparison to standard techniques.

Table 2. Computation of recall

Number of Leaf Samples (LS)	Traditional DNN With SURF	MDNN With DSURF
Leaf sample1	0.75	0.79
Leaf sample2	0.74	0.81
Leaf sample3	0.76	0.83
Leaf sample4	0.8	0.85
Leaf sample5	0.78	0.81
Leaf sample6	0.81	0.86
Leaf sample7	0.75	0.81
Leaf sample8	0.81	0.86
Leaf sample9	0.8	0.84
Leaf sample10	0.83	0.87

Figure 6. Computation of F-measure graph

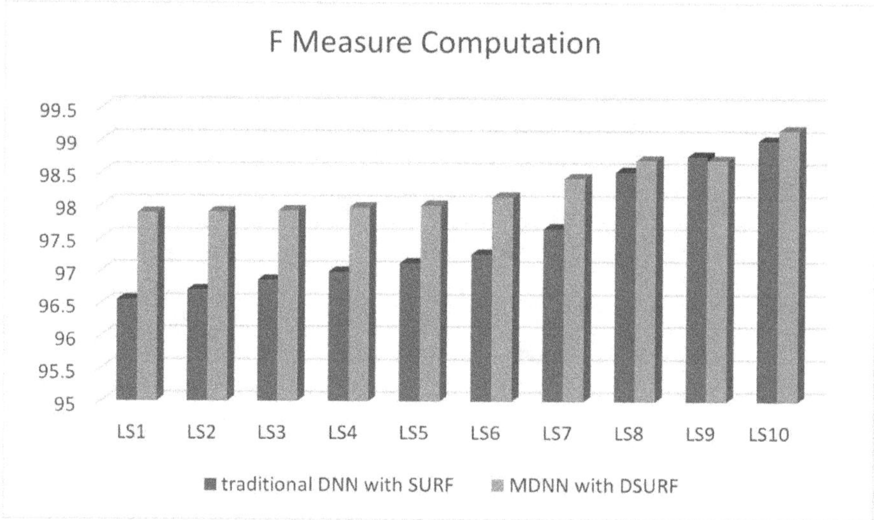

CONCLUSION

In this paper, performed an analytical analysis on a novel to effectively detect and discriminate diseases in apple tree leaf images using the MDNN idea as a deep neural network technique. Early detection of plant illnesses is very vital studies work, due to the fact agriculture has a big problem, and due to this hassle, manufacturing and the rate of farming is declining. In this research assignment, we designed and evolved techniques primarily based on the innovative idea of figuring out apple plant illnesses with the help of their leaf traits. The proposed MDNN is applied in numerous ranges prior to implementing MDNN consisting of picture enhancement and ROI classification within the preprocessing section, function extraction with the help of DSURF descriptor, and schooling and fragmentation using MDNN. Through a lot of these strategies, the accuracy of the proposed became advanced. the usage of MDNN and DSURF improvement measures, we concluded from the survey that the accuracy of the proposed model elevated by 18.03% in size over the available instant and differentiated techniques. Soon, we can examine and compare procedures to different real-world items with multi-leaf infestations and develop plant-based diagnostics primarily based on deep learning strategies.

Table 3. Computation of F-measure

Number of Leaf Samples (LS)	Traditional DNN With SURF	MDNN With DSURF
L.S1	0.81	0.77
L.S2	0.83	0.77
L.S3	0.85	0.8
L.S4	0.87	0.82
L.S5	0.85	0.83
L.S6	0.89	0.85
L.S7	0.86	0.81
L.S8	0.88	0.84
L.S9	0.88	0.85
L.S10	0.88	0.85

Future Scope

This study presents an improved method to improve energy efficiency in wireless networks by implementing Multi Model learning integration. A probability distribution function (PDF) is used to detect dead nodes over the lifetime of a neuron. When dead notes and dead notes are received, the built-in alarm notifies the server of the relevant information. The use of Kmeans clustering and neural networks (NNs) is for the integration and differentiation of diseases affecting the leaves of plants. Accurate and efficient diagnosis of disease is the main goal of the given approach. Results of the experiments show that this proposed technique is an important method that can be of great help in diagnosing leaf diseases with minimal computational cost.

Figure 7. Error computation graph

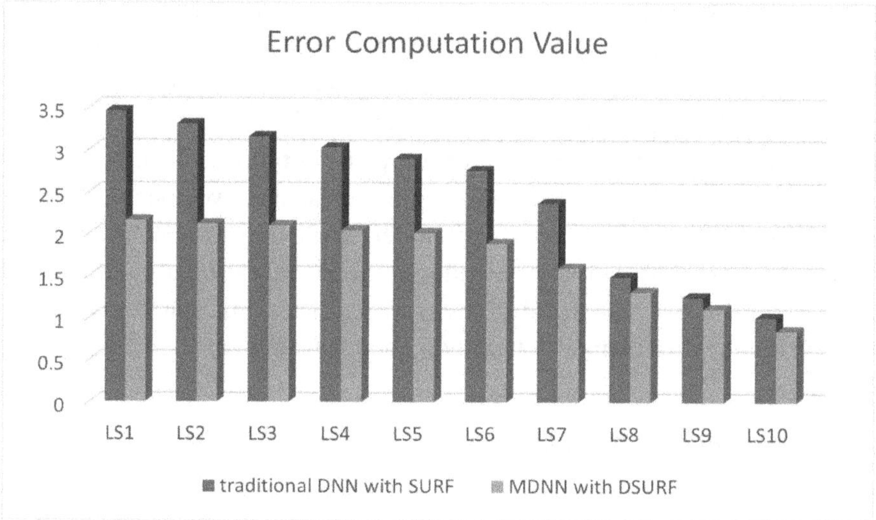

Table 4. Computation of error

Number of Leaf Samples (LS)	Traditional DNN With SURF	MDNN With DSURF
LS1	2.15	3.45
LS2	2.11	3.30
LS3	2.09	3.15
LS4	2.04	3.02
LS5	2.01	2.89
LS6	1.88	2.75
LS7	1.59	2.36
LS8	1.31	1.49
LS9	1.11	1.25
LS10	0.85	1.01

Simulations are performed using MATLAB tools. The obtained results showed the efficiency of dead and half-dead node detection of the proposed research method. Future work may be improved by extending the lifespan of the network using better data collection methods, further development of research is to create open multimedia (audio/video) about diseases and automatically solve them after disease detection.

Figure 8. Accuracy computation graph

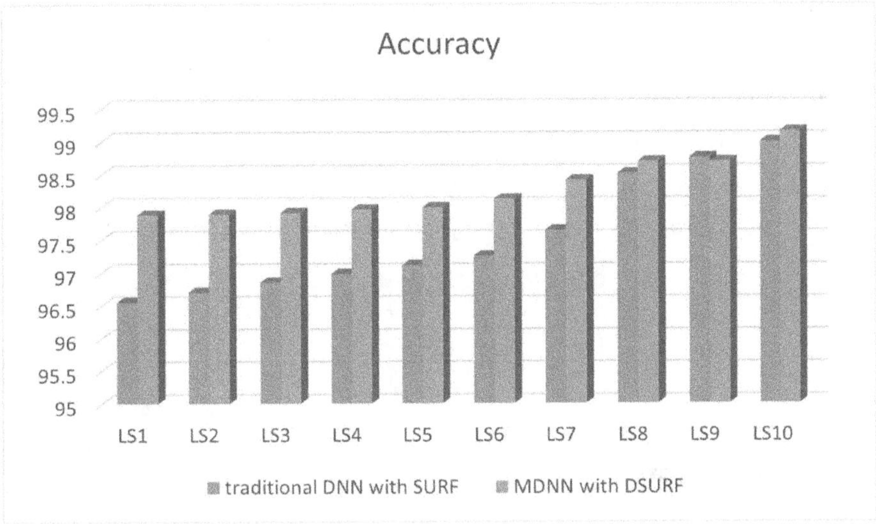

Table 5. Computation of accuracy

Number of Leaf Samples (LS)	Traditional DNN With SURF	MDNN With DSURF
LS1	96.55	97.88
LS2	96.7	97.89
LS3	96.85	97.91
LS4	96.98	97.96
LS5	97.11	97.99
LS6	97.25	98.12
LS7	97.64	98.41
LS8	98.51	98.69
LS9	98.75	98.69
LS10	98.99	99.15

Availability of Data

The data which is being used to support the result of this study may be obtained from the corresponding author.

Figure 9. Comparison of proposed and existed work for accuracy

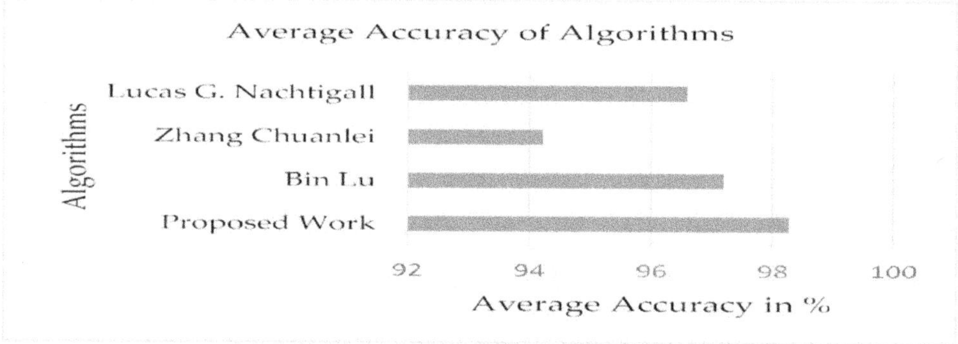

Conflict of Interests

There is no conflict of interests in regards to the publication of this paper.

Funding Statement

This study did not receive any funding.

REFERENCES

Alava, J., Kusme, K., Loor, A., & Santander, E. (2019). RoCoLe: A robusta coffee leaf images dataset for evaluation of machine learning based methods in plant diseases recognition. *Data in Brief, 25*, 104414. doi:10.1016/j.dib.2019.104414 PMID:31516934

Alfian, G., Syafrudin, M., Farooq, U., Ma'arif, M. R., Syaekhoni, M. A., Fitriyani, N. L., Lee, J., & Rhee, J. (2019). Improving efficiency of RFID-based traceability system for perishable food by utilizing IoT sensors and machine learning model. *Food Control, 110*, 107016. doi:10.1016/j.foodcont.2019.107016

Ali & Abdin. (2021). Impact of Activation Functions and Number of Layers on the Classification of Fruits using CNN. *2021 8th International Conference on Computing for Sustainable Global Development (INDIA Com)*, 227-231.

Almadhor, A., Rauf, H., Lali, M., Damasevicius, R., Alouffi, B., & Alharbi, A. (2021). Ai-driven framework for recognition of guava plant diseases through machine learning from dslr camera sensor based high resolution imagery. *Sensors (Basel), 21*(11), 3830. doi:10.339021113830 PMID:34205885

Annabel, L., Annapoorani, T., & Deepalakshmi, P. (2019). Machine Learning for Plant Leaf Disease Detection and Classification–A Review. In *2019 International Conference on Communication and Signal Processing (ICCSP)*. IEEE. 10.1109/ICCSP.2019.8698004

Banus, N., Boada, I., Xiberta, P., Toldra, P., & Bustins, N. (2021). Deep learning for the quality control of thermoforming food packages. *Scientific Reports, 11*(1), 1–15. doi:10.103841598-021-01254-x PMID:34750436

Ben, L., Gaudreau, H., Dallaire, L., Tessier, M., & Fliss, I. (2019). Bioprotective culture: A new generation of food additives for the preservation of food quality and safety. Industrial Biotechnology Journal, 15(3), 138-147.

Bhole & Kumar. (2020). Mango quality grading using deep learning technique: perspectives from agriculture and food industry. *Proceedings of the 21st Annual Conference on Information Technology Education,* 180-186.

Brooks, S., Klomchit, A., Chimthai, S., Jaidee, W., & Bastian, A. C. (2022). SRNE2BP a Fungal Endophyte with Biocontrol Properties to Control Early Blight and Fusarium Wilt Disease in Tomato and Plant Growth Promotion Activity. *Current Microbiology, 79*(4), 1–15. doi:10.100700284-022-02803-x PMID:35175450

Chakrabarti, S. K., Sharma, S., & Shah, M. A. (2022). Potato Pests and Diseases: A Global Perspective. In *Sustainable Management of Potato Pests and Diseases* (pp. 1–23). Springer. doi:10.1007/978-981-16-7695-6_1

Choudhary, R. C., Bairwa, H. L., Kumar, U., Javed, T., Asad, M., Lal, K., Mahawer, L. N., Sharma, S. K., Singh, P., Hassan, M. M., Abo-Shosha, A. A., Rajagopal, R., & Abdelsalam, N. R. (2022). Influence of organic manures on soil nutrient content, microbial population, yield and quality parameters of pomegranate (Punica granatum L.) cv. Bhagwa. *PLoS One, 17*(4), e0266675. doi:10.1371/journal.pone.0266675 PMID:35436290

Chowdhury, M., Rahman, T., Khandakar, A., Ayari, M., Khan, A., Islam, M., Emadi, N., Reaz, M., & Ali, S. (2021). Automatic and reliable leaf disease detection using deep learning techniques. *Agricultural Engineering, 3*(2), 294–312.

Debauche, O., Mahmoudi, S., Elmoulat, M., & Sidi Mahmoudi, P. (2020). Edge AI-IoT pivot irrigation, plant diseases, and pests' identification. *Procedia Computer Science*, *177*, 40–48. doi:10.1016/j.procs.2020.10.009

Gobalakrishnan, N., Pradeep, K., Raman, C. J., Ali, L., & Gopinath, M. P. (2020). A systematic review on image processing and machine learning techniques for detecting plant diseases. In *International Conference on Communication and Signal Processing (ICCSP)*. IEEE. 10.1109/ICCSP48568.2020.9182046

Guine, & Raquel. (2019). The use of artificial neural networks (ANN) in food process engineering. *International Journal of Food Engineering*, *5*(1), 15–21.

Gungor, V. C., Lu, B., & Hancke, G. P. (2010). Opportunities and challenges of wireless sensor networks in smart grid. *IEEE Transactions on Industrial Electronics*, *57*(10), 3557–3564. doi:10.1109/TIE.2009.2039455

Hasan, R., Yusuf, S., & Alzubaidi, L. (2020). Review of the state of the art of deep learning for plant diseases: A broad analysis and discussion. *Plants*, *9*(10), 1302. doi:10.3390/plants9101302 PMID:33019765

Katiyar, S., Khan, R., & Kumar, S. (2021). Artificial Bee Colony Algorithm for Fresh Food Distribution without Quality Loss by Delivery Route Optimization. *Journal of Food Quality*, *2021*, 1–9. doi:10.1155/2021/4881289

Khan, R., Kumar, S., Dhingra, N., & Bhati, N. (2021). The Use of Different Image Recognition Techniques in Food Safety: A Study. *Journal of Food Quality*, *2021*, 1–10. doi:10.1155/2021/7223164

Khan, R., Tyagi, N., & Chauhan, N. (2021). Safety of food and food warehouse using VIBHISHAN. *Journal of Food Quality*, *2021*, 1–12. doi:10.1155/2021/1328332

Khan, R., Zhang, C., Albattah, W., & Qamar, A. (2021). *Image-Based Detection of Plant Diseases: From Classical Machine Learning to Deep Learning Journey*. Wireless Communications and Mobile Computing.

Lam, M., Nguyen, T., & Chung, W. (2020). Deep learning-based food quality estimation using radio frequency-powered sensor mote. *IEEE Access : Practical Innovations, Open Solutions*, *8*, 88360–88371. doi:10.1109/ACCESS.2020.2993053

Liu, Y., Pu, H., & Sun, D. (2021). Efficient extraction of deep image features using convolutional neural network (CNN) for applications in detecting and analysing complex food matrices. *Trends in Food Science & Technology*, *113*, 193–204. doi:10.1016/j.tifs.2021.04.042

Loey, M., ElSawy, A., & Afify, M. (2020). Deep learning in plant diseases detection for agricultural crops: A survey. *International Journal of Service Science, Management, Engineering, and Technology, 11*(2), 41–58. doi:10.4018/IJSSMET.2020040103

Medus, Saban, Frances, Bataller, & Rosadao. (2021). Hyperspectral image classification using CNN: Application to industrial food packaging Food Control. *Food Control, 125*.

Meng, W., Yuan, J., Zhang, C., Bai, Z., Zhou, W., Yan, J., & Li, X. (2013). Parenteral analgesics for pain relief in acute pancreatitis: A systematic review. *Pancreatology, 13*(3), 201–206. doi:10.1016/j.pan.2013.02.003 PMID:23719588

Misra, N. N. (2020). IoT, big data and artificial intelligence in agriculture and food industry. IEEE Internet of Things Journal, 35.

Nachtigall, L. G., Araujo, R. M., & Nachtigall, G. R. (2016). Classification of apple tree disorders using convolutional neural networks. In *2016 IEEE 28th International Conference on Tools with Artificial Intelligence (ICTAI)* (pp. 472-476). IEEE. 10.1109/ICTAI.2016.0078

Nayak, J., Vakula, K., Dinesh, P., Naik, B., & Pelusi, D. (2020). Intelligent food processing: Journey from artificial neural network to deep learning. *Computer Science Review, 38*, 100297. doi:10.1016/j.cosrev.2020.100297

Ozdemir, A., & Polat, K. (2020). Deep learning applications for hyperspectral imaging: A systematic review. *Journal of the Institute of Electronics and Computer, 2*(1), 39–56. doi:10.33969/JIEC.2020.21004

Sahni, V., Srivastava, S., & Khan, R. (2021). Modelling techniques to improve the quality of food using artificial intelligence. *Journal of Food Quality, 2021*, 1–10. doi:10.1155/2021/2140010

Salim, N., Zeebaree, S., Sadeeq, M., Radie, A. H., Shukur, H., & Rashid, Z. (2021). Study for Food Recognition System Using Deep Learning. *Journal of Physics: Conference Series, 1963*(1), 012014. doi:10.1088/1742-6596/1963/1/012014

Selvaraj, M., Vergara, A., Ruiz, H., Safari, N., Elayabalan, S., Ocimati, W., & Blomme, G. (2019). AI-powered banana diseases and pest detection. *Plant Methods, 15*(1), 1–11. doi:10.118613007-019-0475-z

Simon, Y. L. (2020). Artificial intelligence (AI) in agriculture. Academic Press.

Stangierski, J., Weiss, D., & Kaczmerek, A. (2019). Multiple regression models and Artificial Neural Network (ANN) as prediction tools of changes in overall quality during the storage of spreadable processed Gouda cheese. *European Food Research and Technology, 245*(11), 2539–2547. doi:10.100700217-019-03369-y

Xie, W., Wei, S., Zheng, Z., & Yang, D. (2021). A CNN-based lightweight ensemble model for detecting defective carrots. *Biosystems Engineering, 208*, 287–299. doi:10.1016/j.biosystemseng.2021.06.008

Xing, T., Cai, A., Lu, C., Ye, H., Wu, H., Huai, S., Wang, J., Xu, M., & Lin, Q. (2022). Increasing soil microbial biomass nitrogen in crop rotation systems by improving nitrogen resources under nitrogen application. *Journal of Integrative Agriculture, 21*(5), 1488–1500. doi:10.1016/S2095-3119(21)63673-0

Yan, Z. Z., Chen, Q. L., Li, C. Y., Thi Nguyen, B. A., He, J. Z., & Hu, H. W. (2022). Contrasting ecological processes shape the Eucalyptus phyllosphere bacterial and fungal community assemblies. *Journal of Sustainable Agriculture and Environment, 1*(1), 73–83. doi:10.1002ae2.12007

Zhang, X., Yang, J., Lin, T., & Ying, Y. (2021). Food and agro-product quality evaluation based on spectroscopy and deep learning: A review. *Trends in Food Science & Technology, 112*, 431–441. doi:10.1016/j.tifs.2021.04.008

Zhou, L., Zhang, C., Qiu, F., & He, Y. (2019). Application of deep learning in food: A review. *Comprehensive Reviews in Food Science and Food Safety, 18*(6), 1793–1811. doi:10.1111/1541-4337.12492 PMID:33336958

Zhu, L., Spachos, P., Pensini, E., & Plataniotis, K. N. (2021). Deep learning and machine vision for food processing: A survey. *Current Research in Food Science, 4*, 233–249. doi:10.1016/j.crfs.2021.03.009 PMID:33937871

Chapter 15
Optimal Control Strategies for Mitigating Frequency Deviation in Microgrid-Based Smart Village Systems

Sudheeksha Misra
Govind Ballabh Pant Institute of Engineering and Technology, India

Bhola Jha
Govind Ballabh Pant Institute of Engineering and Technology, India

V. M. Mishra
Govind Ballabh Pant Institute of Engineering and Technology, India

Surjeet Singh Patel
Govind Ballabh Pant Institute of Engineering and Technology, India

ABSTRACT

Frequency deviation is a common issue in microgrids, especially in smart village systems where the energy demand varies significantly due to changing loads and intermittent renewable energy sources. This chapter aims to propose optimal control strategies that effectively mitigate frequency deviation in microgrid-based smart village systems. The research will explore the variation of frequency by using different control techniques such as conventional PI, PI plus fuzzy, and PI plus ANFIS. The proposed strategies will be evaluated using simulation models. The results shows that the PI plus ANFIS shows the best results, which will provide valuable insights for designing reliable and efficient microgrid-based smart village systems, addressing the problem of frequency deviation and improving overall system performance.

DOI: 10.4018/978-1-6684-6418-2.ch015

INTRODUCTION

Microgrid-based smart village systems are innovative energy solutions that combine local renewable energy sources, energy storage, and distribution systems to provide reliable and sustainable electricity to remote and rural areas. These systems are typically designed to operate independently or in conjunction with the main grid, depending on the availability and reliability of the grid in the area.

Microgrids are small-scale, localized power grids that can operate autonomously or in parallel with the main grid. They can generate electricity from renewable energy sources such as solar panels, wind turbines, or small-scale hydropower systems, and store excess energy in batteries or other energy storage technologies. Microgrids also include energy distribution systems that can manage the flow of electricity among different energy sources, loads, and storage devices within the system.

The concept of microgrid-based smart village systems has gained traction due to their numerous benefits in providing reliable and sustainable electricity to remote and rural areas. Some of the key advantages of microgrid-based smart village systems are:

- **Reliable Electricity Supply:** Microgrids can provide reliable electricity supply to remote and rural areas that may have limited or no access to the main grid. By utilizing local renewable energy sources and energy storage, microgrids can ensure a stable and consistent power supply, reducing dependency on uncertain grid availability.
- **Energy Independence:** Microgrids allow villages and communities to become more self-sufficient in terms of electricity generation. By generating electricity from local renewable sources, microgrid-based smart village systems can reduce dependence on fossil fuels and centralized power generation, contributing to energy independence and sustainability.
- **Cost-Effective Energy Solutions:** Microgrids can offer cost-effective energy solutions to remote and rural areas by utilizing local renewable energy sources, reducing reliance on expensive fossil fuels, and avoiding the need for costly grid extension. This can lead to economic benefits for the communities by reducing energy costs and creating local job opportunities in the renewable energy sector.
- **Environmental Sustainability:** Microgrid-based smart village systems can contribute to environmental sustainability by utilizing renewable energy sources, reducing greenhouse gas emissions, and promoting a low-carbon energy transition. This can help in mitigating climate change impacts and promoting sustainable development in remote and rural areas.
- **Enhanced Energy Access and Socio-Economic Development:** Reliable access to electricity can have significant socio-economic impacts on remote

and rural communities. Microgrid-based smart village systems can enable access to electricity for lighting, education, healthcare, and livelihood opportunities, improving the quality of life and fostering socio-economic development in these areas.

While microgrids offer numerous benefits, they also face several technical challenges that need to be addressed for successful implementation. The frequency deviation is a critical issue in microgrid systems that can impact their performance and reliability. Microgrids, being smaller and localized power grids, are more susceptible to frequency fluctuations compared to larger interconnected power grids. Frequency deviation refers to the variation in the nominal frequency of the electricity generated by the microgrid from its normal operating frequency of 50 Hz or 60 Hz, depending on the region.

There are several causes of frequency deviation in microgrids. Some of the common causes include:

- **Variability of Renewable Energy Sources:** Microgrids often rely on renewable energy sources such as solar and wind, which are inherently variable in nature. Changes in solar irradiation or wind speed can result in fluctuations in the power output of these sources, leading to frequency deviations in the microgrid.
- **Imbalance Between Generation and Load:** Microgrids may experience frequency deviation when there is an imbalance between the generation capacity and the load demand within the system. If the generation capacity is not adequately matched with the load demand, it can result in frequency fluctuations.
- **Lack of Grid Connection:** In microgrids that operate in islanded mode, i.e., disconnected from the main grid, there is no external source of power or frequency regulation. This can result in frequency deviation, as the microgrid has to rely solely on its local generation and storage resources to meet the load demand.

The impacts of frequency deviation in microgrids can be significant and can affect the overall performance and reliability of the system. Some of the impacts of frequency deviation in microgrids include:

- **Unstable System Operation:** Frequency deviations can lead to unstable operation of the microgrid, affecting the performance of connected loads and causing disruptions in the normal operation of electrical devices and equipment.

- **Reduced Power Quality:** Frequency deviations can result in poor power quality, including fluctuations in voltage levels, increased harmonic distortion, and other power quality issues, which can negatively impact the performance of sensitive loads.
- **Risk of Equipment Damage:** Frequent and prolonged frequency deviations can pose risks to the health and lifespan of electrical equipment within the microgrid, including generators, inverters, and energy storage systems, leading to increased maintenance costs and equipment damage.
- **Safety Concerns:** Frequency deviations can also pose safety concerns, particularly for critical loads such as medical equipment, as they may not function optimally or may fail to operate during frequency fluctuations, posing risks to human health and safety.

Frequency deviation is a significant problem in microgrid systems, caused by various factors such as renewable energy source variability, load demand imbalance, and lack of grid connection. It can have detrimental impacts on the performance, reliability, power quality, and safety of microgrids, necessitating the need for optimal control strategies to mitigate frequency deviation and ensure stable and reliable operation of microgrid-based smart village systems.

The main objective of this research paper is to propose optimal control strategies for mitigating frequency deviation in microgrid-based smart village systems. The research aims to address the problem of frequency deviation in microgrids by developing effective control strategies that can regulate the frequency within acceptable limits and ensure stable and reliable operation of microgrid-based smart village systems. The proposed control strategies will be designed to optimize the operation of renewable energy sources, energy storage systems, and loads within the microgrid, taking into consideration the dynamic nature of renewable energy sources, load demand fluctuations, and other system constraints. The research will utilize fuzzy and ANFIS. The effectiveness of the proposed control strategies will be evaluated through simulations. The ultimate goal of this research is to contribute to the development of reliable and sustainable microgrid systems for smart villages, ensuring stable and high-quality electricity supply to remote and rural areas, thereby improving the overall energy access and socio-economic development of these communities.

Hirsch et al. (2018) provides a comprehensive review of microgrids, including their components, operation modes, and control strategies. It highlights the challenges and opportunities in microgrid technologies, covering topics such as microgrid architectures, power electronics interfaces, energy management systems, and grid integration issues. Kumar and Tyagi (2016) presents a review of microgrid control strategies aimed at enhancing the integration of renewable energy sources and managing frequency deviation. The paper discusses various control strategies, including droop control, virtual inertia control, and frequency-based load shedding, which are commonly used in microgrid systems to regulate the frequency of the microgrid during normal and abnormal operating conditions. Kumar and Bhimasingu (2017) presents a control strategy based on virtual inertia and droop control for mitigating frequency deviation in autonomous microgrids. It considers the variability of renewable energy sources and load demand fluctuations in microgrid systems and proposes a virtual inertia control scheme that emulates the behavior of synchronous generators to provide frequency stability. The droop control strategy is used to regulate the active power output of the distributed generation units in the microgrid. It presents simulation results demonstrating the effectiveness of the proposed control strategy in managing frequency deviation and maintaining the stability of the microgrid system. Rezaei et al. (2020) proposes a hierarchical frequency control strategy for microgrids, incorporating virtual inertia and adaptive droop control, to enhance the stability and reliability of microgrid systems. A two-layer control architecture is used, where the higher layer implements virtual inertia control to provide frequency stability, and the lower layer uses adaptive droop control to regulate the distributed generation units active power output. The virtual inertia control is designed to emulate the behavior of synchronous generators, and the adaptive droop control adjusts the droop gain based on the frequency deviation. Ma et al. (2022) proposes a novel approach for optimal frequency control in an islanded microgrid with distributed generation units using reinforcement learning. The authors develop a reinforcement learning-based controller that can learn the optimal frequency control actions from interactions with the microgrid system. The proposed controller takes into account the state of the microgrid, including generation and load levels, and makes control decisions to maintain the frequency within a desired range. The effectiveness of the proposed approach is demonstrated through simulation studies, showing that the reinforcement learning-based controller can achieve better frequency control performance compared to traditional control methods. Abubakr et al. (2021) proposes an adaptive virtual inertia control strategy for frequency stabilization in microgrids with high renewable energy penetration. The authors propose a virtual inertia concept that emulates the inertia of conventional synchronous generators in microgrid systems, which helps to stabilize the frequency under high renewable energy penetration. Xu et al. (2021) addresses the issue of communication delays in dynamic frequency control of islanded microgrids. The authors propose a distributed frequency control algorithm that considers communication delays among distributed generation units in a microgrid. The algorithm uses a consensus-based approach to achieve frequency synchronization and adjusts the output power of distributed generation units based on local frequency measurements and neighbor information. Simulation results show that the proposed algorithm can effectively mitigate the adverse effects of communication delays on frequency control in islanded microgrids. Rathnayake et al. (2021) provides a comprehensive review of frequency-based load shedding techniques for microgrids. The authors present an overview of existing methods for frequency-based load shedding, which is an important strategy to maintain power balance and stability in microgrids during abnormal conditions. The paper discusses various load shedding techniques, including frequency-based methods based on static and dynamic thresholds, as well as advanced techniques using intelligent algorithms. The authors also identify future research directions and challenges in frequency-based load shedding for microgrids. Kotsampopoulos et al. (2018) presents a survey of droop control methods for frequency regulation in microgrids. The authors provide an overview of different droop control strategies, which are widely used in microgrids to regulate the output power of distributed energy resources (DERs) and maintain system frequency. The paper discusses various types of droop control, such as conventional droop control, enhanced droop control, and virtual impedance-based droop control, and compares their advantages and limitations. The authors also highlight recent developments and future research directions in droop control methods for microgrid frequency regulation. Ding et al. (2019) proposes an adaptive frequency control strategy for inverter-based microgrids with time-delayed communication. The authors address the issue of communication delays in microgrid control systems, which can adversely affect frequency stability. The proposed control strategy uses an adaptive mechanism to estimate the communication delays and adjust the control action of distributed generation units accordingly. The paper presents simulation results to demonstrate the effectiveness of the proposed adaptive frequency control strategy in mitigating the impact of communication delays and improving frequency stability in inverter-based microgrids.

Test System Description

Renewable energy sources most surely have an impact on the dynamic character of the power system. The grid's incorporation of renewable energy sources is growing nowadays. In order to conduct the study, a standard test system suggested by Bevrani et al. (2012) is used, where all the parameters are the same. The block diagram of the same is given below:

1. **PV Model:**

A photovoltaic (PV) model is a mathematical representation of the behavior of a PV system, which is a type of solar energy system that converts sunlight directly into electricity using semiconductor materials. PV models are used in the field of solar energy research, engineering, and design to analyze the performance of PV systems, optimize their efficiency, and predict their output under different conditions. The solar cell circuit is shown in Figure 2.

Figure 1. Block diagram of test system

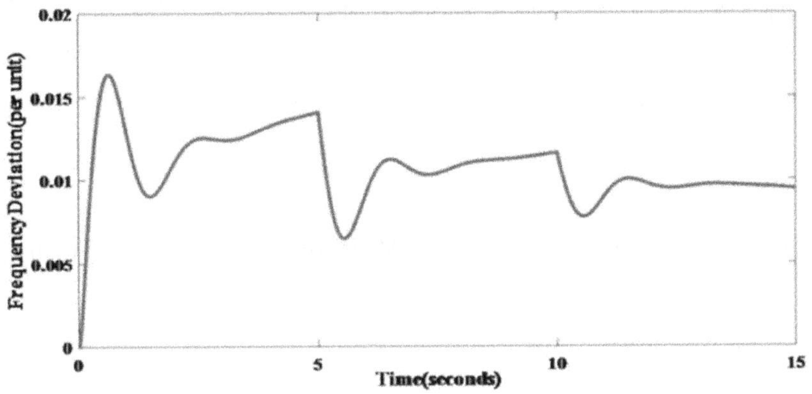

Figure 2. Circuit of solar cell

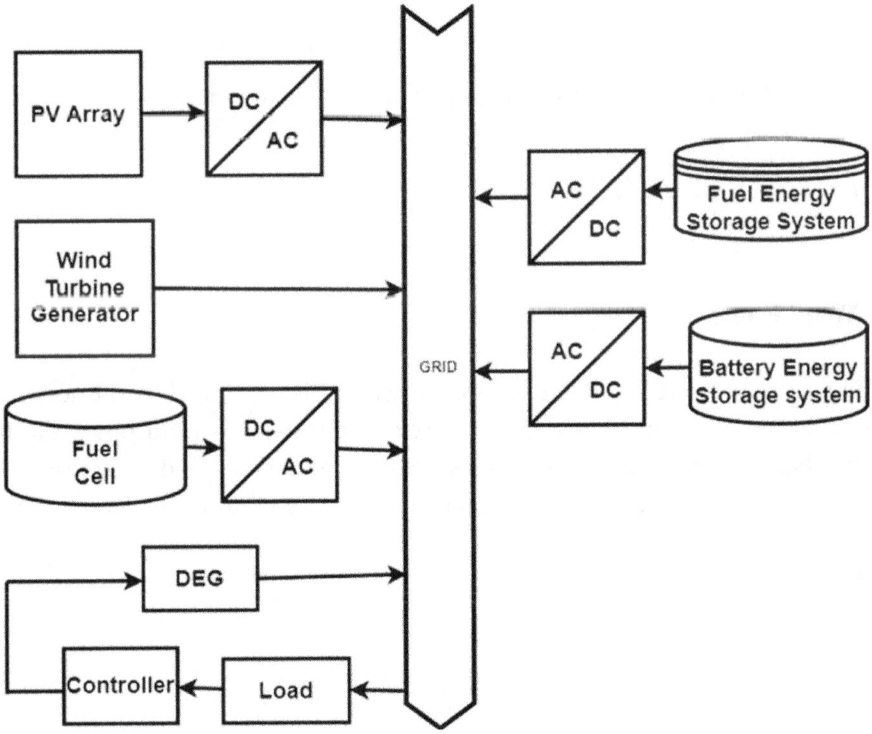

Here the current I will be formulated as:

$$I = I_L - I_D - I_{sh} \tag{1}$$

Here, I stands for the current produced by light in the cell, I_D the voltage-dependent current lost to recombination, and the current I_{sh} lost because of shunt resistances. Using the Shockley equation for an ideal diode, a single diode is represented in this model.

$$I_D = I_0 \left[\exp\left(\frac{V + IR_s}{nV_T} \right) - 1 \right] \tag{2}$$

Here n is the diode ideality factor and V_T is the thermal voltage given by:

$$V_T = \frac{kT_c}{q} \tag{3}$$

And the shunt current can be written as:

$$I_{sh} = (V+IR_S)/R_{sh} \tag{4}$$

Now from the above equations the current I can be drawn as:

$$I = I_L - I_0 \left[\exp\left(\frac{V + IR_s}{nV_T} \right) - 1 \right] - \frac{V + IR_s}{R_{sh}} \tag{5}$$

2. Wind Energy:

Wind energy is a renewable form of energy that is generated by harnessing the kinetic energy of wind and converting it into electricity using wind turbines. The basic principle behind wind energy is that the kinetic energy of the wind is captured by the blades of a wind turbine, which then rotates a generator to produce electrical power.

$$P = 0.5 \times \rho \times A \times v^3 \tag{6}$$

3. Diesel Engine Generator (DEG):

The transfer function of a diesel engine generator system describes the relationship between the input (typically the control signal or the load demand) and the output (typically the electrical power output) of the system in the frequency domain model is given by (Papathanassiou & Papadopoulos, 2001)

$$P_d = -\left(\frac{1}{R}+\frac{K_I}{s}\right)\left(\frac{1}{T_{sg}s+1}\right)\left(\frac{K_{deg}}{T_{sg}s+1}\right)\Delta f_e \tag{7}$$

4. Fuel cell, BESS, and FESS:

Fuel Cells: Fuel cells are devices that generate electricity through an electrochemical process using hydrogen and oxygen (or another oxidizing agent) as fuel. Fuel cells are highly efficient and emit only water and heat as byproducts, making them a clean and environmentally-friendly source of power. They can be used in various applications, including stationary power generation, transportation (such as fuel cell vehicles), and portable power systems.

Battery Energy Storage Systems (BESS): Battery energy storage systems store electrical energy in batteries for later use. They typically consist of one or more batteries, power electronics for managing the charging and discharging of the batteries, and control systems for optimizing the system performance. Battery energy storage systems can be used for a wide range of applications, including smoothing out intermittent renewable energy generation, providing backup power during grid outages, and improving power quality in electric grids.

Fuel Energy Storage Systems: Fuel energy storage systems store energy in the form of a fuel that can be converted into electricity when needed. Examples of fuel energy storage systems include pumped hydro storage (where water is pumped to a higher elevation and then released to generate electricity), compressed air energy storage (where compressed air is stored in underground caverns and then released to generate electricity), and thermal energy storage (where heat is stored and then used to generate electricity).

The transfer function FC, BESS, FESS are given below:

$$\frac{1}{1+T_{FC}},\ \frac{1}{1+T_{BESS}},\ \frac{1}{1+T_{FESS}}$$

Where T_{FC}, T_{BESS}, T_{FESS} are time constant of FC, BESS and FESS respectively.

5. Fuzzy Logic Controller (FLC):

Fuzzy logic controller (FLC) is a type of control system that uses fuzzy logic, a mathematical framework for dealing with uncertainty and ambiguity, to make decisions and control processes. FLCs are particularly useful in systems where traditional control methods may not be suitable due to the presence of imprecise or vague information.

FLCs consist of three main components: the fuzzifier, the fuzzy inference engine, and the defuzzifier.

Fuzzifier: The fuzzifier converts crisp input data (such as sensor measurements) into fuzzy values, which represent the degree of membership of the input data to different linguistic terms or fuzzy sets. Fuzzy sets are defined using membership functions, which describe the shape and characteristics of the fuzzy sets.

Fuzzy Inference Engine: The fuzzy inference engine uses fuzzy rules to infer the appropriate output fuzzy sets based on the fuzzy values obtained from the fuzzifier. Fuzzy rules typically have the form "IF (antecedent) THEN (consequent)", where the antecedent is a combination of fuzzy sets from the input variables, and the consequent is a fuzzy set from the output variable. The inference engine uses fuzzy logic operators (such as AND, OR, and NOT) to combine the fuzzy sets and determine the degree of membership of the output fuzzy sets.

Defuzzifier: The defuzzifier converts the fuzzy output values obtained from the inference engine into crisp output values. This is done by using a defuzzification method, such as centroid, mean of maximum (MOM), or weighted average defuzzification, which calculates a single crisp value based on the shape and position of the fuzzy output sets.

6. **ANFS:** Adaptive Network-Based Fuzzy Inference System

ANFIS was introduced by Jang in 1993 as a way to improve the accuracy and adaptability of fuzzy logic systems. It uses a hybrid approach that combines the learning capabilities of neural networks with the interpretability and linguistic reasoning of fuzzy logic systems. It consists of a five-layer structure, with each layer performing a specific function:

Input Layer: The input layer receives the crisp input data and passes it through the membership function, which fuzzifies the input data and converts it into fuzzy values.

Fuzzification Layer: The fuzzification layer calculates the membership grades for each input data ADApoint based on the shape and characteristics of the membership functions. These membership grades represent the degree of membership of the input data to different fuzzy sets.

Rule Layer: The rule layer calculates the firing strength of each fuzzy rule by multiplying the membership grades of the input variables. This layer combines the fuzzy sets from the fuzzification layer according to the fuzzy rules defined in the system.

Defuzzification Layer: The defuzzification layer calculates the weighted average of the fuzzy sets from the rule layer, taking into account the firing strength of each rule. This process results in a crisp output value.

Output Layer: The output layer produces the final output value of the ANFIS system based on the crisp output value obtained from the defuzzification layer.

ANFIS uses a learning algorithm, such as the back-propagation or gradient descent algorithm, to adaptively adjust the parameters of the membership functions and the fuzzy rules based on a training dataset. This allows ANFIS to learn from data and automatically adjust its parameters to improve its accuracy and adaptability to changing conditions.

Result and Discussion

The load profile is shown in Figure 3 which clearly illustrate that load change after 5 sec by 0.02 p.u. The reason of taking this type of load is to show the load changing nature of the microgrid. The Figure 3, Figure 4 and Figure 5 show the frequency deviation of system with load and it can be noted that the frequency deviation in case of PI+FLC+ANFIS is much lesser than the other two method.

Figure 3. Load profile of a microgrid

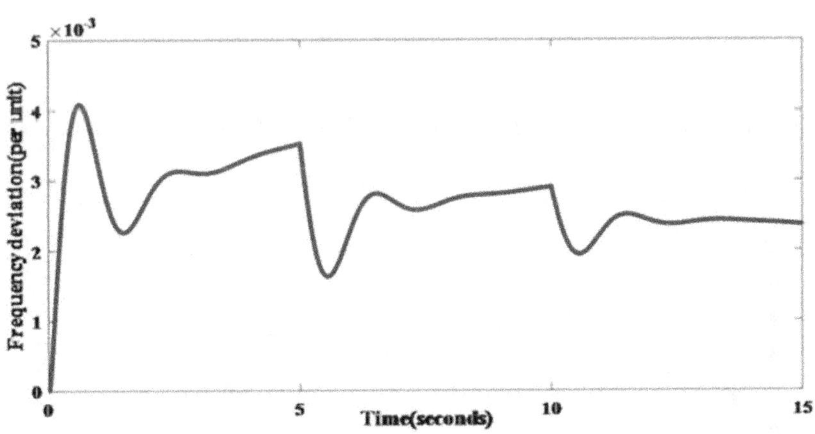

Figure 4. Frequency deviation in case of conventional PI controller

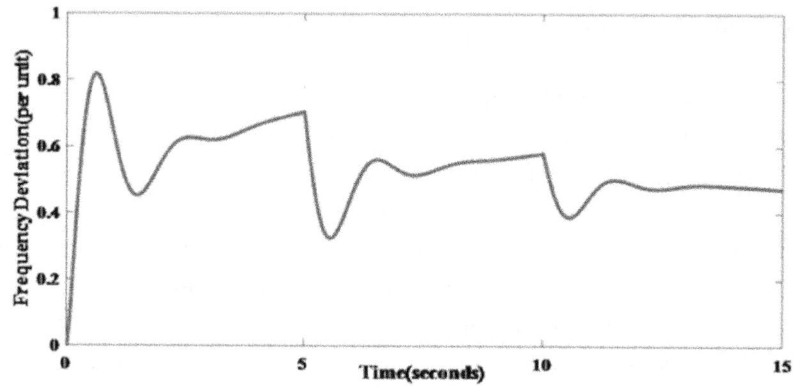

Figure 5. Frequency deviation in case of PI+FLC

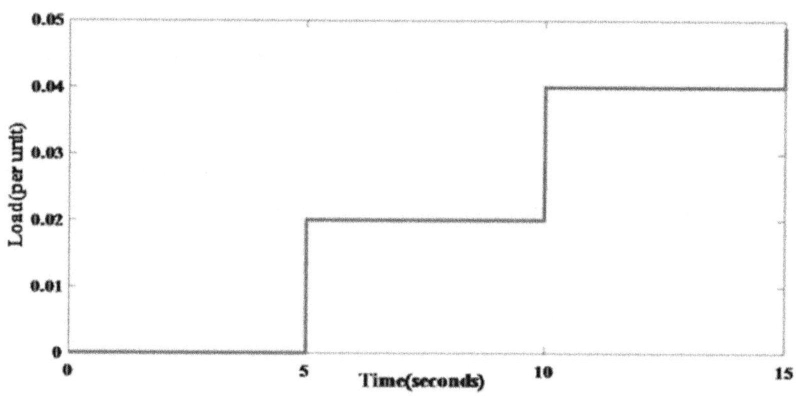

Figure 6. Frequency deviation in case of PI+FLC+ANFIS

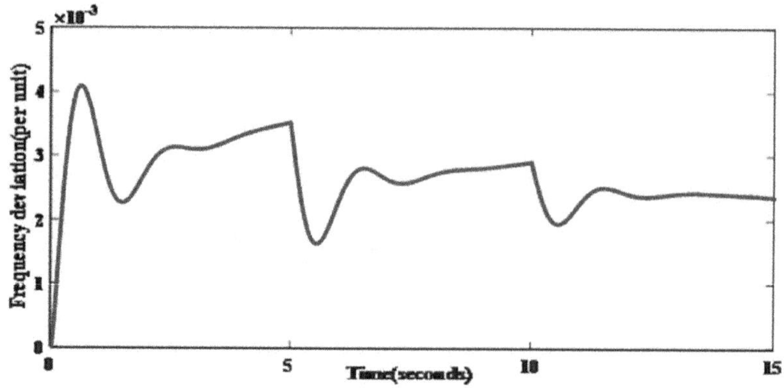

CONCLUSION

Frequency regulation is a critical aspect of microgrid operation and is essential for ensuring reliable and stable power supply within the microgrid. It is a measure of the cyclic variation in the electrical signal, and it needs to be kept within a narrow range to ensure that electrical devices and equipment connected to the microgrid operate correctly. Deviations from the nominal frequency can lead to issues such as equipment malfunction, increased energy consumption, and potential damage to sensitive electronic devices. Here a microgrid system is chosen where PV, Wind, Fuel cell, FESS, BESS are there and the frequency deviation is studied in three cases, first is conventional PI, second is PI+FLC, third is PI+FLC+ANFIS controller from which it is concluded that the third case shows the best result.

Future Scope

Microgrids have significant potential for the future development of smart villages, enabling sustainable and reliable electricity supply, enhancing energy access, optimizing energy consumption, leveraging digital technologies, and improving community resilience. Continued advancements in renewable energy technologies, smart grid solutions, business models, and policy frameworks are expected to drive the adoption and growth of microgrids in smart villages, contributing to rural electrification, economic development, and social empowerment.

REFERENCES

Abubakr, H., Mohamed, T. H., Hussein, M. M., Guerrero, J. M., & Agundis-Tinajero, G. (2021). Adaptive frequency regulation strategy in multi-area microgrids including renewable energy and electric vehicles supported by virtual inertia. *International Journal of Electrical Power & Energy Systems*, *129*, 106814. doi:10.1016/j.ijepes.2021.106814

Bevrani, H., Habibi, F., Babahajyani, P., Watanabe, M., & Mitani, Y. (2012). Intelligent frequency control in an AC microgrid: Online PSO-based fuzzy tuning approach. *IEEE Transactions on Smart Grid*, *3*(4), 1935–1944. doi:10.1109/TSG.2012.2196806

Ding, D., Han, Q. L., Wang, Z., & Ge, X. (2019). A survey on model-based distributed control and filtering for industrial cyber-physical systems. *IEEE Transactions on Industrial Informatics*, *15*(5), 2483–2499. doi:10.1109/TII.2019.2905295

Hirsch, A., Parag, Y., & Guerrero, J. (2018). Microgrids: A review of technologies, key drivers, and outstanding issues. *Renewable & Sustainable Energy Reviews*, *90*, 402–411. doi:10.1016/j.rser.2018.03.040

Kotsampopoulos, P., Lagos, D., Hatziargyriou, N., Faruque, M. O., Lauss, G., Nzimako, O., Forsyth, P., Steurer, M., Ponci, F., Monti, A., Dinavahi, V., & Strunz, K. (2018). A benchmark system for hardware-in-the-loop testing of distributed energy resources. *IEEE Power and Energy Technology Systems Journal*, *5*(3), 94–103. doi:10.1109/JPETS.2018.2861559

Kumar, M., & Tyagi, B. (2016, November). A state of art review of microgrid control and integration aspects. In *2016 7th India International Conference on Power Electronics (IICPE)* (pp. 1-6). IEEE. 10.1109/IICPE.2016.8079411

Kumar, Y. P., & Bhimasingu, R. (2017, November). Fuzzy logic based adaptive virtual inertia in droop control operation of the microgrid for improved transient response. In 2017 IEEE PES Asia-Pacific Power and Energy Engineering Conference (APPEEC) (pp. 1-6). IEEE. doi:10.1109/APPEEC.2017.8309006

Ma, G., Li, J., & Zhang, X. P. (2022). A review on optimal energy management of multimicrogrid system considering uncertainties. *IEEE Access : Practical Innovations, Open Solutions*, *10*, 77081–77098. doi:10.1109/ACCESS.2022.3192638

Papathanassiou, S. A., & Papadopoulos, M. P. (2001). Dynamic characteristics of autonomous wind–diesel systems. *Renewable Energy*, *23*(2), 293–311. doi:10.1016/S0960-1481(00)00180-4

Rathnayake, D. B., Akrami, M., Phurailatpam, C., Me, S. P., Hadavi, S., Jayasinghe, G., Zabihi, S., & Bahrani, B. (2021). Grid forming inverter modeling, control, and applications. *IEEE Access : Practical Innovations, Open Solutions*, *9*, 114781–114807. doi:10.1109/ACCESS.2021.3104617

Rezaei, N., Mazidi, M., Gholami, M., & Mohiti, M. (2020). A new stochastic gain adaptive energy management system for smart microgrids considering frequency responsive loads. *Energy Reports*, *6*, 914–932. doi:10.1016/j.egyr.2020.04.021

Xu, Y., Dong, Z., Li, Z., Liu, Y., & Ding, Z. (2021). Distributed optimization for integrated frequency regulation and economic dispatch in microgrids. *IEEE Transactions on Smart Grid*, *12*(6), 4595–4606. doi:10.1109/TSG.2021.3089421

Chapter 16
Solar-Powered Pesticide-Spraying RFID Robot

K. Raju
St. Joseph Engineering College, India

Roshan Rajeshwar Tulaskar
St. Joseph Engineering College, India

Canute Sherwin
St. Joseph Engineering College, India

ABSTRACT

In this chapter, a solar powered pesticide-spraying RFID robot has been designed and developed for its use in different types of fruits and vegetable fields. The robot is equipped with DC motors, DC pump, a solar panel, and an ultrasonic sensor. The robot's navigation has been automated using RFID (radio frequency identification) technology. An ultrasonic sensor is used for sensing any obstacle in the path of the robot's navigation. A solar panel of 30W is used for charging the batteries that are used for driving the DC motors and the DC pump. The wheels of the robot are powered by DC motors, and the DC pump is used for spraying pesticides through the nozzles. The height of the nozzles can be varied depending on the size of the crop. The robot has been designed for spraying pesticides automatically thereby controlling pests, improving productivity, and reducing the health hazards of farmers posed by pesticides.

DOI: 10.4018/978-1-6684-6418-2.ch016

INTRODUCTION

The history of agriculture in India dates back to Indus valley civilization era and even before that in some regions of Southern India. Today India ranks second worldwide in farm output, so there is always a need for research and development and innovation in the field of agricultural technology. Agriculture plays a vital role in the Indian economy. Over 70 percent of the rural people depend on agriculture as their principal means of livelihood. Many people have agricultural lands where they use equipment that operate on fossil fuels and this causes harm to ozone layer and also results in the depletion of fossil fuels for the future generation. Also the attack of pests on the crops is one of the major problems of reduction in the productivity of agricultural land. The pesticides are used in agriculture in order to protect the plants against all harmful organisms such as fungicides, insecticides, and molluscicides. They are also used for influencing the life processes of plants. They might be useful in destroying undesired plants or parts of plants and preserving plant's product from organisms. Pesticides are the substances meant to control pests. They are chemical or biological agents and are applied on both the crops and the fruits and the vegetable plants such as grapes, watermelon, strawberry, carrot, potato etc. Strawberries remained at the top of the list of consumption of pesticides with at least twenty different kinds of pesticides are being sprayed and become one of the difficult tasks for the farmer. The currently available techniques of spraying pesticides are time consuming, laborious and hazardous. Due to the migration of workers in modern times to different other sectors, it has created a huge shortage of labor in the rural areas. Few pesticide sprayers are available in the market but no sprayer could work automatically without any human intervention. The main concern is that the pesticides are harmful to the human health and they spread diseases and also the worker has to work under the hot sun for long hours. Many projects were also carried out to solve the problems of the farmers. It is important to concentrate on some of the aspects like how to improve the productivity, how to reduce the human effort and how to solve and ease the problems of workers. In order to overcome the problems faced by the farmers in spraying the pesticides on the crops, a solar powered wireless operated all terrain eco-friendly pesticide spraying robot has been designed and fabricated and the details of the robot have been discussed in this chapter.

BACKGROUND

The following literature review has been carried out to identify the possible gaps found in the available literature on the pesticide spraying robots. Madhu et al. (2023)

have worked on an Arduino based pesticide spraying robot for spraying pesticide for cucumber plantation. The design proposed by them ensures the successful navigation of a robot along the rows of plantation in a glasshouse and spraying the pesticide uniformly. The system also ensures the optimization of carbon dioxide enrichment, temperature, humidity, moisture, fertilizer feed, pest and fungus control. The main components of the design include Arduino UNO, motor driver, integrated circuit (IC), motor, spray and robot mechanism. Navya and Sudha (2023) have proposed an artificial intelligence (AI) based robot for harvesting and pesticide spraying and maintaining water management systems in agriculture using internet of things (IOT). The design basically consists of a programmed robot which performs the functions of seed planting, water pouring, pesticide and compost splashing. The system works on a forecast algorithm that waters agricultural areas by monitoring the weather and real-time soil moisture sensor. Ashish et al. (2022) have analyzed the state-of-the-art reviews in the development of autonomous pesticide spraying robots that are divided into four main categories; platform mobility and steering, localization and navigation control, sensing and target detection, and pesticide spraying arrangement. The article also focused on the efficient methodology that can be used for future perspectives to enhance the pesticide spraying operations in agriculture. Sushma Priya et al. (2022) have developed an Arduino based pesticide spraying robot for agricultural purposes. The spraying mechanism, in particular, includes an Arduino UNO microcontroller. It additionally includes a tank of 250 ml capability to which a water pump is hooked up. A splitter is connected to this water pump along with two nozzles at the two opposite ends and through this nozzle spraying is performed. Rohit Sandikar et al. (2022) have designed and fabricated an Arduino based Bluetooth controlled pesticide spraying robot. The robot was developed by integrating the control mode, movement, spraying, and sensor. The system is compact, portable and a well-founded platform that can help farmers for pest control of crops with precision. This approach will help farmers using the fundamental principles of sensors and robotics technology. Yash et al. (2022) have developed an Arduino based autonomous robot for spraying pesticide. They have proposed a remotely operated spraying robot designed to lessen the time, reduce the labor costs and prevent the human hazards involved in spraying potentially toxic chemicals. The main parts of the system include an input unit, spray and control processing unit and an output unit. Mounika et al. (2022) have developed an Arduino based pesticide spraying robot for agricultural purpose. The robot sprays pesticides, cuts the grass, and performs seeding on its own with the help of a mobile application with an aim to avoid harm to the environment and the human operators. The bot sprinkles the pesticides, covering all plants in the farm uniformly. Sunny and Badashah (2022) have worked on Bluetooth based agrobot for spraying the pesticide. A cell phone can be used to control agricultural equipment including machines for spreading urea, sowing seeds, and pulling weeds with the use of a Bluetooth device connected to an embedded system. This straightforward technique is made to spray liquid pesticides. The main components of the system include; control circuit, power supply, rechargeable batteries, DC motors, Bluetooth device, 89C51 microprocessor, integrated circuit and pumping motors. Abhishek More et al. (2022) have designed and developed a solar operated pesticide spraying semi-automatic vehicle. It is designed to spray pesticides directly on the crops with minimum wastage, making it cost effective and environment friendly. With an adjustable chassis it can be used for different types of crop fields. It uses solar power to charge a battery, which powers the pump. The semi-automatic vehicle can be controlled through a mobile application and Bluetooth connectivity. Ahamed Mustak et al. (2022) have adopted an embedded design of fuzzy PID variable spraying control method and proposed a novel spray flow control method based on Programmable Logic Controller (PLC). The experimental results show that the overshoot of Proportional–Integral–Derivative (PID) control is10.76%, and the overshoot of fuzzy PID control is7.17% which can meet the requirements of an advanced spray control flow system. Yuichi Hasuda and Keigo Ozaki (2022) have developed a pesticide spraying robot moving in a greenhouse with a line-trace that sprays pesticides while line-tracing between the ridges in a narrow greenhouse by analyzing the image of the packing plastic tape installed on the ridges. The proposed design ensures successful navigation of a robot along the rows of plantation in a glasshouse and spraying the pesticide uniformly. Kavyashree and Vishwanath (2022) have developed a prototype of a pesticide spraying robot using a microcontroller for greenhouse application. Microcontroller PIC16F877A is used to control the spraying and movement of the robot. The robot is controlled using a mobile application and Bluetooth connectivity. The designed robot is expected to solve some of the major problems associated with the traditional pesticide spraying techniques. Mustafid et al. (2022) have conducted research to design and analyze an autonomous robot that can precisely spray the pesticide on cabbage with appropriate spraying position, droplets diameter, and application dosage. The robot can run straight along the cabbage beds using a wall-follower method and successfully detect and spray the pesticide solution on each cabbage by using machine vision with an accuracy of 88% and 80% accuracy of spraying position.

Andre et al. (2021) have developed a smart and a novel electric sprayer that can be assembled on a robot for spraying the pesticide. The sprayer has a crop perception system that calculates the leaf density based on a support vector machine (SVM) classifier using image histograms (local binary pattern (LBP), vegetation index, average, and hue). This density can then be used as a reference value to feed a controller that determines the air flow, the water flow rate, and the water density of the sprayer. The results of the leaf density classifier show an accuracy score that varies between 80% & 85%. Punit Kanse et al. (2021) have developed an Arduino based pesticide spraying robot for spraying pesticides on the crops. The designed system has an ability to avoid obstacles, spray, track motion, monitor orientation and compensate for path errors for better stability and reliability. It is an integration of multi-sensor module design. Kassim et al. (2020) have designed and developed an Arduino based autonomous pesticide sprayer robot for fertigation farms. The robot sprays pesticide under the crops leaves and is also used for spraying pesticide for crops such as tomato, rockmelon, papaya, pineapple, vegetables etc. The study involves the development of unmanned pesticide sprayer that can be mobilized autonomously. It is because that the pesticide is a hazardous component that can affect human health in the future if it is exposed during manual spraying, especially in a closed area such as in the greenhouse. The flexible sprayer boom also can be flexibly controlled in the greenhouse and outdoor environment such as open space farms. Rajesh Kanna and Vikram (2020) have proposed a microcontroller based pesticide spraying robot. The gadget incorporates a solitary splash syphon engine with a consequently separate flexible spraying utilizing ultrasonic sensors, all mounted on a pan tilt unit. The site-explicit spraying gadget plans to splash explicit targets while diminishing the utilization of pesticides. Ashutosh et al. (2020) have proposed an agricultural pesticide spraying robotic system using android application. The evolved system includes plotting a prototype that uses simple cost-effective components like microprocessors, solar panels, wireless, different motors and terminal equipment which help the farmers in numerous crop field activities. The most attractive feature of this design is electrostatic spraying. Chaitanya et al. (2020) have developed an Arduino based pesticide spraying robot. This works by taking input images using a camera, analyzing them using a machine learning process. The system displays thc disease presented on the leaf, stem or plant. It also displays the exposed area of the disease and also predicts the remedies, turnson the pesticide sprayer which sprays the respective pesticide on the exposed area of disease.

George Adamides et al. (2017) have presented the design aspects and development processes to transform a general-purpose mobile robotic platform into a semi-autonomous agricultural robot sprayer focusing on the user interfaces for teleoperation. Argade et al. (2017) have developed a remote controlled solar powered automatic grass cutter and pesticide spraying robot that is intended to reduce the human effort and be helpful to the farmer. As it is operated on the solar energy it does not have any adverse effects on the environment. Ahalya et al. (2017) have developed a solar powered micro controller based semi-automatic pesticide sprayer vehicle with three wheels for vineyards but its performance is not on par with the performance of a petrol based pesticide sprayer. Shubhangi and Sujata (2017) have proposed a pesticide sprayer robot for remote applications using a mobile or laptop. Amruta Sulakhe and Karanjkar (2015) have worked on the design and operation of agricultural based pesticide spraying robot that can complete the work automatically and meet the spraying requirement in a greenhouse but the cost of the robot is high as it is controlled by a camera for navigation and a skilled operator is required for operating the computer which is used for the navigation. Swati & Belsare (2015) have developed a robot for seed sowing. Pandurang Lad et al. (2015) have proposed a solar operated pesticide sprayer wherein they have used solar energy for operating the pump. Mitul Raval et al. (2015) have developed an all-terrain mobile robot using ARM 7 controller for spraying pesticides. Peng Jian-sheng (2014) has proposed the design of an intelligent Wi-Fi wireless controlled spraying pesticide robot. A microcontroller core, a wireless router for the network connection point, is used for controlling the sprayer. The system is controlled by an android mobile application through video capturing.

Keeping in view the problems of the farmers and the gaps identified in the available literature, an attempt has been made in this chapter to design and develop an ecofriendly fully automated high speed solar powered pesticide spraying robot using RFID technology.

DESIGN DETAILS OF RFID ROBOT

Selection of DC Motor

When selecting the drive wheel motors for mobile vehicles, a number of factors must be taken into account to determine the maximum torque required. One method of computing the torque is as follows:

- **Rolling Resistance**

Rolling Resistance (RR) is the force necessary to propel a vehicle over a particular surface.

$$\begin{aligned}\text{Rolling Resistance} &= \text{Coefficient of rolling resistance} \times \text{Gross vehicle weight}\\ &= 0.01 \times (25 \times 9.81)\\ &= 2.45\ \text{N}\end{aligned}$$

Where Coefficient of rolling resistance = 0.01
Gross vehicle weight = 25 kg

- **Grade Resistance**

Grade Resistance (GR) is the amount of force necessary to move a vehicle at a slope or grade.

Grade Resistance = Gross vehicle weight × sin (θ)
= (25 × 9.81) × sin (3)
where 'θ' = 3°
= 12.84 N

- **Acceleration Force**

Acceleration Force (AF) is the force necessary to accelerate from a stop to maximum speed in a desired time.

Acceleration Force = Gross vehicle weight ×

$$\left(\frac{\text{Maximum speed}}{\text{Time required to achieve maximum speed}}\right)$$

$$= 25 \times \frac{2.77}{1}$$

$$= \mathbf{69.25\ N}$$

Where maximum speed = 2.77 m/s
Time for maximum speed = 1s

- **Total Tractive Effort**
  Total Tractive Effort = Rolling Resistance + Grade Resistance + Acceleration Force
  = 2.45 + 12.84 + 69.25
  = 84.54 N

- **Torque**
  Torque = Total Tractive Effort × Radius of wheel
  = 84.54 × 0.08
  Where radius of wheel = 0.08 m
  = 6.76 N-m

- **Power**
  Power = $(2\pi NT)/60$
  (Where 'N' = 45 rpm)
  $= (2\pi \times 45 \times 6.76)/60$
  $= 31.86$ W

Based on the power calculation, two D C motors of 35 W have been selected.

- **Battery Capacity Particulars**
  Power consumed by two D.C Motors = $35 \times 2 = 70$W
  Power consumed by the D.C Pump = 60W
  Total Power consumed 'P' = 70 + 60 = 130W
  Current required per hour is 'I'= (P/V)x 1 hr
  = (130/12) x 1
  = 10.83 Ah

Based on the design calculation, a battery of 12 V &14 Ah has been selected.

- **Run-Time of the Battery**
  - Current drawn by D.C Motor
    $P = V \times I \times \eta$
    $70 = 12 \times I \times 0.8$
  where $\eta_{motor} = 80\%$
    $I = 7.29A$
  - Current drawn by D.C Pump
    $P = V \times I \times \eta$
  where $\eta_{pump} = 80\%$
    $60 = 12 \times I \times 0.8$
    $I = 6.25A$

- **Battery Stand By-Time**
  Stand by- time (hour) = Battery Capacity (Ah) / total current drawn (A)
  = 14 / (7.3+6.25)
  = 1Hr

- **Solar Panel Details**

Depending on the size of robot, a solar panel of 30 W has been selected.

Total current supplied by the solar panel,

$$I = P/V$$
$$= 30/12$$
$$I = 2.5A$$

Time taken by the solar panel to charge the battery

Time = Battery capacity/Total current supplied by the solar panel
= 14/2.5
= 5.6 Hr

DETAILS OF COMPONENTS OF RFID ROBOT

Mechanical Components

Chassis Structure

The chassis of the robot is made of galvanized iron (G.I) of 1" square hollow box section (Figure 1a).The main reason for selection of GI square box hollow section, is to reduce the weight of the machine to a considerable extent.

Shaft for Sprocket Mounting

The solid circular shafts are used for sprocket mountings (Figure 1b). The shafts are supported by the bearings. The material used for the shaft is "Bright drawn mild steel", which is an improved quality material, free of scale, and has been cold worked (drawn or roll) to size. It is produced to close dimensional tolerances. Straightness and flatness are better than black steel. It is more suitable for repetitive precision machining. Bright drawn steel has more consistent hardness, and increased tensile strength. The diameter of the shaft is 25 mm and its length is 150 mm.

Bearing Block

A bearing block is a pedestal used to provide support for a rotating shaft with the help of compatible bearings and various accessories (Figure 1c). Housing material for a block bearing is typically made of mild steel. The bearing's bolt hole diameter is 10 mm, bearing height is 55 mm and bore size is 32 mm.

Bearing

A bearing is a machine element that constrains the relative motion to the desired motion, and reduces friction between moving parts. The design of the bearing may, for example, provide for free linear movement of the moving part or for free rotation around a fixed axis; or, it may prevent a motion by controlling the vectors of normal forces that bear on the moving parts. The movement may be of sliding or rotating type (Figure 1d).

Sprocket

A sprocket is a profiled wheel with teeth that mesh with a chain (Figure 1e). Sprocket is mounted on the axle. Sprocket is used to transmit rotary motion between two shafts. In the RFID robot sprockets are used to transmit power from driving shaft to the driven shaft by means of a chain. Two sprockets with same number of teeth have been used to maintain the same speed between the driver and driven shaft. The number of teeth on the sprocket is 16 and its width is 3.2 mm.

Wheel

Proper selection of wheel is very important in order to overcome the total tractive force and enable the robot to move freely. In this robot there are four wheels of 6" each. The wheels are provided with proper gripping material i.e. rubber for the purpose of effective turning. The drum of the wheel is made up of nylon (Figure 1f).

Chain Drive

The chain drive is a way of transmitting mechanical power from one place to another (Figure 1g). For the RFID robot, roller chain drive is used because the distance between the sprockets is short and is very efficient with minimum frictional loss. Finite numbers of links form the chain drive. The pitch of the chain is the distance between its drive links. It is determined by measuring the distance between any three consecutive drive links and dividing by two. The chain's width is 3.2 mm, pitch is 12.7 mm and the number of links is 110.

Axle

An axle is a central shaft for a rotating wheel. At one end of the axle wheel is fixed using M10 locking nut, the other end is directly coupled to the DC Motor. Axle is mounted on the chassis through the bearing blocks in order to provide the relative motion. Each axle is fixed with a sprocket in order to transmit power to the adjacent driven wheel (Figure 1h).

Sheet Metal Body

The sheet metal body is made up of galvanized iron of 1 mm thickness (Figure 1i). The main purpose of sheet metal body is to enclose the complete robot to safeguard all the electrical & electronic components from the pesticides. The skeleton of the body is made of mild steel in order to provide sufficient strength and also acts as a support to solar panel and pesticides spraying attachment.

Figure 1. Mechanical components of RFID robot

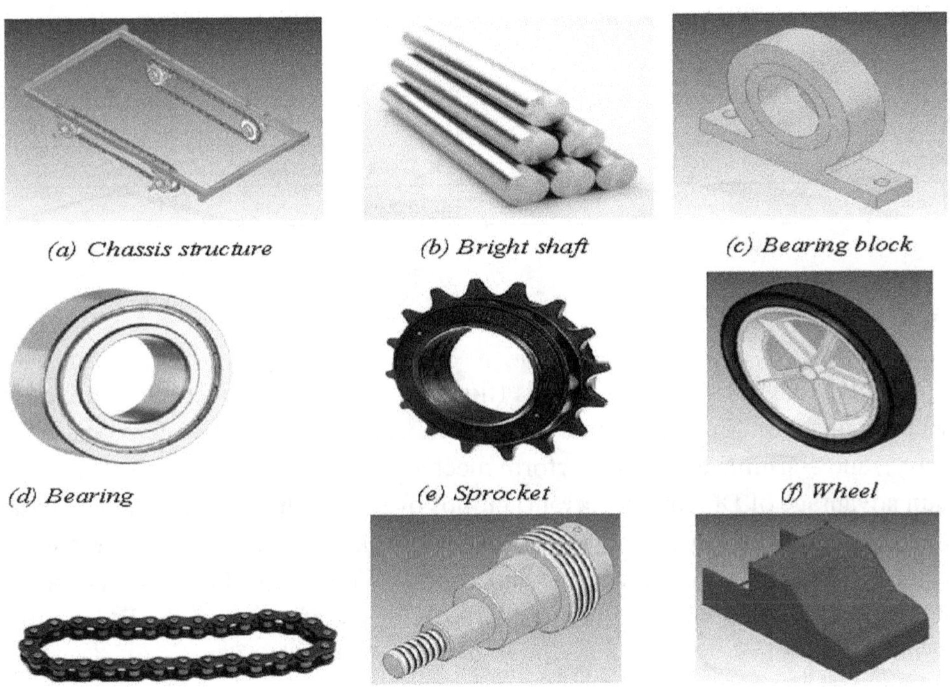

(a) Chassis structure *(b) Bright shaft* *(c) Bearing block*

(d) Bearing *(e) Sprocket* *(f) Wheel*

(g) Chain *(h) Axle* *(i) Sheet metal body*

ELECTRICAL COMPONENTS

DC Motor

A DC motor is any of a class of rotary electrical machines that converts direct current (DC) electrical energy into mechanical energy (Figure 2a). Nearly all types of DC motors have some internal mechanism, either electromechanical or electronic; to periodically change the direction of current flow in part of the motor. A DC motor's speed can be controlled over a wide range, using either a variable supply voltage or by changing the strength of current in its field windings. In the RFID robot a DC motor is used to run the wheels of the robot. The specification of DC Motor is as shown in Table 1.

Table 1. Specification of DC motor

Voltage	12 V
Speed	45 rpm
Power	35 Watts
Torque	6.76 N-m

DC Pump

A pump is a device that moves fluids (liquids or gases) or sometimes slurries by mechanical action. Pumps operate by some mechanism (typically reciprocating or rotary), and consume energy to perform mechanical work by moving the fluid. The main advantage of DC (direct current) pumps over AC (alternating current) pumps is that they can operate directly from a battery, making them more convenient and portable. They are easier to operate and control (Figure 2b). In the RFID robot a DC pump is used for spraying the pesticides. The specification of the DC pump is as shown in Table 2.

Table 2. Specification of DC pump

Voltage	12 V
Power	60 W
Rate of flow	5 lit/min
Weight	0.628 kg

Battery

An electric battery is a device consisting of one or more electrochemical cells with external connections provided to power electrical devices such as flashlights, smart phones, and electric cars (Figure 2c). When a battery is supplying electric power, its positive terminal is the cathode and its negative terminal is the anode. The terminal marked negative is the source of electrons that when connected to an external circuit will flow and deliver energy to an external device. When a battery is connected to an external circuit, electrolytes are able to move as ions within, allowing the chemical reactions to be completed at the separate terminals and so deliver energy to the external circuit. It is the movement of those ions within the battery which allows current to flow out of the battery to perform work. In the RFID robot a rechargeable battery has been used which stores the energy received from the solar panel. The battery is of 12 V, 7.2 Ah and weighs 1.4 kg.

Solar Panel

The photovoltaic solar panels absorb sunlight as a source of energy to generate electricity (Figure 2d). A photovoltaic (PV) module is a packed; connect assembly of typically 6x10 photovoltaic solar cells. The photovoltaic modules constitute the photovoltaic array of a photovoltaic system that generates and supplies solar electricity in commercial and residential applications. Solar panels are devices that convert light into electricity. They are called "solar" panels because most of the time, the most powerful source of light available is the sun. In the RFID robot solar panel is used to charge the battery and use that power to drive the motor and the pump. The specification of the solar panel is as shown in Table 3.

Table 3. Specification of solar panel

No of Cells	36
Weight	1.2 kg
Voltage	12V
Power	30W

Charge Controller

A charge controller or battery regulator limits the rate at which electric current is added to or drawn from electric batteries. It prevents overcharging and may protect against overvoltage, which can reduce battery performance or lifespan, and may pose a safety risk. It may also prevent completely draining ("deep discharging") a battery, or perform controlled discharges, depending on the battery technology, to protect battery life. The terms "charge controller" or "charge regulator" may refer to either a stand-alone device, or to control circuitry integrated within a battery pack, battery-powered device, or battery charger. The charge controller is of 12V and 10A (Figure 2e).

Relay

A relay is a simple electromechanical switch made up of an electromagnet and a set of contacts. Relays are found hidden in all sorts of devices. In fact, some of the first computers ever built used relays to implement Boolean gates. An electromagnetic relay is a type of electrical switch controlled by an electromagnet. The electromagnetic relay is used in a variety of applications, including alarms and sensors, signal switching, and the detection and control of faults in electrical distribution lines. In general, the point of a relay is to use a small amount of power in the electromagnet, say, from a small dashboard switch or a low power electronic circuit to move an armature that is able to switch a much larger amount of power. The relay used in RFID robot is of 12 V (Figure 2f).

Figure 2. Electrical components of RFID robot

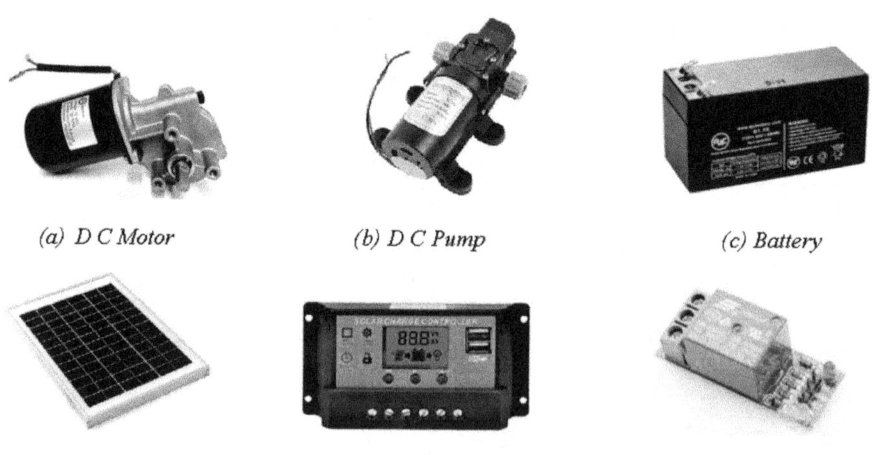

(a) D C Motor *(b) D C Pump* *(c) Battery*

(d) Solar panel *(e) Charge controller* *(f) Relay*

ELECTRONIC COMPONENTS

Arduino

Arduino is an open source computer hardware and software company, project, and user community that designs and manufactures single-board microcontrollers and micro-controller kits for building digital devices and interactive objects that can sense and control objects in the physical and digital world. Arduino board designs use a variety of microprocessors and controllers. The boards are equipped with sets of digital and analogue input/output (I/O) pins that may be interfaced to various expansion boards or bread boards (*shields*) and other circuits. The boards feature serial communications interfaces, including Universal Serial Bus (USB) on some models, which are also used for loading programs from personal computers. In this RFID robot Arduino is used to send signals to the RFID reader for turning purpose and an Ultrasonic sensor is also connected to Arduino for obstacle detection. GSM module is also interacted with Arduino for sending message to farmer for any obstacle (Figure 3a).

RFID Reader

The RFID Reader is a small electronic device used to transmit and receive radio signals between two devices (Figure 3b). In an embedded system it is often desirable to communicate with another device wirelessly. This wireless communication may be accomplished through radio frequency communication. RF communications incorporate a transmitter and a receiver. They are of various types and ranges. Some can transmit up to 500 feet. RF modules are widely used in electronic design owing to the difficulty of designing radio circuitry. Good electronic radio design is notoriously complex because of the sensitivity of radio circuits and the accuracy of components and layouts required for achieving operation on a specific frequency. RFID technology uses digital data in an RFID tag, which is made up of integrated circuits containing a tiny antenna for transferring information to an RFID transceiver. The majority of RFID tags contain at least an integrated circuit for modulating and demodulating radio frequency and an antenna for transmitting and receiving signals. Wavelengths in the 2.4 GHz range are limited because they can be absorbed by water. In this RFID robot it has been used for the navigation of the robot.

Figure 3. Electronic components of RFID robot

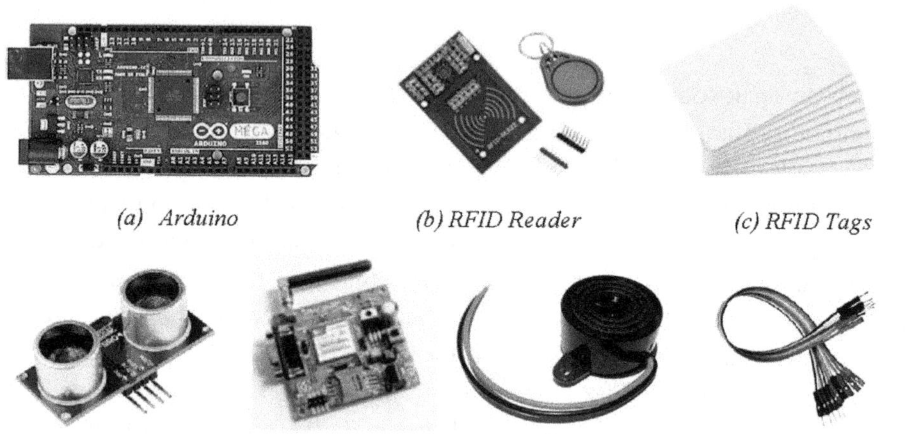

(a) Arduino *(b) RFID Reader* *(c) RFID Tags*

(d) Ultrasonic sensor *(e) GSM Module* *(f) Bumper* *(g) Jump wires*

RFID Tags

The RFID (Radio Frequency Identification) tags are made as simple as possible and they contain comparatively little in the way of electronics (Figure 3c). Fundamentally an RFID tag consists of two main elements:

Electronic Circuit: The electronic circuits within an RFID tag are kept to the minimum to ensure that the costs are minimized and power levels are kept as low as possible.

Antenna: The antenna within the RFID tag is the element that takes the largest amount of space. It must be able to operate satisfactorily at the frequency of operation.

Types of RFID Tags

Passive Tags: Passive RFID tags are by far the most common. They do not contain any power and receive this from the RFID reader. This is sufficient to power any device in the RFID tag and reply with the required data.

Semi-Passive Tags: This form of RFID tag uses a battery to supply the internal operation of the tag, but relies on the RFID reader to supply the power to transmit the signal to the reader.

Active Tags: An active RFID tag is one in which battery power is used to supply power to the electronics. This enables greater distances to be achieved as the tag is not dependent on the received power to provide a reflected signal.

Ultrasonic Sensor

An Ultrasonic sensor is a device that can measure the distance from an object by using sound waves. It measures distance by sending out a sound wave at a specific frequency and listening for that sound wave to bounce back. The Ultrasonic sensor sends out a high-frequency sound pulse and then times how long it takes for the echo of the sound to reflect back. The sensor has two openings on its front. One opening transmits ultrasonic waves, (like a tiny speaker), the other receives them, (like a tiny microphone). In this RFID robot it is used to detect the obstacle (Figure 3d).

GSM Module

A GSM modem is a device which is either a mobile phone or a modem that can be used to make a computer or any other processor communicate over a network. A GSM modem requires a SIM card to operate and operates over a network range subscribed by the network operator. It can be connected to a computer through serial, USB or Bluetooth connection (Figure 3e). A GSM modem can also be a standard GSM mobile phone with the appropriate cable and software driver to connect to a serial port or USB port on the computer. GSM modem is usually preferable to a GSM mobile phone. The GSM modem has wide range of applications in transaction terminals, supply chain management, security applications, weather stations and GPRS mode remote data logging. In this RFID robot it is used in order to send message to the farmer in case of any disturbance in the medium.

Buzzer

A buzzer or beeper is an audio signaling device, which may be mechanical, electromechanical, or piezoelectric type. In a buzzer, the simplest sort of doorbell, an electromagnet is used to operate a self-interrupting circuit. It requires a DC voltage to operate; it should generally be between 3V and 28V, depending on the model (Figure 3f).

Jumper Wires

A jump wire (also known as jumper, jumper wire, jumper cable, DuPont wire, or DuPont cable – named for one manufacturer of them) is an electrical wire, or group of them in a cable, with a connector or pin at each end (or sometimes without them – simply "tinned"), which is normally used to interconnect the components of a breadboard or other prototype or test circuit, internally or with other equipment or components, without soldering. Jumper wires are used for making connections between items on breadboard and Arduino's header pins (Figure 3g).

METHODOLOGY

Block Diagram of the Proposed RFID Robot

Figure 4. Block diagram of the proposed RFID robot

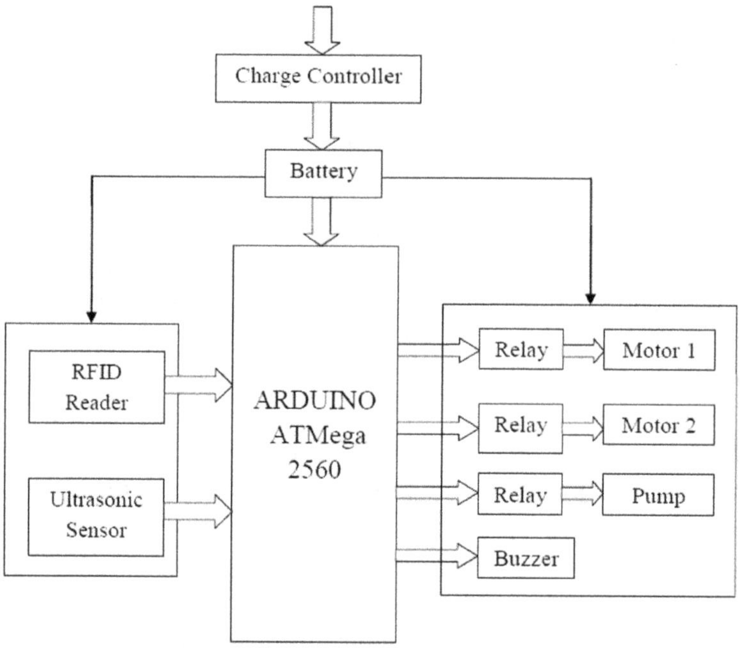

The block diagram of the proposed robot is shown in Figure 4. The control unit consists of an Arduino AT Mega 2560. The Arduino processes the data from the RFID reader module and the ultrasonic sensor to provide sufficient output. The DC motors used in the robot and the pump used for spraying are controlled based on the data processed by the Arduino. The system makes use of the RFID technology for automated navigation and an ultrasonic sensor for detecting the obstacles. In this proposed RFID robot ultrasonic sensor is placed at the front side of the robot.

The sensor at the front will be focused and if it detects any obstacle in the forward path of the robot, then the robot will be stopped. Normally the robot is designed to move on a straight path. The RFID reader module is mounted at the bottom of the robot. The RFID tags will be placed at different locations in the field where the robot has to take a turn. Each tag has a different UID (unique code of identification). While programming the Arduino, all the UID's of every tag are stored in Arduino to pre-define the path as per the need. The typical RFID technology has two main components namely RFID reader and RFID tags. The reader continuously produces radio frequency waves. As the waves hit the tags, the tags will be energized and will emit the data to the reader contained in it. The reader decodes the data encoded in the provided tags and the signal is passed on to the controller i.e. Arduino for further data processing.

Figure 5. The CAD modeling of the designed RFID robot: (a) front view, (b) side view, (c) top view, (d) 3D view (all dimensions are in 'mm')

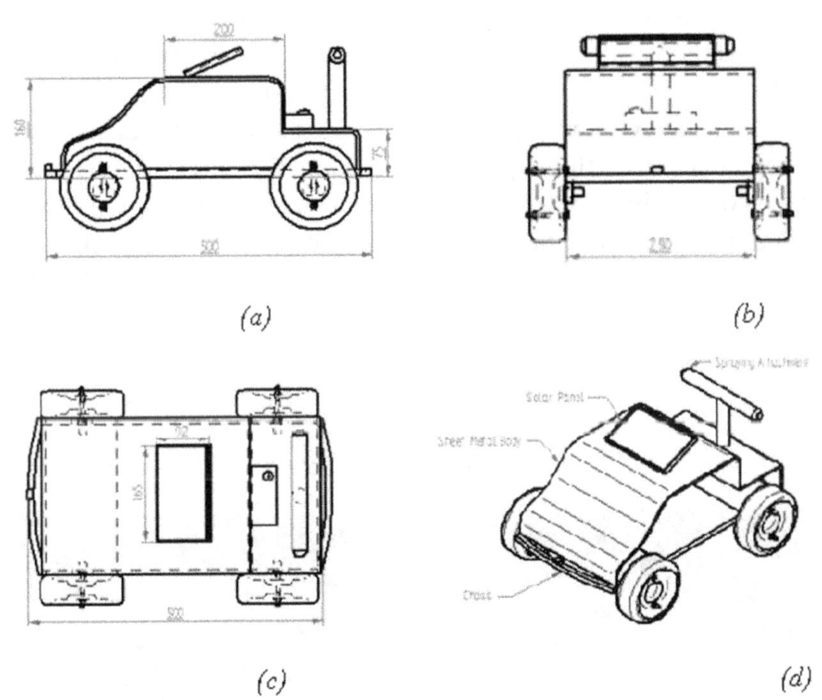

CAD Modeling of RFID Robot

Figure 5 depicts the CAD modeling of the designed RFID robot. Figure 5a-c shows the front, side and top views of the robot respectively and the Figure5d shows the three-dimensional view of the robot.

The Working Mechanism of RFID Robot

The RFID robot system consists of a solar panel which delivers an output of 12V and 30W power to the charging unit. The charging unit is used to strengthen the signal from the solar panel. A charge controller of 10A is used as a charging unit in the system which charges the battery. The power from the charged batteries is utilized for driving the motors and the pump. The chassis of the robot is made up of 1" G.I. hollow square pipe (Figure 6a). The layout of the RFID robot is shown in Figure 6b. The RFID robot is a four wheeled vehicle with drive mechanism being powered by DC motors to the front wheels. Since there is no steering mechanism, the drive is transmitted to the rear wheels with the help of the chain drive mechanism. The chassis of the robot is enclosed with sheet metal body to protect the electronic circuit board, batteries, motors, pump from the pesticides and the atmosphere. The pesticides are stored in a PVC tank. The proposed system can carry 10 liters of pesticide. Since the robot will be used in different kinds of fruits and vegetable fields, it is essential to vary the height of the nozzles. The height of the nozzles can be varied by incorporating a rack and pinion mechanism to the spraying attachment. The drive is provided to the pinion through the use of 12V, 60 rpm DC motor. A two-way switch is used to change the direction of the motor which will allow the spraying attachment to move accordingly. The navigation is automated with the use of RFID technology.

Figure 6. CAD design of (a) chassis, (b) layout of robot

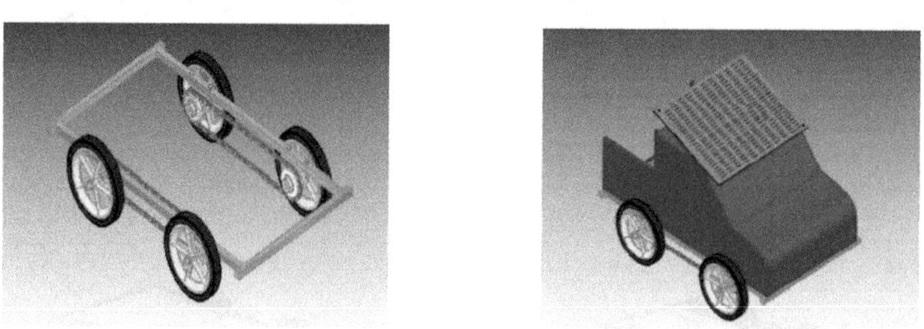

(a) (b)

RESULTS AND DISCUSSION

The designed and fabricated (manufactured)RFID robot is shown in Figure 7. It is fully an automated robot developed using RFID technology for outdoor applications mainly for spraying pesticides in the fruits and the vegetable fields like vineyards, strawberry, cucumber, bell pepper etc. The height of the nozzles used for spraying pesticides can be varied depending on the size of the crop.

Argade et al. (2017) have developed a solar powered robot for spraying pesticides but the robot is navigated by a remote controller. Ahalya et al. (2017) have developed a solar powered robot for spraying pesticides but it is a semiautomatic robot having three wheels and controlled by a microcontroller. Pandurang Lad et al. (2015) have proposed a solar operated pesticide sprayer wherein they have used solar energy for operating the pump. Mitul Raval et al. (2015) have developed a mobile robot using the ARM7 controller for spraying the pesticides. Peng Jian-Sheng (2014) had developed an intelligent robot for spraying pesticides but the robot is not powered by the solar energy and a mobile is used for its navigation. Surya Shankar & Valarmathi (2017) also have developed a robot based on the RFID technology but it is for indoor applications and more over the accuracy of the navigation of the robot was very poor. Tsukiyama (2011) had developed a mobile robot based on the same RFID technology but it is for indoor applications. The developed robot in the present chapter is a battery operated and doesn't require a remote controller or mobile for its navigation and the batteries are being recharged by using the solar energy.

Figure 7. Manufactured RFID robot

SCOPE FOR THE FUTURE WORK

For further improvement the following features can be incorporated in the proposed RFID robot:

- The spraying mechanism can be closely observed by using a camera which could be mounted near the robotic arm giving the farmer live feedback about the spraying.
- Integrated GSM (Global System for Mobile communication) module which could control the start/stop and run operation of the robot.
- SMS (Short Message Service) based system to start and stop the service.
- Pre-programmed GUI (Graphical User Interface) based navigation system.
- Programming based on the type of the crop and the quantity.
- Combination of both RFID Technology and GPS system can be used for the effective navigation purpose.

CONCLUSION

The following conclusions may be drawn from this chapter on RFID robot:

- A fully automated battery-operated pesticide spraying robot has been developed by using RFID technology.
- Solar energy is used for recharging the batteries.
- RFID reader and tags are used for the navigation of robot.
- Height of the nozzles used for spraying the pesticides can be varied depending on the size of the crop.
- An ultrasonic sensor is used for detecting any obstacle in the path of robot.
- The developed automated solar powered robot is suitable for spraying the pesticides in the fruits and the vegetable fields.

ACKNOWLEDGMENT

This project was supported by The Institution of Engineers (India) – IE(I), India [RDUG2017020].

REFERENCES

Adamides, Katsanos, Constantinou, Christou, Xenos, Hadzilacos, & Edan. (2020). Design and development of a semi-autonomous agricultural vineyard sprayer: Human–robot interaction aspects. *Journal of Field Robotics, 34*(8), 1407–1426.

Ahalya, M., Muktha, A., Veena, M., Vidyashree, G., & Rehna, V. J. (2017). Solar Powered Semi-Automatic Pesticides Sprayer for use in Vineyard. *SSRG International Journal Electronics and Communication Engineering*, *4*(4), 55–58.

Ashish, T., Meshram, Anil, V. V., Kavita, B. K., & Avinash, M. B. (2022).Pesticide spraying robot for precision agriculture: A categorical literature review and future trends. *Journal of Field Robotics, 39*(2), 153 – 171.

Ashutosh, B. A., Vivek, D. J., Rushabh, R. S., & Ganesh, K. G. (2020). Agricultural pesticide spraying Robotic system controlled using Android application. *International Journal of Innovations in Engineering Research and Technology*, *7*(4), 62–67.

Baltazar, A. R., Neves dos Santos, F., Moreira, A. P., Valente, A., & Cunha, J. B. (2021). Smarter Robotic sprayer system for precision agriculture. *Electronics (Basel)*, *10*(17), 2061–2070.

Chaitanya, P. V. R., Dileep Kotte, A., Srinath, A., & Kalyan, K. B. (2020). Development of smart pesticide spraying Robot. *International Journal of Recent Technology and Engineering*, *8*(5), 2193–2202.

Hasuda, Y., & Ozaki, K. (2022). The pesticide spraying Robot moving in a greenhouse with a Line-Trace. *International Journal of Engineering Research & Technology (Ahmedabad)*, *11*(1), 304–307.

Jiwtode, Ukani, Sonaskar, & Saurabh. (2022). Smart pesticide spraying robot. *International Research Journal of Engineering and Technology*, *99*(7), 706–710.

Kanse, Masekar, & Pokale, & Patil,. (2021). Smart pesticide spraying Robot. *InternationalJournal of Creative Research Thoughts*, *9*(4), 5683–5695.

Kassim, A. M., Termezai, M. F. N. M., Jaya, A. K. R. A., Azahar, A. H., Sivarao, S., Jafar, F. A., Jaafar, H. I., & Aras, M. S. M. (2020). Design and development of autonomous pesticide sprayer Robot for fertigation farm. *International Journal of Advanced Computer Science and Applications*, *11*(2), 545–551.

Kavyashree, N., & Vishwanath, N. P. (2022). Operation of agriculture based pesticide spraying Robot. *EPRA International Journal of Multidisciplinary Research*, *8*(7), 180–185.

Lad, P., Patil, V., Patil, P., Patil, T., & Patil, P. (2015). Solar operated pesticide sprayer. *International Journal of Advanced Research in Science & Engineering, 4*(01), 122–126.

Madhu, S., Sai Teja, U., Mahender, B., & Vamshi, D. (2023). Pesticide spraying robot. *International Journal of Advances in Engineering and Management, 5*(4), 772–777.

More, Kemble, Javir, Jadhav, & Kamble, & Sawant. (2022). Design and development of solar operated pesticide spraying vehicle. *Journal of Emerging Technologies and Innovative Research, 9*(8), 223–228.

Mounika, Thota, & Vallamdas. (2022). Agricultural pesticide spraying Robot. *International Journal of Innovative Research in Technology, 9*(4), 356–359.

Mustafid, M. A., Subrata, I. D. M., Pramuhadi, G., & Harahap, I. S. (2022). Design and performance test of autonomous precision spraying Robot for cabbage cultivation. *IOP Conference Series: Earth &Environmental Science, 1038.*

Mustak, A., Ma, H., Song, L., & Jin, Y. (2022). Embedded design of automatic pesticide spraying Robot control system. *10th International Symposium on Computational Intelligence and Industrial Applications (ISCIIA2022)*, 1–6.

Navya, P., & Sudha, D. (2023). Artificial intelligence based robot for harvesting, pesticide spraying and maintaining water management system in agriculture using IoT. *AIP Conference Proceedings, 2523*, 020025.

Peng, J.-S. (2014). An intelligent robot system for spraying pesticides. *The Open Electrical & Electronic Engineering Journal, 8*(1), 435–444. doi:10.2174/1874129001408010435

Prahlad, A. P., Bhagwan, B. S., Subhash, K. S., Vijay, P. N., & Kale, R. U. (2017). Solar powered automatic grass cutter and pesticides spreading robot. *International Research Journal of Engineering and Technology, 4*(5), 3372–3374.

Rajesh Kanna, P., & Vikram, R. (2020). Agricultural Robot – A pesticide spraying device. *International Journal of Future Generation Communication and Networking, 13*(1), 150–160.

Raval, M., Dhandhukia, A., & Mohile, S. (2015). Development and Automation of Robot with Spraying Mechanism for Agricultural Applications. *International Journal of Research in Emerging Science and Technology, 2*(8), 60–66.

Sandikar, Sawant, Jadhav, & Misal. (2022). Design and fabrication of pesticides spray Robot. *International Research Journal of Engineering and Technology, 9*(5), 1258–1261.

Shubhangi, Londhe, B. K., & Sujata, K. (2017). Remotely operated pesticide sprayer robot in agricultural field. *International Journal of Computers and Applications, 167*(3), 26–29.

Sulakhe, & Karanjkar. (2015). Design and operation of agriculture based pesticide spraying robot. *International Journal of Scientific Research, 6*(4), 1286–1289.

Sunny, M., & Badashah, S. J. (2022). Bluetooth Based Agrobot used to spray the pesticides. *International Journal for Research in Applied Science and Engineering Technology, 10*(6), 2449–2452.

Surya Shankar, R. P. & Valarmathi, K. (n.d.). *RFID based tag finding mobile Robot for indoor navigation.* www.slideshare.net/divakark3/ieee-confrence

Sushma Priya, K., Praneetha Reddy, R., & Pradeep, Y. (2022). Agricultural pesticide spraying Robot. *International Research Journal of Modernisation in Engineering Technology and Science., 4*(7), 800–804.

Swati, D. (2015). Seed sowing using robotics technology. *International Journal of Scientific Research and Management, 3*(5), 2889–2892.

Tsukiyama, T. (2011). RFID based navigation system for indoor mobile robots. *IFAC Proceedings, 44*(1), 1084 – 1089.

Compilation of References

| E-Agriculture. (n.d.). *Can artificial intelligence help improve agricultural productivity?* Can Artificial Intelligence help improve agricultural productivity? FAO. https://www.fao.org/e-agriculture/news/can-artificial-intelligence-help-improve-agricultural-productivity

Abdel-Fattah, M. K., Mokhtar, A., & Abdo, A. I. (2021). Application of neural network and time series modeling to study the suitability of drain water quality for irrigation: A case study from Egypt. *Environmental Science and Pollution Research International, 28*(1), 898–914. doi:10.100711356-020-10543-3 PMID:32822008

Abubakr, H., Mohamed, T. H., Hussein, M. M., Guerrero, J. M., & Agundis-Tinajero, G. (2021). Adaptive frequency regulation strategy in multi-area microgrids including renewable energy and electric vehicles supported by virtual inertia. *International Journal of Electrical Power & Energy Systems, 129*, 106814. doi:10.1016/j.ijepes.2021.106814

Adafruit. (n.d.). https://cdn-learn.adafruit.com/downloads/pdf/pir-passive-infrared-proximity-motion-sensor.pdf

Adamides, Katsanos, Constantinou, Christou, Xenos, Hadzilacos, & Edan. (2020). Design and development of a semi-autonomous agricultural vineyard sprayer: Human–robot interaction aspects. *Journal of Field Robotics, 34*(8), 1407–1426.

Aditya, S., & Kulkarni, S. C. (2016). Adoption and Utilization of Drones for Advanced Precision Farming: A Review. *International Journal on Recent and Innovation Trends in Computing and Communication, 4*(5), 563-565.

Agarwal, S., & Agarwal, N. (2017). *An algorithm based low-cost automated system for irrigation with soil moisture sensor*. International Conference on Computer Communication and Informatics (ICCCI -2017), Coimbatore, India. 10.1109/ICCCI.2017.8117741

Aggarwal. (2022). COVID-19 image classification using deep learning: Advances, challenges and opportunities. *Computers in Biology and Medicine*, 105350.

Agrahari, R. K., Kobayashi, Y., Tanaka, T. S. T., Panda, S. K., & Koyama, H. (2021). Smart fertilizer management: The progress of imaging technologies and possible implementation of plant biomarkers in agriculture. *Soil Science and Plant Nutrition, 67*(3), 248–258. doi:10.1080/00380768.2021.1897479

Ahalya, M., Muktha, A., Veena, M., Vidyashree, G., & Rehna, V. J. (2017). Solar Powered Semi-Automatic Pesticides Sprayer for use in Vineyard. *SSRG International Journal Electronics and Communication Engineering*, *4*(4), 55–58.

Ahsan. (2021). COVID-19 detection from chest X-ray images using feature fusion and deep learning. *Sensors, 21*(4), 1480.

Ahsan, M. M., Gupta, K. D., Islam, M. M., Sen, S., Rahman, M. L., & Shakhawat Hossain, M. (2020). Covid-19 symptoms detection based on nasnetmobile with explainable ai using various imaging modalities. *Machine Learning and Knowledge Extraction*, *2*(4), 490–504. doi:10.3390/make2040027

Alagumariappan, P., Dewan, N. J., Muthukrishnan, G. N., Raju, B. K., Bilal, R. A. A., & Sankaran, V. (2020). Intelligent plant disease identification system using machine learning. *Engineering Proceedings*, *2*, 49. doi:10.3390/ecsa-7-08160

Alakus. (2020). Comparison of deep learning approaches to predict COVID-19 infection. *Chaos, Solitons & Fractals*, *140*, 110120.

Alava, J., Kusme, K., Loor, A., & Santander, E. (2019). RoCoLe: A robusta coffee leaf images dataset for evaluation of machine learning based methods in plant diseases recognition. *Data in Brief*, *25*, 104414. doi:10.1016/j.dib.2019.104414 PMID:31516934

Al-Bahadly, I., & Thompson, J. (2015). Garden Watering System Based on Moisture Sensing. *International Conference on Sensing Technology*. 10.1109/ICSensT.2015.7438404

Alfian, G., Syafrudin, M., Farooq, U., Ma'arif, M. R., Syaekhoni, M. A., Fitriyani, N. L., Lee, J., & Rhee, J. (2019). Improving efficiency of RFID-based traceability system for perishable food by utilizing IoT sensors and machine learning model. *Food Control*, *110*, 107016. doi:10.1016/j.foodcont.2019.107016

Alhari, M. I. (2022). Developing IT Strategic Transformation of Smart Village Concept for Indonesian Village Model. *2022 International Conference Advancement in Data Science, E-learning and Information Systems (ICADEIS)*. IEEE. 10.1109/ICADEIS56544.2022.10037570

Ali & Abdin. (2021). Impact of Activation Functions and Number of Layers on the Classification of Fruits using CNN. *2021 8th International Conference on Computing for Sustainable Global Development (INDIA Com)*, 227-231.

Ali, M. M., Bachik, N. A., Muhadi, N. A., Yusof, T. N. T., & Gomes, C. (2019). Non-destructive techniques of detecting plant diseases: A review. *Physiological and Molecular Plant Pathology*, *108*, 101426–101437. doi:10.1016/j.pmpp.2019.101426

Almadhor, A., Rauf, H., Lali, M., Damasevicius, R., Alouffi, B., & Alharbi, A. (2021). Ai-driven framework for recognition of guava plant diseases through machine learning from dslr camera sensor based high resolution imagery. *Sensors (Basel)*, *21*(11), 3830. doi:10.339021113830 PMID:34205885

Alonso, J., Castañón, Á. R., & Bahamonde, A. (2013). Support Vector Regression to predict carcass weight in beef cattle in advance of the slaughter. *Computers and Electronics in Agriculture, 91*, 116–120. doi:10.1016/j.compag.2012.08.009

Alonso, J., Villa, A., & Bahamonde, A. (2015). Improved estimation of bovine weight trajectories using Support Vector Machine Classification. *Computers and Electronics in Agriculture, 110*, 36–41. doi:10.1016/j.compag.2014.10.001

Alpaydin, E. (2010). *Introduction to Machine Learning*. MIT Press.

Al-Saddik, H., Simon, J.-C., & Cointault, F. (2017). Development of spectral disease indices for 'Flavescence Doree' grapevine disease identification. *Sensors (Basel), 17*(12), 2772. doi:10.339017122772 PMID:29186057

Anagnostis, A., Benos, L., Tsaopoulos, D., Tagarakis, A., Tsolakis, N., & Bochtis, D. (2021). Human activity recognition through recurrent neural networks for human-robot interaction in agriculture. *Sensors (Basel), 11*, 2188.

Anagnostis, A., Papageorgiou, E., & Bochtis, D. (2020). Application of Artificial Neural Networks for Natural Gas Consumption Forecasting. *Sustainability (Basel), 12*(16), 6409. doi:10.3390u12166409

Anil, K. (1989). *Fundamentals of Digital Image Processing*. Prentice-Hall.

Annabel, L., Annapoorani, T., & Deepalakshmi, P. (2019). Machine Learning for Plant Leaf Disease Detection and Classification–A Review. In *2019 International Conference on Communication and Signal Processing (ICCSP)*. IEEE. 10.1109/ICCSP.2019.8698004

Anvarjon, T., Mustaqeem, & Kwon, S. (2020). Deep-Net: A Lightweight CNN-Based Speech Emotion Recognition System Using Deep Frequency Features. *Sensors (Basel), 20*(18), 5212. doi:10.339020185212 PMID:32932723

Ashish, T., Meshram, Anil, V. V., Kavita, B. K., & Avinash, M. B. (2022).Pesticide spraying robot for precision agriculture: A categorical literature review and future trends. *Journal of Field Robotics, 39*(2), 153 – 171.

Ashutosh, B. A., Vivek, D. J., Rushabh, R. S., & Ganesh, K. G. (2020). Agricultural pesticide spraying Robotic system controlled using Android application. *International Journal of Innovations in Engineering Research and Technology, 7*(4), 62–67.

AWMA. (1992). *Air & Waste Management Association, Air Pollution Engineering Manual*. Van Nostrand Reinhold.

Ayman, H. A. E., & Ayman, A. A. K. Understanding color image processing by machine vision for biological materials. In *Structure and function of food engineering* (pp. 227–274). INTEHC.

Badia Melis, R., Qian, J. P., Fan, B. L., Hoyos-Echevarria, P., Ruiz-García, L., & Yang, X. T. (2016). Artificial neural networks and thermal image for temperature prediction in apples. *Food and Bioprocess Technology, 9*(7), 1089–1099. doi:10.100711947-016-1700-7

Balleda, K. (2014). Agpest: An efficient rule-based expert system to prevent pest diseases of rice & wheat crops. In *Proc. Intelligent Systems and Control (ISCO)-*. IEEE. (Original work published 2014) doi:10.1109/ISCO.2014.7103957

Baltazar, A. R., Neves dos Santos, F., Moreira, A. P., Valente, A., & Cunha, J. B. (2021). Smarter Robotic sprayer system for precision agriculture. *Electronics (Basel)*, *10*(17), 2061–2070.

Banus, N., Boada, I., Xiberta, P., Toldra, P., & Bustins, N. (2021). Deep learning for the quality control of thermoforming food packages. *Scientific Reports*, *11*(1), 1–15. doi:10.103841598-021-01254-x PMID:34750436

Barbedo, J. G. A. (2013). Digital image processing techniques for detecting, quantifying and classifying plant diseases. *SpringerPlus*, *2*(1), 660–671. doi:10.1186/2193-1801-2-660 PMID:24349961

Baun, D. L., Bergfeldt, B., Mogensen, E. P. B., Vehlow, J., & Christensen, T. H. (2004). Thermal treatment of stabilised APC-residue in a waste incinerator pilot plant. Part 2: Leaching characteristics of bottom ashes. *Waste Management & Research*, *22*, 58–68. doi:10.1177/0734242X04042284 PMID:15113115

Bayar. (2018). Constrained convolutional neural networks: A new approach towards general purpose image manipulation detection. *IEEE Transactions on Information Forensics and Security*, *13*(11), 2691-2706.

Beachler, D. S., Joseph, J., & Pompelia, M. (1995). *Lesson 2: Fabric Filter Bag Cleaning*. North Carolina State University.

Bednarczyk, A. T. (2022). *From good to great: the next step in embedded navigation*. https://www.tomtom.com/newsroom/explainers-and-insights/future-of-embedded-navigation/

Belasque, J. Jr, Gasparoto, M. C. G., & Marcassa, L. G. (2008). Detection of mechanical and disease stresses in citrus plants by fluorescence spectroscopy. *Applied Optics*, *47*(11), 1922–1926. doi:10.1364/AO.47.001922 PMID:18404192

Ben, L., Gaudreau, H., Dallaire, L., Tessier, M., & Fliss, I. (2019). Bioprotective culture: A new generation of food additives for the preservation of food quality and safety. Industrial Biotechnology Journal, 15(3), 138-147.

Bencini, L., Chiti, F., Collodi, G., Di Palma, D., Fantacci, R., Manes, A., & Manes, G. (2009). Agricultural Monitoring Based on Wireless Sensor Network Technology: Real Long-life Deployments for Physiology and Pathogens Control. In *Third International Conference on Sensor Technologies and Applications*. IEEE. 10.1109/SENSORCOMM.2009.63

Benos, L., Tagarakis, A. C., Dolias, G., Berruto, R., Kateris, D., & Bochtis, D. (2021). Machine Learning in Agriculture: A Comprehensive Updated Review. *Sensors (Basel)*, *21*(11), 3758. doi:10.339021113758 PMID:34071553

Bernhard, A. (2019). *The invention that saved a million ships*. BBC.

Bevrani, H., Habibi, F., Babahajyani, P., Watanabe, M., & Mitani, Y. (2012). Intelligent frequency control in an AC microgrid: Online PSO-based fuzzy tuning approach. *IEEE Transactions on Smart Grid*, *3*(4), 1935–1944. doi:10.1109/TSG.2012.2196806

Bhagat, P. R., Naz, F., & Magda, R. (2022). Artificial intelligence solutions enabling sustainable agriculture: A bibliometric analysis. *PLoS One*, *17*(6), e0268989. doi:10.1371/journal.pone.0268989 PMID:35679287

Bhole & Kumar. (2020). Mango quality grading using deep learning technique: perspectives from agriculture and food industry. *Proceedings of the 21st Annual Conference on Information Technology Education,* 180-186.

Bing, C. (2014). The Design and Operation of City Agricultural Sightseeing System with the Internet of Things Technology. In *Sixth International Conference on Measuring Technology and Mechatronics Automation.* IEEE.

Bonnet, P., Joly, A., Goëau, H., Champ, J., Vignau, C., Molino, J.-F., Barthélémy, D., & Boujemaa, N. (2016). Plant identification: Man vs. machine. *Multimedia Tools and Applications*, *75*(3), 1647–1665. doi:10.100711042-015-2607-4

Borges, E., Matos, A. P., Cardoso, J. M., Correia, C., Vasconcelos, T., & Gomes, N. (2012). *Early detection and monitoring of plant diseases by Bioelectric Impedance Spectroscopy.* In *Proceedings of the 2012 IEEE 2nd Portuguese Meeting in Bioengineering (ENBENG)*, Coimbra, Portugal. 10.1109/ENBENG.2012.6331377

Brooks, S., Klomchit, A., Chimthai, S., Jaidee, W., & Bastian, A. C. (2022). SRNE2BP a Fungal Endophyte with Biocontrol Properties to Control Early Blight and Fusarium Wilt Disease in Tomato and Plant Growth Promotion Activity. *Current Microbiology*, *79*(4), 1–15. doi:10.100700284-022-02803-x PMID:35175450

Business Bliss Consultants. (2018). *Importance of Image Processing Technology in Agriculture.* UK DISS. https://ukdiss.com/examples/image-processing-for-agriculture-sector.php?vref=1

Butz, P., Hofmann, C., & Tauscher, B. (2005). Recent developments in NoninvasiveTechniques for freshfruit and vegetable internalquality analysis. *Journal of Food Science*, *70*(9), 131–141. doi:10.1111/j.1365-2621.2005.tb08328.x

Camargo, A., & Smith, J. S. (2009). An imageprocessing based algorithm to automatically. identify plant disease visual. symptoms. *Biosystems Engineering*, *102*(1), 9–21. doi:10.1016/j.biosystemseng.2008.09.030

Cen, H., Weng, H., Yao, J., He, M., Lv, J., Hua, S., Li, H., & He, Y. (2017). Chlorophyll fluorescence imaging uncovers photosynthetic fingerprint of citrus Huanglongbing. *Frontiers in Plant Science*, *8*, 1509–1519. doi:10.3389/fpls.2017.01509 PMID:28900440

Chaitanya, P. V. R., Dileep Kotte, A., Srinath, A., & Kalyan, K. B. (2020). Development of smart pesticide spraying Robot. *International Journal of Recent Technology and Engineering*, *8*(5), 2193–2202.

Chakrabarti, S. K., Sharma, S., & Shah, M. A. (2022). Potato Pests and Diseases: A Global Perspective. In *Sustainable Management of Potato Pests and Diseases* (pp. 1–23). Springer. doi:10.1007/978-981-16-7695-6_1

Chandler, S. (2019, February 7). *What are the advantages of a smart irrigation system? TechTarget.* IoT Agenda. Retrieved April 26, 2023, from https://www.techtarget.com/iotagenda/answer/What-are-the-advantages-of-a-smart-irrigation-system

Chaudhary, P., & Chaudhari, A. (2012). Color Transform Based Approach for Diseases SpotDetection on Plant Leaf. *International Journal of Computer Science and Telecommunications, 3*(6).

Cheng, Q., Zhang, S., Bo, S., Chen, D., & Zhang, H. (2020). Augmented Reality Dynamic Image Recognition Technology Based on Deep Learning Algorithm. *IEEE Access : Practical Innovations, Open Solutions*, *8*, 137370–137384. doi:10.1109/ACCESS.2020.3012130

Chicco. (2021). The Matthews correlation coefficient (MCC) is more reliable than balanced accuracy, bookmaker informedness, and markedness in two-class confusion matrix evaluation. *BioData Mining, 14*(1), 1-22.

Chlingaryan, A., Sukkarieh, S., & Whelan, B. (2018). Machine learning approaches for crop yield prediction and nitrogen status estimation in precision agriculture: A review. *Computers and Electronics in Agriculture*, *151*, 61–69. doi:10.1016/j.compag.2018.05.012

Choudhar, G. M., & Gulati, V. (2015). *Advances in Image Processing for Detection of Plant Diseases*, *2*(2), 135–141.

Choudhary, R. C., Bairwa, H. L., Kumar, U., Javed, T., Asad, M., Lal, K., Mahawer, L. N., Sharma, S. K., Singh, P., Hassan, M. M., Abo-Shosha, A. A., Rajagopal, R., & Abdelsalam, N. R. (2022). Influence of organic manures on soil nutrient content, microbial population, yield and quality parameters of pomegranate (Punica granatum L.) cv. Bhagwa. *PLoS One*, *17*(4), e0266675. doi:10.1371/journal.pone.0266675 PMID:35436290

Chowdhury. (2020). Can AI help in screening viral and COVID-19 pneumonia? *IEEE Access, 8*, 132665-132676.

Chowdhury, M., Rahman, T., Khandakar, A., Ayari, M., Khan, A., Islam, M., Emadi, N., Reaz, M., & Ali, S. (2021). Automatic and reliable leaf disease detection using deep learning techniques. *Agricultural Engineering*, *3*(2), 294–312.

Cohn, R. D., Arbes, S. J. Jr, Jaramillo, R., Reid, L. H., & Zeldin, D. C. (2006). National prevalence and exposure risk for cockroach allergen in U.S. households. *Environmental Health Perspectives*, *114*(4), 522–526. doi:10.1289/ehp.8561 PMID:16581539

Collewet, C., & Marchand, E. (2009). Colorimetry-based visual servoing. In *2009 IEEE/RSJ Interna-tional Conference on Intelligent Robots and Systems.* IEEE. .10.1109/IROS.2009.5354416

Colomina, I., & Molina, P. (2014). Unmanned aerial systems for photogrammetry and remote sensing: A review. *ISPRS Journal of Photogrammetry and Remote Sensing*, *92*, 79–97. doi:10.1016/j.isprsjprs.2014.02.013

Costa, C., Antonucci, F., Pallattino, F., Aguzzi, J., Sun, D. W., & Menesatti, P. (2011). Shapeanalysis of agriculturalproducts: A review of recentresearch advances andpotential application to computer vision. *Food and Bioprocess Technology*, *4*(5), 673–692. doi:10.100711947-011-0556-0

Courtenay, J., & Bryant, M. (2008). *Pleated cartridges provide increased baghouse capacity and improved filter performance.* Academic Press.

Crane-Droesch, A. (2018). Machine learning methods for crop yield prediction and climate change impact assessment in agriculture. *Environmental Research Letters*, *13*(11), 114003. doi:10.1088/1748-9326/aae159

Crop Protection Innovations. (2022, June 30). Retrieved October 10, 2022, from https://www.bayer.com/en/agriculture/crop-

Cvar, N. (2020). *The Use of IoT Technology in Smart Cities and Smart Villages: Similarities, Differences, and Future Prospects*. MDPI. www.mdpi.com/journal/sensors

Damadam, S., Zourbakhsh, M., Javidan, R., & Faroughi, A. (2022). An Intelligent IoT Based Traffic Light Management System: Deep Reinforcement Learning. *Smart Cities*, *5*(4), 1293–1311. doi:10.3390martcities5040066

DataFlair. (2019). *Importance of AI in Healthcare Sector*. https://data-flair.training/blogs/ai-in-healthcare-sector/amp

De Paul Obade, V., & Rattan, L. (2016). Towards a standard technique for soil quality assessment. *Geoderma*, *265*(8), 96–102. doi:10.1016/j.geoderma.2015.11.023

Debauche, O., Mahmoudi, S., Elmoulat, M., & Sidi Mahmoudi, P. (2020). Edge AI-IoT pivot irrigation, plant diseases, and pests' identification. *Procedia Computer Science*, *177*, 40–48. doi:10.1016/j.procs.2020.10.009

Degada, A., Thapliyal, H., & Mohanty, S. P. (2021). Smart Village: An IoT Based Digital Transformation In *2021 IEEE 7th World Forum on Internet of Things (WF-IoT)*. IEEE. 10.1109/WF-IoT51360.2021.9594980

Deisenroth, M. P., Faisal, A. A., & Ong, C. S. (2020). *Mathematics for Machine Learning*. Cambridge University Press. doi:10.1017/9781108679930

Deshusses, M., & Cox, H. H. J. (2003). *Biotrickling Filters for Air Pollution Control Chapter*. doi:10.1002/0471263397.env105

Dharmaraj, V. & Vijayanand, C. (2018). Artificial Intelligence (AI) in Agriculture. *International Journal of Current Microbiology and Applied Sciences, 7*(12)

Dhingra, G., Kumar, V., & Joshi, H. D. (2017). Study of digital image processing techniques for leaf disease detection and classification. *Multimedia Tools and Applications*, *77*(15), 19951–20000. doi:10.100711042-017-5445-8

Digiteum Team. (2022, November 25). *How to build an IOT solution for an agricultural irrigation system*. Digiteum. Retrieved April 26, 2023, from https://www.digiteum.com/iot-solutions-agricultural-irrigation-system/

DIN EN 1948: Teil 1-3. (1996). *Bestimmung der Massenkonzentration von PCDD/F*. Determination of Massconcentration of PCDD/F. Deutsche Norm. VDI/DIN Handbuch Reinhaltung der Luft, Band 5, Beuth Verlag.

Ding, D., Han, Q. L., Wang, Z., & Ge, X. (2019). A survey on model-based distributed control and filtering for industrial cyber-physical systems. *IEEE Transactions on Industrial Informatics*, *15*(5), 2483–2499. doi:10.1109/TII.2019.2905295

Divani, D., & Patil, P. (2016). Automated Plant Watering System. *International Conference on Computation of Power, Energy Information and Communication (ICCPEIC)*. 10.1109/ICCPEIC.2016.7557245

Dutta, R., Smith, D., Rawnsley, R., Bishop-Hurley, G., Hills, J., Timms, G., & Henry, D. (2015). Dynamic cattle behavioural classification using supervised ensemble classifiers. *Computers and Electronics in Agriculture*, *111*, 18–28. doi:10.1016/j.compag.2014.12.002

El Mezouari, El Fazziki, & Sadgal. (2022). Smart Irrigation System. *IFCA Papers Online*, *55*(10), 3298-3303.

Elavarasan, D., & Vincent, P. D. (2020). Crop yield prediction using deep reinforcement learning model for sustainable agrarian applications. *IEEE Access : Practical Innovations, Open Solutions*, *8*, 86886–86901. doi:10.1109/ACCESS.2020.2992480

Elavarasan, Vincent, D. R., Sharma, V., Zomaya, A. Y., & Srinivasan, K. (2018). Forecasting yield by integrating agrarian factors and machine learning models: A survey. *Computers and Electronics in Agriculture*, *155*, 257–282. doi:10.1016/j.compag.2018.10.024

Ella, S., & Andari, R. N. (2019). Utilization of ICT in Building a Smart Village Model for Village Development in Indonesia. *In 2019 International Conference on ICT for Smart Society (ICISS)*. IEEE. 10.1109/ICISS48059.2019.8969820

EPA. (1992). *Control Techniques for Volatile Organic Emissions from Stationary Sources*. EPA-453/R-92-018.

Espinoza, C. Z., Khot, L. R., Sankaran, S., & Jacoby, P. W. (2017). High Resolution Multispectral and Thermal Remote Sensing Based Water Stress assessment. In Subsurface Irrigated Grapevines. *Remote Sensing (Basel)*, *9*(9), 961. doi:10.3390/rs9090961

Evers, M. (2022). Urban Planning Goes Digital in Spanish 'Smart City'. *Spiegel*. https://www.spiegel.de/international/world/santander-a-digital-smart-city-prototype-in-spain-a-888480.html

Fabio, V., Athina, K., Rene, B. & Martin, R. (2017). Assessing accuracy and geographical transferability of machine learning algorithms for wind speed modelling, 297-310. *Lecture Notes in Geoinformation and Cartography*. Springer. doi:10.1007/978-3-319-56759-4_17

Facts, R. S. (2022). *How Can Automated and Connected Vehicles Improve Road Safety*. https://roadsafetyfacts.eu/how-can-automated-and-connected-vehicles-improve-road-safety/

Farnoud, A. (2008). *Electrostatic Removal of Diesel Particulate Matter*. ProQuest.

Feng, Y.-Z., & Sun, D.-W. (2012, November). Application of Hyperspectral imaging in food safety inspection and control: A review. *Critical Reviews in Food Science and Nutrition, 52*(11), 1039–1058. doi:10.1080/10408398.2011.651542 PMID:22823350

Freixenet, J., Munoz, X., Raba, D., Marti, J., & Cuff, X. (2002). Yet another survey on imagesegmentation: region andboundary information integration. In A. Heyden (Ed.) ECCV, LNCS 2352. Springer.

Fresnel. (1821). Note sur le calcul des teintes que la polarisation développe dans les lames cristallisées. *Annales de Chimie et de Physique, 17*, 102–11. doi:10.5281/zenodo.4058004

Fujiyoshi, H., Hirakawa, T., & Yamashita, T. (2019). Deep learning-based image recognition for autonomous driving. *IATSS Research, 43*(4), 244–252. doi:10.1016/j.iatssr.2019.11.008

Gammage, R. B., & Berven, B. A. (1996). *Indoor Air and Human Health* (2nd ed.). Lewis Publishers.

Gangavarapu, T., Jaidhar, C. D., & Chanduka, B. (2020). Applicability of machine learning in spam and phishing email filtering: Review and approaches. *Artificial Intelligence Review, 53*(7), 5019–5081. doi:10.100710462-020-09814-9

Gang, S., & Qiao, X.-L. (2008). AdaptiveColor Image Enhancement basedon Human Visual Properties. *Proceedings of International Congress on Image and Signal Processing*.

Gao, H., Zhu, F., & Cai, J. (2010). A review of Non-destructiveDetection forfruit Quality" *IFIP Advances in Information and Communication Technology, 317*, 133–140. doi:10.1007/978-3-642-12220-0_21

Gobalakrishnan, N., Pradeep, K., Raman, C. J., Ali, L., & Gopinath, M. P. (2020). A systematic review on image processing and machine learning techniques for detecting plant diseases. In *International Conference on Communication and Signal Processing (ICCSP)*. IEEE. 10.1109/ICCSP48568.2020.9182046

Goel, D., Gulati, P., Jha, S. K., Kumar, N., Khan, A., & Kumari, P. (2022). *Evaluation of Machine Learning Techniques for Crop Yield Prediction, Artificial Intelligence Applications in Agriculture and Food Quality Improvement*. IGI Global. doi:10.4018/978-1-6684-5141-0.ch007

Gonzalez-de-Soto, M., Emmi, L., Benavides, C., Garcia, I., & Gonzalez-de-Santos, P. (2016). Reducing air pollution with hybrid-powered robotic tractors for precision agriculture. *Biosystems Engineering, 143*, 79–94. doi:10.1016/j.biosystemseng.2016.01.008

Gonzalez, R. C. (1992). *DigitalImage Processin Reading*. Wesley Addison.

Gracia, L., Parra, L., Jimenez, J. M., Lloret, J., & Lorenz, P. (2020). IoT-Based Smart Irrigation Systems: An Overview on the Recent Trends on Sensors and IoT Systems for Irrigation in Precision Agriculture. *Surveys of Sensor Networks and Sensor Systems Deployments*, *20*(4), 1042. doi:10.339020041042 PMID:32075172

Guilmartin, J. F. (2022). *Unmanned aerial vehicle*. Encyclopedia Britannica. https://clearpathrobotics.com/blog/2021/01/husky-ugv-drives-smart-precision-agriculture-for-steep-slope-vineyards/

Guine, & Raquel. (2019). The use of artificial neural networks (ANN) in food process engineering. *International Journal of Food Engineering*, *5*(1), 15–21.

Gungor, V. C., Lu, B., & Hancke, G. P. (2010). Opportunities and challenges of wireless sensor networks in smart grid. *IEEE Transactions on Industrial Electronics*, *57*(10), 3557–3564. doi:10.1109/TIE.2009.2039455

Gutierrez, A., Ansuategi, A., Susperregi, L., Tub-o, C., T, I., & Da, L. (2019). A Benchmarking of Learning Strategies for Pest Detection and Identification on Tomato Plants for Autonomous Scouting Robots Using Internal Databases. Journal of Sensors.

Hasan, R., Yusuf, S., & Alzubaidi, L. (2020). Review of the state of the art of deep learning for plant diseases: A broad analysis and discussion. *Plants*, *9*(10), 1302. doi:10.3390/plants9101302 PMID:33019765

Hasuda, Y., & Ozaki, K. (2022). The pesticide spraying Robot moving in a greenhouse with a Line-Trace. *International Journal of Engineering Research & Technology (Ahmedabad)*, *11*(1), 304–307.

Hengl, T., Leenaars, J. G., Shepherd, K. D., Walsh, M. G., Heuvelink, G., Mamo, T., Tilahun, H., Berkhout, E., Cooper, M., Fegraus, E., Wheeler, I., & Kwabena, N. A. (2017). Soil nutrient maps of Sub-Saharan Africa: Assessment of soil nutrient content at 250 m spatial resolution using machine learning. *Nutrient Cycling in Agroecosystems*, *109*(1), 77–102. doi:10.100710705-017-9870-x PMID:33456317

Heredity. (2013). *The peppered moth and industrial melanism: Evolution of a natural selection case study*. Academic Press.

Hill, M. G., Connolly, P. G., Reutemann, P., & Fletcher, D. (2014). The use of data mining to assist crop protection decisions on kiwi fruit in New Zealand. *Computers and Electronics in Agriculture*, *108*, 250–257. https://cropforlife.com/list-of-agricultural-practices/ https://www.kdnuggets.com/2019/05/machine-learning-agriculture-applications-techniques.html. doi:10.1016/j.compag.2014.08.011

Hilmani, A. (2020). *Automated Real-Time Intelligent Traffic Control System for Smart Cities Using Wireless Sensor Networks*. https://www.hindawi.com/jounarls/wcmc/2020/8841893/

Hirsch, A., Parag, Y., & Guerrero, J. (2018). Microgrids: A review of technologies, key drivers, and outstanding issues. *Renewable & Sustainable Energy Reviews*, *90*, 402–411. doi:10.1016/j.rser.2018.03.040

Hopfield, J. J. (1982). Neural Networks and Physical Systems with Emergent Collective Computational Abilities. *Proceedings of the National Academy of Sciences of the United States of America*, *79*(8), 2554–2558. doi:10.1073/pnas.79.8.2554 PMID:6953413

Houlden, G., McNeill, S., Aminu-Kano, M., & Bell, J. N. B. (n.d.). *Air pollution and agricultural aphid pests. I: Fumigation experiments with SO_2 and NO_2*. Academic Press.

Hsu, S-L. (Ed.). (1999). *Brownfields and Property Values*. United States Environmental Protection Agency.

Huang, Y., Reddy, K. N., Fletcher, R. S., & Pennington, D. (2018). 2017. UAV Low-Altitude Remote Sensing for Precision Weed Management. *Weed Technology*, *32*(1), 2–6. doi:10.1017/wet.2017.89

Hua, Y., Zhang, N., Yuan, X., Quan, L., Yang, J., Nagasaka, K., & Zhou, X. (2019). Recent Advances in Intelligent Automated Fruit Harvesting Robots. *The Open Agriculture Journal*, *13*(1), 101–106. doi:10.2174/1874331501913010101

Imperial College at Silwood Park. (1990). Environmental Pollution. Academic Press.

India.gov. (n.d.). https://www.india.gov.in/spotlight/saansad-adarsh-gram-yojana

Industrial Melanism. (2012). College of William and Mary. doi:10.1002/9780470015902.a0001788.pub3

IoT Solutions World Congress. (2019, July 10). *IoT transforming the future of Agriculture: IoT Solutions World Congress: May 21 – 23 Barcelona*. IOT Solutions World Congress. Retrieved April 26, 2023, from https://www.iotsworldcongress.com/iot-transforming-the-future-of-agriculture/

Ismail, S., Dawoud, D. W., Ismail, N., Marsh, R., & Alshami, A. S. (2022). Iot-Based Water Management Systems: Survey and Future Research Direction. *IEEE Access : Practical Innovations, Open Solutions*, *10*, 35942–35952. Advance online publication. doi:10.1109/ACCESS.2022.3163742

Iyer. (2021). *AI enabled applications towards intelligent transportation*. doi:10.1016/j.treng.2021.100083

Jain, P. (2021, July 23). *Artificial Intelligence in agriculture: Using modern day AI to solve traditional farming problems*. Analytics Vidhya. https://www.analyticsvidhya.com/blog/2020/11/artificial-intelligence-in-agriculture-using-modern-day-ai-to-solve-traditional-farming-problems/

Jain, T., Garg, P., Tiwari, P. K., Kuncham, V. K., Sharma, M., & Verma, V. K. (2021). Performance prediction for crop irrigation using different machine learning approaches. In Examining the Impact of Deep Learning and IoT on Multi-Industry Applications (pp. 61-79). IGI Global. doi:10.4018/978-1-7998-7511-6.ch005

Java. (n.d.). *Artificial Intelligence in agriculture - javatpoint*. Java. https://www.javatpoint.com/artificial-intelligence-in-agriculture#:~:text=AI%20enables%20better%20decision%2Dmakin g&text=Moreover%2C%20AI%2Dpowered%20machines%20can,decisions%20and%20do%20efficient%20farming

Jayalakshmi, R., & Balasubramanian, T. (2015). A brief survey of classification techniques applied to soil fertility prediction. *International Journal of Scientific Research in Computer Science Applications and Management Studies*. http://www.ijsrcsams.com/images/stories/Past_Issue_Docs/ijsrcsamsv8i2p121.pdf

Jha, K., & Shah, M. (2019, June 17). *A comprehensive review on automation in agriculture using artificial intelligence*. Artificial Intelligence in Agriculture. https://www.sciencedirect.com/science/article/pii/S2589721719300182#bbb0025

Jia, H. Y. (2010). Soil fertility grading with Bayesian Network transfer learning. *IEEE International Conference on Machine Learning and Cybernetics*. IEEE. 10.1109/ICMLC.2010.5580915

Jiwtode, Ukani, Sonaskar, & Saurabh. (2022). Smart pesticide spraying robot. *International Research Journal of Engineering and Technology*, *99*(7), 706–710.

Johann, A. L., de Araújo, A. G., Delalibera, H. C., & Hirakawa, A. R. (2016). Soil moisture modeling based on stochastic behavior of forces on a no-till chisel opener. *Computers and Electronics in Agriculture*, *121*, 420–428. doi:10.1016/j.compag.2015.12.020

John, K., Abraham Isong, I., Michael Kebonye, N., Okon Ayito, E., Chapman Agyeman, P., & Marcus Afu, S. (2020). Using machine learning algorithms to estimate soil organic carbon variability with environmental variables and soil nutrient indicators in an alluvial soil. *Land (Basel)*, *9*(12), 487. doi:10.3390/land9120487

Johnny, M., Gerardo, B., & Tanguilig, B. (2020). Pest Detection & Extraction UsingI mage Processing Techniques. *International Journal of Computer and Communication Engineering, 3*.

Joshi, S. R. (2019). Emerging Model Villages in India: A Study of Punsari Village from the State of Gujarat (India). *Croatian and Comparative Public Administration, 19*(2). doi:10.31297/hkju.19.2.3

Kalis, B., Collier, M., & Fu, R. (2018). *10 Promising AI Applications in Health Care*. https://hbr.org/2018/05/10-promising-ai-applications-in-heath-care

Kamyshova, G., Osipov, A., Gataullin, S., Korchagin, S., Ignar, S., Gataullin, T., Terekhova, N., & Suvorov, S. (2022). Artificial neural networks and computer vision's-based phytoindication systems for variable rate irrigation improving. *IEEE Access : Practical Innovations, Open Solutions*, *10*, 8577–8589. doi:10.1109/ACCESS.2022.3143524

Kansara, K. (2015). Sensor Based Automated Irrigation System with IOT: A Technical Review. *International Journal of Computer Science and Information Technologies*, *6*(6), 5331–5333.

Kanse, Masekar, & Pokale, & Patil,. (2021). Smart pesticide spraying Robot. *InternationalJournal of Creative Research Thoughts*, *9*(4), 5683–5695.

Kapoor, A., Bhat, S. I., Shidnal, S., & Mehra, A. (2016). Implementation of IoT (Internet of Things) and Image processing in smart agriculture. *International Conference on Computation System and Information Technology for Sustainable Solutions*. IEEE. DOI:10.1109/CSITSS.2016.7779434

Karadağ, K., Tenekeci, M. E., Tasaltın, R., & Bilgilic, A. (2018). Detection of pepper fusarium disease using machine learning algorithms based on spectral reflectance. *Sustain Comput, 8*. Advance online publication. doi:10.1016/j.suscom.2019.01.001

Karasekreter, N., Başçiftçi, F., & Fidan, U. (2013). A new suggestion for an irrigation schedule with an artificial neural network. *Journal of Experimental & Theoretical Artificial Intelligence, 25*(1), 93–104. doi:10.1080/0952813X.2012.680071

Karmokar, B. C. (2015). Tea leaf diseases recognition using neural network ensemble. *International Journal of Computer Applications, 114*(17), 27–30. doi:10.5120/20071-1993

Karthika, J., Mathankumar, K., Santhose, M., Sharan, T., & Shariharan, S. (2021). Disease Detection in Cotton-Leaf Spot Using. Image-Processing. *Journal of Physics*.

Kassim, A. M., Termezai, M. F. N. M., Jaya, A. K. R. A., Azahar, A. H., Sivarao, S., Jafar, F. A., Jaafar, H. I., & Aras, M. S. M. (2020). Design and development of autonomous pesticide sprayer Robot for fertigation farm. *International Journal of Advanced Computer Science and Applications, 11*(2), 545–551.

Katiyar, S., Khan, R., & Kumar, S. (2021). Artificial Bee Colony Algorithm for Fresh Food Distribution without Quality Loss by Delivery Route Optimization. *Journal of Food Quality, 2021*, 1–9. doi:10.1155/2021/4881289

Kavyashree, N., & Vishwanath, N. P. (2022). Operation of agriculture based pesticide spraying Robot. *EPRA International Journal of Multidisciplinary Research, 8*(7), 180–185.

Khaki, S., & Wang, L. (2019). Crop yield prediction using deep neural networks. *Frontiers in Plant Science, 10*, 621. doi:10.3389/fpls.2019.00621 PMID:31191564

Khandelwal, P. M., & Chavhan, H. (2019). *Artificial Intelligence in Agriculture: An Emerging Era of Research*. Kavikulguru Institute of Technology and Science.

Khan, R., Kumar, S., Dhingra, N., & Bhati, N. (2021). The Use of Different Image Recognition Techniques in Food Safety: A Study. *Journal of Food Quality, 2021*, 1–10. doi:10.1155/2021/7223164

Khan, R., Tyagi, N., & Chauhan, N. (2021). Safety of food and food warehouse using VIBHISHAN. *Journal of Food Quality, 2021*, 1–12. doi:10.1155/2021/1328332

Khan, R., Zhang, C., Albattah, W., & Qamar, A. (2021). *Image-Based Detection of Plant Diseases: From Classical Machine Learning to Deep Learning Journey*. Wireless Communications and Mobile Computing.

Khirade Sachin, D., & Patil, A. B. (2015). *Plant Disease Detection Using Image Processing. In International Conference on Computing Communication Control and Automation*. IEEE.

Khirade, S. D., & Patil, A. B. (2015). Plant Disease Detection Using Image Processing. *International Conference on Computing Communication Control and Automation*. IEEE. 10.1109/ICCUBEA.2015.153

Kidd, N. A. C. (1991). The implications of air pollution for conifer aphid population dynamics: A simulation analysis. *Journal of Applied Entomology*, *111*(1-5), 166–171. doi:10.1111/j.1439-0418.1991.tb00307.x

Kirchoff, G. (1859). *Kirchoff laws of thermal radiation*. Monatsberichte der Akademie der Wissenschaften zu Berlin.

Klompenburg, T. V., Kassahun, A., & Catal, C. (2020). Crop yield prediction using machine learning: A systematic literature review. *Computers and Electronics in Agriculture*, *177*, 105709. doi:10.1016/j.compag.2020.105709

Klyushin, D., & Tymoshenko, A. (2021). Optimization of drip irrigation systems using artificial intelligence methods for sustainable agriculture and the environment. In *Artificial intelligence for sustainable development: Theory, practice and future applications* (pp. 3–17). Springer. doi:10.1007/978-3-030-51920-9_1

Kokkotis, C., Moustakidis, S., Papageorgiou, E., Giakas, G., & Tsaopoulos, D. E. (2020). Machine learning in knee osteoarthritis: A review. *Osteoarthritis and Cartilage Open*, *2*(3), 100069. doi:10.1016/j.ocarto.2020.100069 PMID:36474688

Kotsampopoulos, P., Lagos, D., Hatziargyriou, N., Faruque, M. O., Lauss, G., Nzimako, O., Forsyth, P., Steurer, M., Ponci, F., Monti, A., Dinavahi, V., & Strunz, K. (2018). A benchmark system for hardware-in-the-loop testing of distributed energy resources. *IEEE Power and Energy Technology Systems Journal*, *5*(3), 94–103. doi:10.1109/JPETS.2018.2861559

Kumar, M., & Tyagi, B. (2016, November). A state of art review of microgrid control and integration aspects. In *2016 7th India International Conference on Power Electronics (IICPE)* (pp. 1-6). IEEE. 10.1109/IICPE.2016.8079411

Kumar, Y. P., & Bhimasingu, R. (2017, November). Fuzzy logic based adaptive virtual inertia in droop control operation of the microgrid for improved transient response. In 2017 IEEE PES Asia-Pacific Power and Energy Engineering Conference (APPEEC) (pp. 1-6). IEEE. doi:10.1109/APPEEC.2017.8309006

Kumar, G., Kumar, S., & Kumar, N. (2015). Comparative study of wavelet and wavelet packet transform for denoising telephonic speech signal. *International Journal of Computer Applications*, *110*(15), 1–8. doi:10.5120/19389-9145

Kumar, N., Alam, K., & Siddiqi, A. H. (2016). Savitzky–Golay and Wavelet Transform Based Rama Spectroscopic data Denoising. *International Journal of Control Theory and Applications*, *9*(21), 297–305.

Kumar, N., Siddiqi, A. H., & Alam, K. (2014). Raman spectral data de-noising based on wavelet analysis. *International Journal of Computer Applications*, *102*(16), 20–22. doi:10.5120/17899-8864

Kumar, N., Siddiqi, A. H., & Alam, K. (2016). Raman spectra denoising using wavelet and wavelet packet transform. *Indian Journal of Industrial and Applied Mathematics*, *7*(01), 1–10. doi:10.5958/1945-919X.2016.00001.3

Kumar, N., Tripathi, P., & Alam, K. (2016). Non-Negative Matrix Factorization-Based EEG Signal Classification. *Indian Journal of Industrial and Applied Mathematics*, *7*(2), 212–219. doi:10.5958/1945-919X.2016.00019.0

Kumar, P. P., Akshay, R., Achha, N., Sagar, K., Thirupathi, V., & Reddy, G. R. (2022, May). An affordable multitasking drone for smart farming with artificial intelligence features. In. AIP Conference Proceedings: Vol. 2418. *No. 1* (p. 020002). AIP Publishing LLC. doi:10.1063/5.0081755

Lad, P., Patil, V., Patil, P., Patil, T., & Patil, P. (2015). Solar operated pesticide sprayer. *International Journal of Advanced Research in Science & Engineering*, *4*(01), 122–126.

Lamb, D.W. & Brown, R.B. (2001). Remote-Sensing& Mapping of Weeds in Crops. *Journal of Agricultural Engineering Research*, *78*(2), 117-125.

Lam, M., Nguyen, T., & Chung, W. (2020). Deep learning-based food quality estimation using radio frequency-powered sensor mote. *IEEE Access : Practical Innovations, Open Solutions*, *8*, 88360–88371. doi:10.1109/ACCESS.2020.2993053

Lateef, Z. E. (2021). *Introduction To Machine Learning: All You Need to Know About Machine Learning*. https://www.edureka.co/blog/introduction-to-machine-learning/amp/

Latha et al. (2014). Image Processing in Agriculture. *International journal of innovative research in electrical, electronics, instrumentation and control engineering, 2*(6), 1562-1565.

Lenniy, D. (2023, March 10). *Smart irrigation in agriculture*. Intellias. Retrieved April 26, 2023, from https://intellias.com/smart-irrigation-in-agriculture/

Liakos, K., Busato, P., Moshou, D., Pearson, S., & Bochtis, D. (2018). Machine Learning in Agriculture: A Review. *Sensors (Basel)*, *18*(8), 2674. doi:10.339018082674 PMID:30110960

Li, D. (2021). Application of artificial intelligence and machine learning based on big data analysis in sustainable agriculture. *Acta Agriculturæ Scandinavica. Section B, Soil and Plant Science*, *71*(9), 956–969. doi:10.1080/09064710.2021.1965650

Light Dependent Resistors. (1997). *NORP12 RS stock number 651-507 NSL19-M51 RS stock number 596-14*. https://components101.com/sites/default/files/component_datasheet/LDR%20 Datasheet.pdf

Liu, Y., Pu, H., & Sun, D. (2021). Efficient extraction of deep image features using convolutional neural network (CNN) for applications in detecting and analysing complex food matrices. *Trends in Food Science & Technology*, *113*, 193–204. doi:10.1016/j.tifs.2021.04.042

Loey, M., ElSawy, A., & Afify, M. (2020). Deep learning in plant diseases detection for agricultural crops: A survey. *International Journal of Service Science, Management, Engineering, and Technology*, *11*(2), 41–58. doi:10.4018/IJSSMET.2020040103

Lopez-Arevalo, I., Aldana-Bobadilla, E., Molina-Villegas, A., Galeana-Zapién, H., Muñiz-Sanchez, V., & Gausin-Valle, S. (2020). A Memory- Efficient Encoding Method for Processing Mixed-Type Data on Machine Learning. *Entropy (Basel, Switzerland)*, *22*(12), 1391. doi:10.3390/e22121391 PMID:33316972

López-López, M., Calderón, R., González-Dugo, V., Zarco-Tejada, P., & Fereres, E. (2016). Early detection and quantification of almond red leaf blotch using high-resolution hyperspectral and thermal imagery. *Remote Sensing (Basel)*, *8*(4), 276. doi:10.3390/rs8040276

Madhu, S., Sai Teja, U., Mahender, B., & Vamshi, D. (2023). Pesticide spraying robot. *International Journal of Advances in Engineering and Management*, *5*(4), 772–777.

Ma, G., Li, J., & Zhang, X. P. (2022). A review on optimal energy management of multimicrogrid system considering uncertainties. *IEEE Access : Practical Innovations, Open Solutions*, *10*, 77081–77098. doi:10.1109/ACCESS.2022.3192638

Mahalakshmi & Latha. (2019). Artificial Intelligence with Internet of Things on Healthcare systems: A Survey. *International Journal of Advanced Trends in Computer Science and Engineering*, 2847-2854. doi:10.30534/ijatcse.2019.27862019

Mahlein, A.-K., Hammersley, S., Oerke, E.-C., Dehne, H.-W., Goldbach, H., & Grieve, B. (2015). Supplemental blue LED lighting array to improve the signal quality in hyperspectral imaging of plants. *Sensors (Basel)*, *15*(6), 12834–12840. doi:10.3390150612834 PMID:26039423

Mahmuod Nasr, M. (2021). Smart Healthcare in the Age of AI: Recent Advances, Challenges and Future Prospects. *IEEE Access : Practical Innovations, Open Solutions*, *9*, 145248–145270. doi:10.1109/ACCESS.2021.3118960

Majumder, A., & Standard, B. (2019). *Goods delivery by drones gets closer to reality, updated policy unveiled*. https://www.google.com/amp/s/wap.business-standard.com/article-amp/current-affairs/goods-deliver-by-drones-get-closer-to-reality-updated-policy-unveiled-119011600056.html

Malarvizhi, K., Sowmithra, M., Gokula, P. D., & Kabila, B. (2021). Machine Learning for Plant Species Classification using Leaf Vein Morphometric. *International Journal of Engineering Research & Technology (Ahmedabad)*, *10*(04).

Maldonado, A., Vigneault, C., & Nakaji, K. (2012). Postharvest Technologies of Fresh Horticulture Produce. In Horticulture (pp. 164-172). InTech.

Manfrida, J. P. (n.d.). Insects, Allergy and Indoor Air Quality. *The Environmental Reporter*, *9*(2).

Medus, Saban, Frances, Bataller, & Rosadao. (2021). Hyperspectral image classification using CNN: Application to industrial food packaging Food Control. *Food Control, 125*.

Meena, H., Nandanwar, H., Pahl, D., & Chauhan, A. (2020). IoT based perceptive monitoring and controlling an automated irrigation system. *11th International conference on computing, communication and networking technologies (ICCCNT)*. 10.1109/ICCCNT49239.2020.9225455

Mehra, J., & Rechenberg, H. (1982). *The Historical Development of Quantum Theory* (Vol. 1). Springer-Verlag. doi:10.1007/978-1-4612-5783-7

Meng, W., Yuan, J., Zhang, C., Bai, Z., Zhou, W., Yan, J., & Li, X. (2013). Parenteral analgesics for pain relief in acute pancreatitis: A systematic review. *Pancreatology, 13*(3), 201–206. doi:10.1016/j.pan.2013.02.003 PMID:23719588

Meshram, V., Patil, K., Meshram, V., Hanchate, D., & Ramkteke, S. D. (2021). Machine learning in agriculture domain: A state-of-art survey. *Artificial Intelligence in the Life Sciences, 1*, 100010. doi:10.1016/j.ailsci.2021.100010

Misra, N. N. (2020). IoT, big data and artificial intelligence in agriculture and food industry. IEEE Internet of Things Journal, 35.

Mohanraj, I., Ashokumar, K., & Naren, J. (2016). Field Monitoring and Automation Using IOT in Agriculture Domain. *Procedia Computer Science, 93*, 931–939. doi:10.1016/j.procs.2016.07.275

Mohapatra, J. B., Jha, P., Jha, M. K., & Biswal, S. (2021). Efficacy of machine learning techniques in predicting groundwater fluctuations in agro-ecological zones of India. *The Science of the Total Environment, 785*, 147319. doi:10.1016/j.scitotenv.2021.147319 PMID:33957597

More, Kemble, Javir, Jadhav, & Kamble, & Sawant. (2022). Design and development of solar operated pesticide spraying vehicle. *Journal of Emerging Technologies and Innovative Research*, 9(8), 223–228.

Morgenstern, J. (2019). Kamikatsu: The Zero Waste Town That Turned Its Whole Economy Sustainable. *Forbes*. https://www.forbes.com/sites/juliemorgenstern/2019/06/27/kamikatsu-the-zero-waste-town-that-turned-its-whole-economy-sustainable/?sh=4b4d31743794

Motia, S., & Reddy, S. R. N. (2021, August). Exploration of machine learning methods for prediction and assessment of soil properties for agricultural soil management: A quantitative evaluation. *Journal of Physics: Conference Series, 1950*(1), 012037. doi:10.1088/1742-6596/1950/1/012037

Mounika, R., & Bharathi, P. (2020). Detection of Plant Leaf Diseases Using Image Processing. *Journal of Critical Reviews, 7*(6).

Mounika, Thota, & Vallamdas. (2022). Agricultural pesticide spraying Robot. *International Journal of Innovative Research in Technology, 9*(4), 356–359.

Mucherino, A., Petraq, P., & Panos, M. P. (2009). A survey of data mining techniques applied to agriculture. *Operations Research, 9*(2), 121–140. doi:10.100712351-009-0054-6

Muneret, L., Mitchell, M., Seufert, V., Aviron, S., Pétillon, J., Plantegenest, M., Thiéry, D., & Rusch, A. (2018). Evidence that organic farming promotes pest control. *Nature Sustainability, 1*(7), 361–368. doi:10.103841893-018-0102-4

Mustafa, M., Abbas, A., & Bsoul, Q. (2021). Smart Irrigation System Based on the Internet of Things and the Cloud. *International Journal for Modern Trends in Science and Technology, 7*(9), 19–24.

Mustafa, N. B. A., Fuad, N. A., Ahmed, S. K., Abidin, A. A. Z., Ali, Z., Yit, W. B., & Sharrif, Z. A. M. (2008). Image processing of an agriculture produce: Determination of size and ripeness of a banana, *International Symposium on Information Technology*. DOI:10.1109/ITSIM.2008.4631636

Mustafid, M. A., Subrata, I. D. M., Pramuhadi, G., & Harahap, I. S. (2022). Design and performance test of autonomous precision spraying Robot for cabbage cultivation. *IOP Conference Series: Earth &Environmental Science, 1038.*

Mustak, A., Ma, H., Song, L., & Jin, Y. (2022). Embedded design of automatic pesticide spraying Robot control system. *10th International Symposium on Computational Intelligence and Industrial Applications (ISCIIA2022)*, 1–6.

Muthuminal, R., & Mohana Priya, R. (2022). *An Overview on Techniques Involved in Recharging Ground Water and Its Impacts*. Intechopen, doi:10.5772/intechopen.105448

Nachtigall, L. G., Araujo, R. M., & Nachtigall, G. R. (2016). Classification of apple tree disorders using convolutional neural networks. In *2016 IEEE 28th International Conference on Tools with Artificial Intelligence (ICTAI)* (pp. 472-476). IEEE. 10.1109/ICTAI.2016.0078

Nagappan, M., Gopalakrishnan, V., & Alagappan, M. (2020). Prediction of reference evapotranspiration for irrigation scheduling using machine learning. *Hydrological Sciences Journal, 65*(16), 2669–2677. doi:10.1080/02626667.2020.1830996

Nahvi, B., Habibi., J, Mohammadi., K, Shamshirband, S., Saleh Al Razgan, O. (2016). Using self-adaptive evolutionary algorithm to improve the performance of an extreme learning machine for estimating soil temperature. *Computer Electron Agriculture, 124*, 0168–1699. http://dx.doi.org/ doi:10.1016/j.compag.2016.03.025

Nair, S. M. (2017). Piplantri: The Village That Celebrates Girl Child with 111 Trees. *The Indian Express*. https://indianexpress.com/article/trending/trending-in-india/piplantri-the-village-that-celebrates-girl-child-with-111-trees-4928185/

Nallani & Berlin Hency. (2015). Low Power Cost Effective Automatic Irrigation System. *Indian Journal of Science and Technology, 8*(23).

Narayanavaram, B., Manoj Kumar Reddy, E., & Rashmi, M. R. (2020). Arduino based Automation of Agriculture A Step towards Modernization of Agriculture. *International Conference on Electronics, Communication and Aerospace Technology (ICECA-2020).* 10.1109/ICECA49313.2020.9297546

National Association of Landscape Professionals, Inc. (n.d.). *Shopping for a smart irrigation system?* Love Your Landscape.org. Retrieved April 26, 2023, from https://www.loveyourlandscape.org/expert-advice/water-smart-landscaping/smart-irrigation/shopping-for-a-smart-irrigation-system/

Navya, P., & Sudha, D. (2023). Artificial intelligence based robot for harvesting, pesticide spraying and maintaining water management system in agriculture using IoT. *AIP Conference Proceedings, 2523*, 020025.

Nayak, J., Vakula, K., Dinesh, P., Naik, B., & Pelusi, D. (2020). Intelligent food processing: Journey from artificial neural network to deep learning. *Computer Science Review*, *38*, 100297. doi:10.1016/j.cosrev.2020.100297

Neményi, M. (2018). Precision crop production and artificial intelligence–the future of sustainable agriculture. *Acta Agraria Debreceniensis*, 47-58.

Niu, H. (2019). Estimating evapotranspiration with UAVs in agriculture: A review. In *2019 ASABE Annual International Meeting*. American Society of Agricultural and Biological Engineers, 2019.

Oimkar, K. (2018). Crop Disease Detection Using Deep Learning. In *Fourth International Conference on Computing Communication Control and Automation (ICCUBEA)*. IEEE.

Oklahoma State University. (2017, February 1). *Smart Irrigation Technology: Controllers and sensors*. Smart Irrigation Technology: Controllers and Sensors. Retrieved April 26, 2023, from https://extension.okstate.edu/fact-sheets/smart-irrigation-technology-controllers-and-sensors.html

Ozdemir, A., & Polat, K. (2020). Deep learning applications for hyperspectral imaging: A systematic review. *Journal of the Institute of Electronics and Computer*, *2*(1), 39–56. doi:10.33969/JIEC.2020.21004

Pantazi, X. E., Moshou, D., Alexandridis, T., Whetton, R. L., & Mouazen, A. M. (2016). Wheat yield prediction using machine learning and advanced sensing techniques. *Computers and Electronics in Agriculture*, *121*, 57–65. doi:10.1016/j.compag.2015.11.018

Papageorgiou, E. I., Aggelopoulou, K., Gemtos, T. A., & Nanos, G. D. (2018). Development and Evaluation of a Fuzzy Inference System and a Neuro-Fuzzy Inference System for Grading Apple Quality. *Applied Artificial Intelligence*, *32*(3), 253–280. doi:10.1080/08839514.2018.1448072

Papathanassiou, S. A., & Papadopoulos, M. P. (2001). Dynamic characteristics of autonomous wind–diesel systems. *Renewable Energy*, *23*(2), 293–311. doi:10.1016/S0960-1481(00)00180-4

Pardon, G., Ladhani, L., Sandstrom, N., & van der Wijngaart, W. (n.d.). Aerosol sampling using an electrostatic precipitator integrated with a microfluidic interface, Sensors and actuators. *B, Chemical, 212*, 344-352. doi:10.1016/j.snb.2015.02.008

Passive Infrared Sensor. (n.d.). In *Wikipedia*. https://en.wikipedia.org/wiki/Passive_infrared_sensor

Pegorini, V., Karam, L. Z., Pitta, C. S. R., Cardoso, R., da Silva, J. C. C., Kalinowski, H. J., Ribeiro, R., Bertotti, F. L., & Assmann, T. S. (2015). In vivo pattern classification of ingestive behavior in ruminants using FBG sensors and machine learning. *Sensors (Basel)*, *15*(11), 28456–28471. doi:10.3390151128456 PMID:26569250

Peng, J.-S. (2014). An intelligent robot system for spraying pesticides. *The Open Electrical & Electronic Engineering Journal*, *8*(1), 435–444. doi:10.2174/1874129001408010435

Photorisistor. (n.d.). In *Wikipedia*. https://en.wikipedia.org/wiki/Photoresistor

Pmagy.gov. (n.d.-a). https://pmagy.gov.in/new-dashboard/Districts-Villages-covered

Pmagy.gov. (n.d.-b). https://pmagy.gov.in/new-dashboard/Docs/PMAGY-CSMC-Meeting-12022020.pdf

Polya, G. (2004). *How to Solve It: A New Aspect of Mathematical Method*. Princeton University Press.

Poojith, L., Reddy, B., Kumar G., & Vittal, K. (2004). *International journal of innovative research in electrical, electronics, instrumentation, and control engineering*. Academic Press.

Prabha, D. & Satheesh, K. (2012a). *Crop Disease Identification using Image processing Methods*. In The National Conference on Green Computing, ST Hindu College, Nagercoil, TamilNadu, India.

Prabha, D. & Satheesh, K. (2012b). *A study on Image Processing Methods for Fruit Classification*. In Proceedings of IPC & IPEeL_ACT & CIIT_CENT & CSPE, Chennai, TamilNadu, India.

Prahlad, A. P., Bhagwan, B. S., Subhash, K. S., Vijay, P. N., & Kale, R. U. (2017). Solar powered automatic grass cutter and pesticides spreading robot. *International Research Journal of Engineering and Technology*, *4*(5), 3372–3374.

Proceedings of the 11th International Symposium on Insect-Plant Relationships. (n.d.). *Series Entomologica*.

Pruet, P., Ang, C. S., Farzin, D., & Chaiwut, N. (2015). Exploring the internet of "educational things" (IOET) in rural underprivileged areas. In *12th International Conference on Electrical Engineering/Electronics, Computer,Telecommunications and Information Technology (ECTI-CON)*. IEEE.

Pudasainee, A., Pudasaini, U., & Chaulagain, B. P. (2020). Prospects of ICT based agricultural technology in the context of Nepal. *Nepalese Journal of Agricultural Sciences*, *19*, 211.

Pujari, J. D., Yakkundimath, R., & Byadgi, A. S. (2015). Image Processing Based Detection of Fungal Diseases in Plants. *Procedia Computer Science*, *46*, 1802–1808. doi:10.1016/j.procs.2015.02.137

Quina, M. J., Bordado, J. C. M., & Quinta-Ferreira, R. M. (2011). Air Pollution Control in Municipal Solid Waste Incinerators, The Impact of Air Pollution on Health, Economy, Environment and Agricultural Sources. Academic Press.

Rahman, M. A. (2014).Enabling drone communications with WiMAX Technology. In *The 5th International Conference on Information, Intelligence, Systems and Applications* (pp 323-328). IEEE. 10.1109/IISA.2014.6878796

Rahman. (2020). *COVID-19 radiography database*. https://www. kaggle. com/tawsifurrahman/covid19-radiography-database

Rai, A. K., & Dwivedi, R. K. (2020). Fraud Detection in Credit Card Data using Unsupervised Machine Learning Based Scheme. In *Proceedings of the International Conference on Electronics and Sustainable Communication Systems (ICESC)*, Coimbatore, India, 2–4 July 2020. Institute of Electrical and Electronics Engineers Inc. 10.1109/ICESC48915.2020.9155615

Rajesh Kanna, P., & Vikram, R. (2020). Agricultural Robot – A pesticide spraying device. *International Journal of Future Generation Communication and Networking*, *13*(1), 150–160.

Rashmi, P., Naik, S., & Marfatia, R. (2013). Image Processing and Machine Learning for Automated Fruit Grading System: A Technical Review. *International Journal of Computer Applications*, *81*(16), 29–39. doi:10.5120/14209-2455

Rathnayake, D. B., Akrami, M., Phurailatpam, C., Me, S. P., Hadavi, S., Jayasinghe, G., Zabihi, S., & Bahrani, B. (2021). Grid forming inverter modeling, control, and applications. *IEEE Access : Practical Innovations, Open Solutions*, *9*, 114781–114807. doi:10.1109/ACCESS.2021.3104617

Raval, M., Dhandhukia, A., & Mohile, S. (2015). Development and Automation of Robot with Spraying Mechanism for Agricultural Applications. *International Journal of Research in Emerging Science and Technology*, *2*(8), 60–66.

Rawal, S. (2017, February). IoT Based Smart Irrigation System. *International Journal of Computers and Applications*, *159*(8).

Raza, S.-A., Prince, G., Clarkson, J. P., & Rajpoot, N. M. (2015). Automatic detection of diseased tomato plants using thermal and stereo visible light images. *PLoS One*, *10*(4), e0123262–e0123281. doi:10.1371/journal.pone.0123262 PMID:25861025

Reddy, D. J., & Kumar, M. R. (2021). Crop Yield Prediction using Machine Learning Algorithm. *2021 5th International Conference on Intelligent Computing and Control Systems (ICICCS)*, 1466-1470. 10.1109/ICICCS51141.2021.9432236

Reinecke, M., & Prinsloo, T. (2017). The influence of drone monitoring on crop health and harvest size. In *IEEE 1st International Conference in Next Generation Computing Applications (Next Comp)* (pp. 5-10). IEEE. 10.1109/NEXTCOMP.2017.8016168

Rezaei, N., Mazidi, M., Gholami, M., & Mohiti, M. (2020). A new stochastic gain adaptive energy management system for smart microgrids considering frequency responsive loads. *Energy Reports*, *6*, 914–932. doi:10.1016/j.egyr.2020.04.021

Rezwan, S., & Choi, W. (2022). Artificial Intelligence Approaches for UAV Navigation: Recent Advances and Future Challenges. *IEEE Access : Practical Innovations, Open Solutions*, *10*, 26320–26339. doi:10.1109/ACCESS.2022.3157626

Ritz, C., Putku, E., & Astover, A. (2015). A practical two-step approach for mixed model-based kriging, with an application to the prediction of soil organic carbon concentration. *European Journal of Soil Science*, *66*(3), 548–554. doi:10.1111/ejss.12238

Robitaille, P-M. (2009). *Kirchhoff's Law of Thermal Emission: 150 Years*. Department of Radiology, The Ohio State University.

Rogerson, S. (2020). *Sintrones fanless computer powers smart city applications*. https://www.iotm2mcouncil.org/iot-library/news/smart-cities-news/sintrones-fanless-computer-powers-smart-city-applications/

Rokhmana, C. A. (2015). The Potential of UAV-based Remote Sensing for Supporting Precision Agriculture in Indonesia. *Procedia Environmental Sciences*, *24*, 245–253. doi:10.1016/j.proenv.2015.03.032

Romero, J. R., Roncallo, P. F., Akkiraju, P. C., Ponzoni, I., Echenique, V. C., & Carballido, J. A. (2013). Using classification algorithms for predicting durum wheat yield in the province of Buenos Aires. *Computers and Electronics in Agriculture*, *96*, 173–179. doi:10.1016/j.compag.2013.05.006

Roopa Ranjani & Sravya. (2018). IoT Based Smart Irrigation System. *International Journal of Research, 5*(12).

Ruiz-Ruiz, S., Ambrós, S., Vives, M. D. C., Navarro, L., Moreno, P., & Guerri, J. (2009). Detection and quantitation of Citrus leaf blotch virus by TaqMan real-time RT-PCR. *Journal of Virological Methods*, *160*(1-2), 57–62. doi:10.1016/j.jviromet.2009.04.012 PMID:19406167

Safdar Munir, Bajwa, Ashraf, Anwar, & Rashid. (2021). Intelligent and Smart Irrigation System Using Edge Computing and IoT. Wiley.

Sahni, V., Srivastava, S., & Khan, R. (2021). Modelling techniques to improve the quality of food using artificial intelligence. *Journal of Food Quality*, *2021*, 1–10. doi:10.1155/2021/2140010

Saleem, M., Sapna, K., Potgieter, J., & Arif, K. (2020). Image-Based Plant Disease Identification by Deep Learning Meta Architectures. MDP Journals, 9.

Salim, N., Zeebaree, S., Sadeeq, M., Radie, A. H., Shukur, H., & Rashid, Z. (2021). Study for Food Recognition System Using Deep Learning. *Journal of Physics: Conference Series*, *1963*(1), 012014. doi:10.1088/1742-6596/1963/1/012014

Sandikar, Sawant, Jadhav, & Misal. (2022). Design and fabrication of pesticides spray Robot. *International Research Journal of Engineering and Technology, 9*(5), 1258–1261.

Sankaran, S., Maja, J. M., Buchanon, S., & Ehsani, R. (2013). Citrus Greening detection using visible, near infrared and thermal imaging techniques. *Sensors (Basel)*, *13*(2), 2117–2130. doi:10.3390130202117 PMID:23389343

Sankaran, S., Mishra, A., Ehsani, R., & Davis, C. (2010). A review of advanced techniques for detecting plant diseases. *Computers and Electronics in Agriculture*, *72*(1), 1–13. doi:10.1016/j.compag.2010.02.007

Sarghini, F., & De Vivo, A. (2017a). Interference analysis of a heavy lift multi rotor drone flow field and transported spraying system. *Chemical Engineering Transactions*, *58*, 631–636.

Sarghini, F., & De Vivo, A. (2017b). Analysis of preliminary design requirements of a heavy lift multi rotor drone for agricultural use. *Chemical Engineering Transactions*, *58*, 625–630.

Sarki. (2022). Automated detection of COVID-19 through convolutional neural network using chest x-ray images. *PloS One, 17*(1).

Saswell. (n.d.). *Smart irrigation controllers have difference between weather-based and soil moisture sensors*. Shenzhen Saswell Technology Inc. Retrieved April 26, 2023, from https://www.saswell.com/smart-irrigation-controllers-have-difference-between-weather-based-and-soil-moisture-sensors_n100

Saxena, L., & Armstrong, L. (2014). A survey of image processing techniques for agriculture. In Proceedings of Asian Federation for Information Technology in Agriculture. (pp. 401-413). Australian Society of Information and Communication Technologies in Agriculture.

Schillaci, C., Acutis, M., Lombardo, L., Lipani, A., Fantappiè, M., Märker, M., & Saia, S. (2017). Spatio-temporal topsoil organic carbon mapping of a semi-arid Mediterranean region: The role of land use, soil texture, topographic indices and the influence of remote sensing data to modelling. *The Science of the Total Environment*, *601*, 821–832. doi:10.1016/j.scitotenv.2017.05.239 PMID:28578240

Searchinger, T., Waite, R., Hanson, C. & Ranganathan, J. (2019). *World Resources Report: Creating a Sustainable Food Future*. United Nations, World Resources Institute, and the World Bank.

Seeland, M., Rzanny, M., Alaqraa, N., Wäldchen, J., & Mäder, P. (2017). Plant species classification using flower images—A comparative study of local feature representations. *PLoS One*, *2017*(12), e0170629. doi:10.1371/journal.pone.0170629 PMID:28234999

Selvaraj, M., Vergara, A., Ruiz, H., Safari, N., Elayabalan, S., Ocimati, W., & Blomme, G. (2019). AI-powered banana diseases and pest detection. *Plant Methods*, *15*(1), 1–11. doi:10.118613007-019-0475-z

Seufert, V., Ramankutty, N., & Foley, J. A. (2012). Comparing the yields of organic and conventional agriculture. *Nature*, *485*(7397), 229–232. doi:10.1038/nature11069 PMID:22535250

Shah, D., Dixit, R., Shah, A., Shah, P., & Shah, M. (2020). A comprehensive analysis regarding several breakthroughs based on computer intelligence targeting various syndromes. *Augment Hum Res*, *5*(1), 14. doi:10.100741133-020-00033-z

Sharma, R. (2021, May). Artificial Intelligence in Agriculture: A Review. In *2021 5th International Conference on Intelligent Computing and Control Systems (ICICCS)* (pp. 937-942). IEEE. 10.1109/ICICCS51141.2021.9432187

Sharma, U., Sharma, D., Saleem, S., Kumar, M., Singh, S.P. (2022). Artificial Intelligence-Based Sustainable Agricultural Practices. In *Artificial Intelligence Applications in Agriculture and Food Quality Improvement.* IGI Global. doi:10.4018/978-1-6684-5141-0.ch001

Sharma, R., Kamble, S. S., Gunasekaran, A., Kumar, V., & Kumar, A. (2020). A systematic literature review on machine learning applications for sustainable agriculture supply chain performance. *Computers & Operations Research*, *119*, 104926. doi:10.1016/j.cor.2020.104926

Shriaarthi, B., Suganya, B., & Sridevi, S. (2018). Brain Controlled Car for Disabled Using Artificial Intelligence. *International Journal of Engineering Research & Technology*, 69-71. doi:10.17577/IJERTCONV1IS03024

Shubhangi, Londhe, B. K., & Sujata, K. (2017). Remotely operated pesticide sprayer robot in agricultural field. *International Journal of Computers and Applications*, *167*(3), 26–29.

Simelli, I., & Tsagaris, A. (2015). The use of unmanned aerial systems (UAS) in agriculture. *7th International Conference on Information and Communication Technologies In Agriculture*, (pp. 730-736).

Simon, Y. L. (2020). Artificial intelligence (AI) in agriculture. Academic Press.

Singh, B. (2022). *India's first association of robotic surgeons launched in Guwahati.* https://www.google.com/amp/s/m.economictimes.com/news/india/indias-first-association-of-robotic-surgeons-formally-launched-in-guwahati/amp_articleshow/91219923.cms

Singha, S., Pasupulet, S., Singha, S. S., Singh, S., & Kumar, S. (2021). Prediction of groundwater quality using efficient machine learning technique. *Chemosphere*, *276*, 130265. doi:10.1016/j.chemosphere.2021.130265 PMID:34088106

Singh, K. B., Sengar, N., Das, D., & Misra, S. C. (2022). Village 5.0: Enabling Technologies and its Applications in Development of Smart Village. In *2022 IEEE International Symposium on Smart Electronic Systems (iSES)* (pp.556-561). IEEE. 10.1109/iSES54909.2022.00122

Sivakumar, D. (2019). Computerized growth analysis of seeds using deep learning method, *International Journal of Recent Technology and Engineering (IJRTE)*, *7*(6S5), 1885-1892.

Sladojevic, S., Arsenovic, M., Anderla, A., Culibrk, D., & Stefanovic, D. (2016). Deep neural networks based recognition of plant diseases by leaf image classification. *Computational Intelligence and Neuroscience*, *2016*, 1–11. doi:10.1155/2016/3289801 PMID:27418923

Slaughter, D. C., Giles, D. K., & Downey, D. (2008). Autonomous robotic weed control systems: A review. *Computers and Electronics in Agriculture*, *61*(1), 63–78. doi:10.1016/j.compag.2007.05.008

Smart Village. (n.d.). https://www.smartvillage.biz/

Smith, J. (2020). *Can Artificial Intelligence Replace the Role of Doctors?* https://readwrite.com/can-artificial-intelligence-replace-the-role-of-doctors/

Söderström, M., Börjesson, T., Roland, B., & Stadig, H. (2014). Modelling within-field variations in deoxynivalenol (DON) content in oats using proximal and remote sensing. *Precision Agriculture*, *16*(1), 1–14. doi:10.100711119-014-9373-6

Sofos, F., & Karakasidis, T. E. (2021). Machine Learning Techniques for Fluid Flows at the Nanoscale. *Fluids (Basel, Switzerland)*, *6*(3), 96. doi:10.3390/fluids6030096

Sofu, M. M., Er, O., Kayacan, M. C., & Cetisli, B. (2016). Design of an automatic apple sorting system using machine vision. *Computers and Electronics in Agriculture*, *127*, 395–405. doi:10.1016/j.compag.2016.06.030

Sonaa, G., Passonia, D., Pintoa, L., Pagliaria, D., Masseroni, D., Ortuani, B., & Facchib, A. (2016). UAV multispectral survey to map soil and crop for precision farming applications. *The International Archives of the Photogrammetry, Remote Sensing and Spatial Information Sciences*, *XLI*(B1), 1023–1029. doi:10.5194/isprs-archives-XLI-B1-1023-2016

Stangierski, J., Weiss, D., & Kaczmerek, A. (2019). Multiple regression models and Artificial Neural Network (ANN) as prediction tools of changes in overall quality during the storage of spreadable processed Gouda cheese. *European Food Research and Technology*, *245*(11), 2539–2547. doi:10.100700217-019-03369-y

Stefan, J. (1879). Über die Beziehung zwischen der Wärmestrahlung und der Temperatur [On the relationship between heat radiation and temperature]. *Sitzungsberichte der Mathematisch-naturwissenschaftlichen Classe der Kaiserlichen Akademie der Wissenschaften*, *79*, 391–428.

Strategic Investment Research Unit (SIRU). (2022, June 17). *Artificial Intelligence in Indian agriculture*. Investment Promotion and Facilitation Agency. https://www.investindia.gov.in/team-india-blogs/artificial-intelligence-indian-agriculture

Suchithra, M. S., & Maya, P. (2020). Improving the prediction accuracy of soil nutrient classification by optimizing extreme learning machine parameters. *Information Processing in Agriculture*, *7*(01), 72–82. https://www.sciencedirect.com/science/article/pii/S2214317318304906. doi:10.1016/j.inpa.2019.05.003

Sujatha, R., Chatterjee, J. M., Jhanjhi, N. Z., & Brohi, S. N. (2021). Performance of deep learning vs machine learning in plant leaf disease detection. *Microprocessors and Microsystems*, *80*, 103615. doi:10.1016/j.micpro.2020.103615

Sujith, Sajja, Mahalakshmi, Nuhmani, & Prasanalakshmi. (2021). Systematic review of smart health monitoring using deep learning and Artificial intelligence. *Neuroscience Informatics*, 1-6. doi:10.1016/j.neuri.2021.100028

Suksawat, B., & Komkum, P. (2015). Pineapple quality grading using image processing and fuzzy logic based on Thai Agriculture Standards. *International Conference on Control, Automation and Robotics*. 10.1109/ICCAR.2015.7166035

Sulakhe, & Karanjkar. (2015). Design and operation of agriculture based pesticide spraying robot. *International Journal of Scientific Research*, *6*(4), 1286–1289.

Sunny, M., & Badashah, S. J. (2022). Bluetooth Based Agrobot used to spray the pesticides. *International Journal for Research in Applied Science and Engineering Technology*, *10*(6), 2449–2452.

Surya Shankar, R. P. & Valarmathi, K. (n.d.). *RFID based tag finding mobile Robot for indoor navigation*. www.slideshare.net/divakark3/ieee-confrence

Surya, DSatheesh, K. (2014). Image processing methods and its Role in agricultural sector – A study'. *International Journal of Business Intelligents*, *03*(01), 366–373.

Sushma Priya, K., Praneetha Reddy, R., & Pradeep, Y. (2022). Agricultural pesticide spraying Robot. *International Research Journal of Modernisation in Engineering Technology and Science.*, *4*(7), 800–804.

Swati, D. (2015). Seed sowing using robotics technology. *International Journal of Scientific Research and Management, 3*(5), 2889–2892.

Taneja, K., & Bhatia, S. (2017). Automatic Irrigation System using Arduino UNO. *International Conference on Intelligent Computing and Control Systems ICICCS.*

Temple, J. (2020, April 2). *How artificial intelligence is helping farmers and babies in the developing world.* MIT Technology Review. https://www.technologyreview.com/s/613210/how-artificial-intelligence-is-helping-farmers-and-babies-in-the-developing-world/

Terhoeven-Urselmans, T., Vagen, T.-G., Spaargaren, O., & Shepherd, K. D. (2010). Prediction of soil fertility properties from a globally distributed soil mid-infrared spectral library. *Soil Science Society of America Journal, 74*(5), 1792–1799. doi:10.2136ssaj2009.0218

Thaher, T., & Ishaq, I. (2020). Cloud-based Internet of Things Approach for Smart Irrigation System: Design and Implementation. *International Conference on Promising Electronic Technologies (ICPET).* 10.1109/ICPET51420.2020.00015

The Medical Futurist. (2021). *The Top 12 Health Chatbots.* https://mediclafuturist.com/top-12-health-chatbots/

Therond, O., Duru, M., Roger-Estrade, J., & Richard, G. (2017). A new analytical framework of farming system and agriculture model diversities. A review. *Agronomy for Sustainable Development, 37*(3), 21. doi:10.100713593-017-0429-7

Thomas, B. A. Strickland, II., & J, R. (1997). Color Image Enhancement using Spatially Adaptive Saturation Feedback. *IEEE International Conference on Image Processing, 3*, 30-33.

Tian, Y., Yang, C., Huang, W., Tang, J., Li, X., & Zhang, Q. (2021). Machine learning-based crop recognition from aerial remote sensing imagery. *Frontiers of Earth Science, 15*(1), 54–69. doi:10.100711707-020-0861-x

Tipan, E. (2020). *Manual vs. Automatic: Which Car Should You Get?* https://www.esquiremag.ph/wheels/drive/manual-vs-automatic-car-a2314-20200109-lfrm

Tripathi, P. (2020). Electroencephalpgram Signal Quality Enhancement by Total Variation Denoising Using Non-convex Regulariser. *International Journal of Biomedical Engineering and Technology, 33*(2), 134–145. doi:10.1504/IJBET.2020.107709

Tripathi, P., Kumar, N., Rai, M., & Khan, A. (2022). Applications of Deep Learning in Agriculture. In *Artificial Intelligence Applications in Agriculture and Food Quality Improvement* (pp. 17–28). IGI Global., doi:10.4018/978-1-6684-5141-0.ch002

Tripathi, P., Kumar, N., & Siddiqi, A. H. (2020). De-noising Raman spectra using total variation de-noising with iterative clipping algorithm. In *Computational Science and its Applications, Taylor and Francis Group* (pp. 225–231). CRC Press. doi:10.1201/9780429288739-14

Tripathi, P., & Siddiqi, A. H. (2016). Solution of Inverse Problem for de-noising Raman Spectral Data with Total variation using Majorization-Minimization Algorithm. Int. J. *Computing Science and Mathematics*, *7*(3), 274–282. doi:10.1504/IJCSM.2016.077855

Tsukiyama, T. (2011). RFID based navigation system for indoor mobile robots. *IFAC Proceedings, 44*(1), 1084 – 1089.

Türkoğlu, M., & Hanbay, D. (2019). Plant disease and pest detection using deep learning-based features. *Turkish Journal of Electrical Engineering and Computer Sciences*, *27*(3), 1636–1651. doi:10.3906/elk-1809-181

Tyagi, A. C. (2016). Towards a second green revolution. *Irrigation and Drainage*, *65*(4), 388–389. doi:10.1002/ird.2076

Tyagi, N., Rai, M., Sahw, P., Tripathi, P., & Kumar, N. (2022). Methods for the Recognition of Human Emotions Based on Physiological Response: Facial Expressions. In *Smart Healthcare for Sustainable Urban Development* (pp. 183–202). IGI Global., doi:10.4018/978-1-6684-2508-4.ch013

Ucat, R. C., & Dela, C. J. C. (2019). Postharvest grading classification of Cavendish banana using deep learning and tensorflow, *International Symposium on Multimedia and Communication Technology (ISMAC)*. IEEE. 10.1109/ISMAC.2019.8836129

Uddin, M., Wang, Y., & Woodbury-Smith, M. (2019). *Artificial intelligence for precision medicine in neurodevelopmental disorders*. https://www.nature.com/articles/s41746-019-0191-0

Unpaprom, Y., Dussadeeb, N., & Ramaraj, R. (2018). Modern Agriculture Drones the Development of Smart Farmers. Academic Press.

Uzal, L. C., Grinblat, G. L., Namías, R., Larese, M. G., Bianchi, J. S., Morandi, E. N., & Granitto, P. M. (2018). Seed-per-pod estimation for plant breeding using deep learning. *Computers and Electronics in Agriculture*, *150*, 196–204. doi:10.1016/j.compag.2018.04.024

Vadapalli, A., & Pervalli, S. (2020). Smart Agriculture System using IoT Technology. *International Journal of Advance Research in Science and Engineering*, *9*(9).

Vardhan, P.H., Deepak, S., Aditya, P.T., & Arul, S. (2014). Development of Automated Aerial Pesticide Sprayer. *International Journal of Engineering Science and Research Technology, 3*(4).

Vasudevan, A., Kumar, D. A., & Bhuvaneswari, N. S. (2016). *Precision farming using unmanned aerial and ground vehicles. IEEE Technological Innovations in ICT for Agriculture and Rural Development*. TIAR.

Verified Market Research. (2021). *Healthcare Chatbots Market Size and Forecast*. https://verifiedmarketresearch.com/product/healthcare-chatbots-market/

Vibhute, A. K., Bodhe, S., Vibhute, A. K., & Bodhe, S. (2012). Applications of Image Processing in Agriculture: A Survey. *International Journal of Computer Applications*, *52*(2), 34–40. doi:10.5120/8176-1495

Vieira, S., Lopez Pinaya, W. H., & Mechelli, A. (2020). Introduction to Machine Learning. Academic Press. doi:10.1016/B978-0-12-815739-8.00001-8

View, D. (2020). *Top 10 Best Unmanned Combat Aerial Vehicle*. https://www.defenceview.in/top-10-best-unmanned-combat-aerial-vehicle/

Wabali, V. C., Esiri, A., & Zitte, L. (2017). A sensory assessment of color and textural quality of refrigerated tomatoes preserved with different concentrations of potassium permanganate. *Food Science & Nutrition*, *5*(3), 434–438. doi:10.1002/fsn3.410 PMID:28572927

Wahabzada, M., Mahlein, A.-K., Bauckhage, C., Steiner, U., Oerke, E.-C., & Kersting, K. (2016). Plant phenotyping using probabilistic topic models: Uncovering the hyperspectral language of plants. *Scientific Reports*, *6*(1), 22482–22492. doi:10.1038rep22482 PMID:26957018

Walch, K. (2019, July 9). How ai is Transforming Agriculture. *Forbes*. https://www.forbes.com/sites/cognitiveworld/2019/07/05/how-ai-is-transforming-agriculture/#5442e7ad4ad1

Wang, B., Waters, C., Orgill, S., Cowie, A., Clark, A., Li Liu, D., Simpson, M., McGowen, I., & Sides, T. (2018). Estimating soil organic carbon stocks using different modelling techniques in the semi-arid rangelands of eastern Australia. *Ecological Indicators*, *88*, 425–438. doi:10.1016/j.ecolind.2018.01.049

Wien, W. (1897). On the division of energy in the emission-spectrum of a black body. *Philosophical Magazine*, *43*(262), 214–220. doi:10.1080/14786449708620983

Wilf, P., Zhang, S., Chikkerur, S., Little, S. A., Wing, S. L., & Serre, T. (2016). Computer vision cracks the leaf code. *Proceedings of the National Academy of Sciences of the United States of America*, *113*(12), 3305–3310. doi:10.1073/pnas.1524473113 PMID:26951664

Xiao, Y. Z. (2021). *The Cultural Dynamics in Water Management from Ancient History to the Present Age*. IWA Publishing.

Xia, T., Kustas, W. P., Anderson, M. C., Alfieri, J. G., Gao, F., McKee, L., Prueger, J. H., Geli, H. M. E., Neale, C. M. U., Sanchez, L., Alsina, M. M., & Wang, Z. (2016). Mapping evapotranspiration with highresolution aircraft imagery over vineyards using one-and two-source modeling schemes. *Hydrology and Earth System Sciences*, *2016*(4), 1523–1545. doi:10.5194/hess-20-1523-2016

Xie, W., Wei, S., Zheng, Z., & Yang, D. (2021). A CNN-based lightweight ensemble model for detecting defective carrots. *Biosystems Engineering*, *208*, 287–299. doi:10.1016/j.biosystemseng.2021.06.008

Xing, T., Cai, A., Lu, C., Ye, H., Wu, H., Huai, S., Wang, J., Xu, M., & Lin, Q. (2022). Increasing soil microbial biomass nitrogen in crop rotation systems by improving nitrogen resources under nitrogen application. *Journal of Integrative Agriculture*, *21*(5), 1488–1500. doi:10.1016/S2095-3119(21)63673-0

Xu, Y., Dong, Z., Li, Z., Liu, Y., & Ding, Z. (2021). Distributed optimization for integrated frequency regulation and economic dispatch in microgrids. *IEEE Transactions on Smart Grid*, *12*(6), 4595–4606. doi:10.1109/TSG.2021.3089421

Yadav, A., Rashmi, A. V., Goyal, V., & Mukherjee, R. (2022). Smart Village: Problems and Hybrid Solutions from IoT Perspective. In *2022 2nd International Conference on Innovative Sustainable Computational Technologies (CISCT)*, (pp.1-5). IEEE. 10.1109/CISCT55310.2022.10046527

Yallappa, D., Veerangouda, M., Maski, D., Palled, V., & Bheemanna, M. 2017. Development and evaluation of drone mounted sprayer for pesticide applications to crops. In *2017 IEEE Global Humanitarian Technology Conference (GHTC)*. IEEE. 10.1109/GHTC.2017.8239330

Yan, Z. Z., Chen, Q. L., Li, C. Y., Thi Nguyen, B. A., He, J. Z., & Hu, H. W. (2022). Contrasting ecological processes shape the Eucalyptus phyllosphere bacterial and fungal community assemblies. *Journal of Sustainable Agriculture and Environment*, *1*(1), 73–83. doi:10.1002ae2.12007

Yueling, S., Zengfeng, Z., & Hongsheng, W. (2011). Agricultural information technology development and innovation path. In *International Conference on Electronics, Communications and Control (ICECC)*. IEEE.

Yu, H., Wen, X., Wu, M., Sheng, D., Wu, J., & Zhao, Y. (2022). Data-based groundwater quality estimation and uncertainty analysis for irrigation agriculture. *Agricultural Water Management*, *262*, 107423. doi:10.1016/j.agwat.2021.107423

Yvon, M., Thébaud, G., Alary, R., & Labonne, G. (2009). Specific detection and quantification of the phytopathogenic agent 'Candidatus Phytoplasma prunorum'. *Molecular and Cellular Probes*, *23*(5), 227–234. doi:10.1016/j.mcp.2009.04.005 PMID:19401232

Yvoz, S., Petit, S., Biju-Duval, L., & Cordeau, S. A. (2020). framework to type crop management strategies within a production situation to improve the comprehension of weed communities. *European Journal of Agronomy*, *115*, 126009. doi:10.1016/j.eja.2020.126009

Zavratnik, V., Kos, A., & Duh, E. S. (2019). Smart Villages in Slovenia: Examples of Good Pilot Practices, In Smart Villages in the EU and Beyond (Emerald Studies in Politics and Technology). Emerald Publishing Limited. doi:10.1108/978-1-78769-845-120191010

Zha, J. (2022). Artificial Intelligence in Agriculture. *Journal of Physics*.

Zhang, Huang, Li, Zhao, Fan, Wu, & Liu. (2014). *Application & developments and of computer vision for external quality inspection of vegetables and fruits: A review*. Food Research International.

Zhang, N. (2022). *Plant Disease Detection*. Encyclopedia. https://encyclopedia.pub/entry/2567

Zhang, S., Huang, W., Huang, Y., & Zhang, C. (2020). Plant species recognition methods using leaf image: Overview. *Neurocomputing*, *408*, 246–272. doi:10.1016/j.neucom.2019.09.113

Zhang, X., Yang, J., Lin, T., & Ying, Y. (2021). Food and agro-product quality evaluation based on spectroscopy and deep learning: A review. *Trends in Food Science & Technology*, *112*, 431–441. doi:10.1016/j.tifs.2021.04.008

Zheng, A., & Casari, A. (2018). *Feature Engineering for Machine Learning: Principles and Techniques for Data Scientists*. O'Reilly Media, Inc.

Zheng, H., Zhou, X., Cheng, T., Yao, X., Tian, Y., Cao, W., & Zhu, Y. 2016. Evaluation of a UAVbased hyperspectral frame camera for monitoring the leaf nitrogen concentration in rice. In *2016 IEEE International Geoscience and Remote Sensing Symposium (IGARSS)*. IEEE. 10.1109/IGARSS.2016.7730917

Zhou, L., Zhang, C., Qiu, F., & He, Y. (2019). Application of deep learning in food: A review. *Comprehensive Reviews in Food Science and Food Safety*, *18*(6), 1793–1811. doi:10.1111/1541-4337.12492 PMID:33336958

Zhu, L., Spachos, P., Pensini, E., & Plataniotis, K. N. (2021). Deep learning and machine vision for food processing: A survey. *Current Research in Food Science*, *4*, 233–249. doi:10.1016/j.crfs.2021.03.009 PMID:33937871

About the Contributors

Mohammad Ayoub Khan received the Ph.D. degree in computer engineering from Jamia Millia Islamia, New Delhi, India, and the Master of Technology degree in computer science and engineering from Guru Gobind Singh Indraprastha, New Delhi. He is currently an Associate Professor with the University of Bisha, Saudi Arabia. His research interests include the Internet of Things, RFID, wireless sensor networks, ad hoc networks, smart cities, industrial IoT, signal processing, NFC, routing in network-on-chip, real time, and embedded systems. He has more than 14 years of experience in his research areas. He has published many research papers and books in reputed journals and international IEEE. conferences. He contributes to the research community by undertaking various volunteer activities in the capacity of editor for many journals and as the conference chair.(

Cathryn J. Peoples received the B.A. degree (with honors and diploma in industrial studies) in business studies with computing, M.Sc. degree in telecommunications and internet systems, and Ph.D. degree in networking from Ulster University, U.K., in 2004, 2005, and 2009, respectively. She is currently employed by The Open University in the School of Computing and Communications as an Associate Lecturer for the following modules: Software Engineering, The Computing & IT Project, The Apprenticeship Computing & IT Project, The Engineering Project, Advanced Work-based Learning, and Change, Strategy & Projects at Work. Cathryn is also employed on a part-time basis as a Research Associate at Ulster University working on Internet of Things (IoT) research. She works in a freelancing role for the University of Essex Online through Kaplan Open Learning as a Tutor and Content Developer on the MSc Cyber Security degree programme, and additionally is a Computer Science Assessment Specialist with Cambridge Assessment International Education. Cathryn achieved Full College membership of EPSRC in August 2018. She became an Editor-in-Chief of the EAI Endorsed Transactions on Cloud Systems in January 2020 and a Topic Editor for MDPI Electronics in January 2021. Cathryn is an Editor of the Springer book on 'Advances in Cybersecurity Management', to be published in 2021.

* * *

Nitendra Kumar received a PhD (Mathematics) from Sharda University, Greater Noida, India and a Master of Science (Mathematics and Statistics) from Dr Ram Manohar Lohia Avadh University, Faizabad, India. Currently, he is working as an Assistant Professor at the Amity Business School, Amity University, Noida with interests in the Wavelets and its Variants, Data Mining, Inverse Problems, Epidemic Modeling, Fractional Derivatives and Statistical Methods. He has more than 10 years of experience in his research areas. Dr Kumar has published many research papers in reputed journals and also published 6 books on engineering mathematics. He contributes to the research community by undertaking various volunteer activities in the capacity of editor for two edited books, Guest Editor of reputed journals of Taylor and Francis.

Mritunjay Rai is working as an Assistant Professor in Department of Electronics and Communication at SRM University, Lucknow-Deva Road, Barabanki, UP. He is pursuingPh.D. from IIT-ISM Dhanbad. He has completed his B.Tech in ECE from Sri Ramswaroop Memorial College of Engineering and Management, Lucknow (SRMCEM) in 2009 and Master of Engineering (distinction) in Instrumentation and Control from Birla Institute of Technology-Mesra, Ranchi in 2013. His areas of interest lie in image processing, speech processing, Artificial Intelligence, machine learning, deep learning, Internet of Things (IoT)and robotics & automation. He has more than 9 years of working experience in research as well as academic. In addition, he has guided several UG and PG projects. He has published many research articles in reputed Journals published by IEEE, Inderscience, etc. He has reviewed many research papers of Journals and international/national conferences.

Sakshi Rajput works in the multidisciplinary area of VLSI and Mobility Management (wireless communication). She has cleared Gate exam in ECE also. She has 10 years of teaching experience and has taught VLSI Design (Analog & Digital), Electronics Instrumentation & Measurement, Digital Electronics, STLD, Embedded System, Analog Electronics in the department of ECE. She has published research papers in several International Journals. She is actively involved to obtain solutions to the problem related to wireless communication between devices with the help of self organizing network, deep neural networking & artificial intelligence. She has always remained a meritorious student and always completed assigned tasks and responsibilities with complete dedication. She is also actively involved in various institutional & departmental activities. She has supervised more than 20 B.Tech projects in the field of VLSI Design & Embedded System.

Preeti Sehrawat is currently pursuing her Ph.D from DCRUST in the area of vehicular networking and wireless communication. She is particularly interested in vehicular communication in 5G. She has taught various subjects like STLD, Digital System Design, Electronic Devices etc. during her teaching career.

Padmesh Tripathi is working as an Associate Professor of Mathematics at IIMT College of Engineering, Greater Noida, UP. He received his Ph.D. degree from Sharda University, Greater Noida (India). He completed his Master's and Bachelor's degrees from University of Allahabad, Prayagraj. He has been teaching since last 20 years. He has published many research papers in reputed journals. He has been granted funds from prestigious organizations like: Cambridge University, UK; University of California, USA; University of Eastern Finland, Finland; INRIA, Sophia Antipolis, France, etc. and he visited these organizations. He has been member of Society of Industrial and Applied Mathematics (SIAM), Philadelphia, USA (2011-19). He is life member of Science and Engineering Institute (SCIEI), Los Angeles, USA; Indian Society of Industrial and Applied Mathematics (ISIAM), India; Ramanujan Mathematical Society, India; International Association of Engineers, UK. He is member of EURO working group on continuous optimization (EUROPT), Italy; Society for Foundations of Computational Mathematics, USA. His research interests are inverse problems, signal processing, optimization, data science, etc.

Nalini Tyagi is a well-known author and researcher in the field of image processing. She obtained her Bachelor degree (B.Tech.) from UP Technical University in Electronics and Instrumentation. Then she completed her master's (M.Tech.) from M.D. University, Rohtak in the Electronics and Communication Engineering. She has published original articles in reputed journals in the area of her studies. Her research interests include image processing, Deep Learning, Machine Learning.

Index

A

Adarsh Gram 228-230
Air Pollution 62, 216-219, 222-227
Aphids 216, 219
Application 21, 23, 25, 28-29, 32, 39, 41, 43, 59-60, 65, 69-70, 73, 85, 91, 95, 97-98, 100, 110-111, 113, 124, 128-129, 131-132, 135, 145, 180, 184-185, 187, 190-192, 195, 201-202, 216, 224-225, 232, 234-235, 239-242, 245, 267-268, 285-288, 305
Arduino 187-190, 195, 197-200, 202, 206, 283, 285-287, 297, 300-301
Arthropods 217, 225
Artificial intelligence (AI) 4, 16-18, 20-34, 45, 47, 62, 66, 99-100, 102-103, 114-117, 119, 121-124, 126-133, 146, 151, 161-162, 181-183, 205, 209, 249, 251, 267, 285, 306
Artificial Neural Networks 33, 112-113, 119, 123, 126-127, 132, 266

B

Battery 37, 42, 179, 241, 277, 286, 290-291, 294-296, 298-299, 302-303
Biofilters 216-217, 224-225
Biotic Environmental 218
Black Body Radiation 236-237, 241
Block chain 1, 27

C

Catalytic 216-217, 222-223
Chest X-Rays 208-210, 214
Classification 30, 34, 51, 55, 57-58, 62-67, 76-77, 88, 98, 103-104, 107-116, 128, 209-210, 212, 214-215, 250-254, 261, 264-265, 267
Cloud 4, 24, 60, 158-159, 163, 184-185, 188, 191-192, 194-195, 197, 199, 203, 205-206
Clustering 57, 82-84, 103-104, 248, 250, 261
CNN 83, 208-214, 248, 250-251, 254, 264, 266-267
Control Strategies 269, 272-274
Conventional Methods 18, 134
COVID-19 208-215
Crop Health Monitoring 35
Crop Yield 62, 110, 114-116, 122-123, 126-127, 132, 134-135, 185, 203, 205

D

Data Augmentation 208, 210
Data Fusion 250
DC Motor 283, 288, 292, 294, 302
DC Pump 283, 294
Decision Making 100, 119, 165
Deep Learning 40, 52, 60, 65-67, 84, 98, 100, 108, 113, 117, 131, 166, 170, 183, 208-209, 215, 248-252, 261, 265-268
Deterministic Probability 248
Disease Detection 30, 58, 62-63, 68, 84, 97-99, 109-111, 117, 163, 250, 253, 262, 265
Drone 22, 28-30, 35-37, 39-46, 101, 132, 146, 161, 177-178

DSURF 248, 254, 257-258, 261

E

Evapotranspiration 35, 45-46, 52, 132, 140-141

F

Frequency Deviation 269, 271-273, 279-281
Fresnel Lens 236-241

I

ICT 1, 5, 14, 67, 132
Image Processing 40, 47-48, 52, 56-58, 60-64, 66-67, 69-71, 73-75, 84-85, 95-98, 110, 128, 266
Image Segmentation 57, 69, 73-74, 77
Integrated Agricultural System 119, 121
Internet of things (IoT) 1, 3-4, 10-11, 14, 23, 27-28, 44, 48, 60, 63, 100, 126, 131, 145, 159, 163, 166, 180, 182, 186-188, 190-191, 199, 206-207, 267, 285
IR Sensor 228
Irrigation Methods 135

L

LDR 228, 230, 232-235, 244, 246
LED 7, 59, 64, 157, 168, 185, 198, 218-219, 228, 243, 245-246

M

Machine learning (ML) 4, 10, 14, 16-18, 20-25, 27-32, 41, 47-48, 50-52, 54-56, 60-64, 70, 84, 99-100, 102-103, 105-108, 110-119, 123-124, 126-127, 131-133, 151, 161-162, 175, 182, 188, 190-191, 205, 215, 248, 250, 252, 257, 264-266, 287
Machine Vision 48, 54, 57-60, 66, 96, 268, 286
MDNN 248, 254, 256-258, 260-261
Microgrid 269, 271-274, 279, 281-282
Moisture Level 140, 155, 184, 187, 189-190, 195, 200

O

Oxidation 217, 222

P

Parasitoids 218
Pesticides 23, 29-30, 35, 42-44, 52, 55, 70, 101, 110, 216-217, 283-288, 293-294, 302-307
PIR Sensor 228, 235-236, 243-244
Plant Disease 57-58, 63, 67-68, 86, 96-98, 111, 113, 248, 250, 253, 256
Processing of Images 85

R

Relay 189, 195, 202-203, 228, 233, 237, 296
Renewable Energy 2, 5-6, 9-10, 13, 269-274, 277, 281-282
ResNet 208
RFID Reader 283, 297-298, 300-301, 304

S

Sensors 4, 9-10, 13-14, 24-25, 27-28, 30, 38-39, 48, 52, 55, 57, 60-61, 64-65, 97, 100-101, 110, 112-113, 115-116, 122, 124, 126-127, 137-140, 142-148, 150, 155-160, 164, 168, 170, 176, 184-186, 188, 190-192, 194, 199-200, 203, 205-206, 215, 227, 234, 242, 264-265, 285, 287, 296
Simulation 195, 226, 269, 273-274
Smart Village 1-10, 12-15, 229-230, 247, 269-272
Smarter Irrigation 134
Soil Management 99, 107, 132
Solar Energy 275, 287-288, 303-304
Solar Panel 283, 290-291, 293, 295, 302
Spraying 23, 26, 35-36, 38, 42-43, 45, 52, 60, 130, 153, 224, 283-288, 293-294, 300, 302-307
Sustainable Agriculture 2, 6, 22, 119, 121-

122, 131-133, 203, 268

V

Validation 83, 256

W

Water Management 10, 22, 52, 99, 108, 118, 158, 185-188, 206-207, 285, 306

Weed Detection 18, 52, 70, 80-81, 99, 109-110

www.ingramcontent.com/pod-product-compliance
Lightning Source LLC
LaVergne TN
LVHW081345300926
849161LV00008B/2076